# Enabling Technology in Optical Fiber Communications

# Enabling Technology in Optical Fiber Communications: From Device, System to Networking

Editors

**Yang Yue**
**Jian Zhao**
**Jiangbing Du**
**Zhaohui Li**

MDPI • Basel • Beijing • Wuhan • Barcelona • Belgrade • Manchester • Tokyo • Cluj • Tianjin

*Editors*
Yang Yue
Institute of Modern Optics
Nankai University
Tianjin
China

Jian Zhao
School of Precision Instruments
and Optoelectronics Engineering
Tianjin University
Tianjin
China

Jiangbing Du
State Key Laboratory of
Advanced Optical
Communication Systems and
Networks
Shanghai Jiao Tong University
Shanghai
China

Zhaohui Li
State Key Laboratory of
Optoelectronic Materials and
Technologies, School of
Electronics and Information
Technology
Sun Yat sen University
Guangzhou
China

*Editorial Office*
MDPI
St. Alban-Anlage 66
4052 Basel, Switzerland

This is a reprint of articles from the Special Issue published online in the open access journal *Sensors* (ISSN 1424-8220) (available at: www.mdpi.com/journal/sensors/special_issues/OFC).

For citation purposes, cite each article independently as indicated on the article page online and as indicated below:

LastName, A.A.; LastName, B.B.; LastName, C.C. Article Title. *Journal Name* **Year**, *Volume Number*, Page Range.

**ISBN 978-3-0365-1183-2 (Hbk)**
**ISBN 978-3-0365-1182-5 (PDF)**

# Contents

# About the Editors

**Yang Yue**

Yang Yue received his B.S. and M.S. degrees in electrical engineering and optics from Nankai University, Tianjin, China, in 2004 and 2007, respectively. He received his Ph.D. degree in electrical engineering from the University of Southern California, Los Angeles, CA, USA, in 2012. He is a professor at the Institute of Modern Optics, Nankai University, Tianjin, China. Dr. Yue's current research interests include intelligent photonics, optical communications and networking, optical interconnect, detection, imaging and display technology. He has published over 200 peer-reviewed journal papers and conference proceedings, 4 edited books, 1 book chapter, >10 invited papers, >50 issued or pending patents, and >100 invited presentations.

**Jian Zhao**

Jian Zhao received a B.Eng. degree in optoelectrical engineering from Tianjin University, Tianjin, China, in 2003; an M.S. degree in optics from Nankai University, in 2006; and a Ph.D. degree in electronics and information engineering from The Hong Kong Polytechnic University, in 2010. From 2016 to 2017, he was a visiting scholar of the CREOL, The College of Optics and Photonics, University of Central Florida. He is currently an associate professor at the School of Precision Instruments and Optoelectronics Engineering, Tianjin University. He has authored or coauthored more than 50 articles in peer-reviewed journals and leading international conferences. His current research interests include optical fiber communications, optical fiber sensing, and nonlinear optics.

**Jiangbing Du**

Jiangbing Du received his B.S. (2002–2005) and M.S. (2005–2008) degrees, respectively, from College of Physics and Institute of Modern Optics, Nankai University, Tianjin, China. He obtained his PhD degree (2008–2011) in electronic engineering from The Chinese University of Hong Kong. He worked at Huawei technologies from 2011 to 2012. He joined Shanghai Jiao Tong University in 2012. Dr. Du is a senior member of both IEEE and OSA. He is the author or coauthor of over 150 journal and conference papers, including 26 OFC/ECOC papers. He has served as a TPC/LOC member in a variety of academic conferences, such as OECC, ACP, etc. His research interests include optical communication, photonic integration, and optical signal processing.

**Zhaohui Li**

Zhaohui Li received a B.S. degree from the Department of Physics and an M.S. degree at the Institute of Modern Optics from Nankai University, Tianjin, China, in 1999 and 2002, respectively, and a Ph.D. degree from Nanyang Technological University, Singapore, in 2007. He is currently a professor with the School of Electronics and Information Technology, Sun Yat-sen University, Guangzhou, China. His research interests include optical communication systems, optical signal processing technology, and ultra-fine measurement systems.

*sensors*

MDPI

*Editorial*

# Special Issue on Enabling Technology in Optical Fiber Communications: From Device, System to Networking

**Yang Yue [1,*], Jian Zhao [2], Jiangbing Du [3] and Zhaohui Li [4]**

1 Institute of Modern Optics, Nankai University, Tianjin 300350, China
2 School of Precision Instruments and Optoelectronics Engineering, Tianjin University, Tianjin 300072, China; enzhaojian@tju.edu.cn
3 State Key Laboratory of Advanced Optical Communication Systems and Networks, Shanghai Jiao Tong University, Shanghai 200240, China; dujiangbing@sjtu.edu.cn
4 State Key Laboratory of Optoelectronic Materials and Technologies, School of Electronics and Information Technology, Sun Yat Sen University, Guangzhou 510275, China; lzhh88@mail.sysu.edu.cn
* Correspondence: yueyang@nankai.edu.cn



**Citation:** Yue, Y.; Zhao, J.; Du, J.; Li, Z. Special Issue on Enabling Technology in Optical Fiber Communications: From Device, System to Networking. *Sensors* **2021**, *21*, 1969. https://doi.org/10.3390/s21061969

Received: 2 March 2021
Accepted: 9 March 2021
Published: 11 March 2021



It is well known that optical fiber communications support the global communication networks nowadays, which originates from Charles K. Kao's proposal of using optical fiber as a light transmission medium in 1966 [1]. By utilizing different degrees of freedom of the photon, society has made tremendous progress over the past half century. Revolutionary technologies debut one after another, including wavelength-division multiplexing (WDM), coherent detection, space-division multiplexing (SDM), and so forth [2,3]. Recently, the emerging 5G, cloud computing, and high-definition video have been driving more bandwidth and power-hungry applications. To better serve these needs, the optical fiber communications community has been escalating the research and development efforts in device, system, and networking to the next level.

This Special Issue aims to explore the enabling technology in optical fiber communications. It focuses on the state-of-the-art advances from fundamental theories, devices, and subsystems, to networking applications, as well as future perspectives of optical fiber communications. The collected papers have well accomplished these goals by contributing leading-edge derivation, analysis, and experiments with significant results. The topics cover integrated photonics, fiber optics, fiber and free-space optical communications, and optical networking. The special issue consists of one review paper, nine research articles, and eight letters.

More specifically, from the integrated device perspective, Fang, Y. et al. proposed an $Si_3N_4/SiO_2$ horizontal-slot-waveguide-based polarization beam splitter (PBS) [4]. Its coupling length can be effectively reduced due to the slot design, and the extinction ratios (ER) of the fundamental modes for two orthogonal polarizations are both >20 dB. Furthermore, it features low nonlinearity, which is critical for on-chip high-power systems. For fiber-based devices, Rong, J. et al. numerically simulated shaping the supercontinuum (SC) using the fiber cascading method to significantly increase the SC spectral width and flatness in silica photonic crystal fiber (PCF) [5]. To characterize linear frequency-modulated continuous-wave (FMCW) lasers, Yang, J. et al. [6] proposed a scheme for measuring the mapping relationship between instantaneous frequency and time of a FMCW laser based on a modified coherent optical spectrum analyzer (COSA) and digital signal processing (DSP) method. The authors demonstrated precisely measuring an FMCW laser at a large fast sweep rate of 5000 THz/s, while maintaining <100 kHz uncertainty.

Regarding the system-level technologies for optical communication, Tseng, S. et al. developed a bipolar optical code division multiple access (Bi-OCDMA) technique based on spectral amplitude coding for the formation and transmission of optical-polarized and coded signals over wireless optical channels [7]. The proposed free-space optics communication system used a dual electro-optical modulator design, which could improve the transmission rate. For direct-detection optical communication systems, Zhang, W. et al. compared the complexity, efficiency, and stability performance of pruned Volterra series-based equalization (VE) and neural network-based equalization (NNE) for 112 Gbps vertical-cavity surface-emitting laser (VCSEL)-enabled optical interconnects [8]. The experimental results showed that NNE has more than one order of magnitude bit error rate (BER) advantage over VE at the same computation complexity, and pruned NNE has around 50% lower computation complexity compared to VE at the same BER level. For coherent-detection optical communication systems, Ding, J. et al. investigated the impact of equalization-enhanced phase noise (EEPN) in Nyquist-spaced dual-polarization quadrature amplitude modulation (DP-QAM) links [9]. It was found that EEPN-induced distortions become more significant with the increase of the local oscillator (LO) laser linewidth, and this results in degradations in BER, achievable information rate (AIR), and AIR-distance product. Moreover, Wu, B. et al. proposed a blind discrete-cosine-transform-based phase noise compensation (BD-PNC) to compensate the intercarrier interference (ICI) in the coherent optical offset-quadrature amplitude modulation (OQAM)-based filter-bank multicarrier (CO-FBMC/OQAM) transmission system [10]. The simulation results showed that its BER performance is improved by more than one order of magnitude through the mitigation of the ICI over traditional blind PNC scheme only aiming for common phase error (CPE) compensation. For quantum communication applications, Wu, B. et al. proposed and experimentally demonstrated a secure key generation and distribution system that is compatible with optical amplifiers and standard WDM transmission systems [11]. The key generation system was tested in a 240 km bidirectional fiber-pair link with multiple optical amplifiers, and 38 WDM channels were transmitted together with the key distribution channel.

On the networking level, Holik, M. et al. created an open-source software-based solution for monitoring traffic transmitted through gigabit passive optical network (GPON) [12]. The work described the issue of writing to the Mongo database system, showing that the high processing speed is too high for Python processing and critical operations must be implemented in the C# programming language. He, C. et al. proposed a FiWi broadband access network, integrating the wireless mesh network (WMN) frontend subnetwork, together with time and wavelength division multiplexed PON (TWDM-PON) optical backhaul and adapting power over fiber (PoF) technology [13]. For elastic optical networks (EONs), He, S. et al. proposed an advanced-reservation-based invalid-spectrum-aware (AR-ISA) resource allocation algorithm to improve the networking performance and the resource alignment [14]. Moreover, Rodrigues, E. et al. proposed a crosstalk-aware routing, modulation, spectrum, and core allocation (RMSCA) algorithm that uses a multipath and mapping scheme for improving resource allocation [15]. Simulation results show that the algorithm decreases the blocking ratio by up to four orders of magnitude compared with the other RMSCA algorithms in the literature. Furthermore, Zong, Y. et al. surveyed the state-of-the-art works for the virtual network embedding (VNE) problem towards multidomain heterogeneous converged optical networks, and discussed the future research issues and challenges [16].

Additionally, there are several collected papers aimed at sensing applications. Wu, Z. et al. demonstrated Bragg-grating-assisted Sagnac interferometer in $SiO_2$-$Al_2O_3$-$La_2O_3$ polarization maintaining fiber for strain–temperature discrimination [17]. Xu, X. et al. proposed and demonstrated a temperature and humidity sensor based on a fluorinated polyimide film and fiber Bragg grating [18]. Han, H. et al. proposed a surface plasmon resonance (SPR) sensor based on a dual-side polished microstructured optical fiber (MOF) with a dual core [19]. Wu, Y. et al. proposed a resolution enhancement and signal-to-noise ratio (SNR) improvement scheme for digital optical frequency comb (DOFC)-based Brillouin optical time-domain analysis (BOTDA) ultrafast distributed sensing employing a pump pulse array [20]. Cheng, Y. et al. analyzed the source of the position deviation and proposed a demodulation recursive compensation algorithm to ensure a submillimeter resolution in improved optical frequency domain reflectometry (OFDR) [21].

It has just been over 50 years since the discovery of optical fiber as a low-loss light transmission medium by Charles K. Kao and his coworkers. Within this fairly short period of time, an extensive research community and industry have been established globally. Nowadays, optical fiber communications is the backbone of our information technology infrastructure, supporting voice, video, and data transmission through global networks. One critical issue in its research and development is the challenge of meeting the needs of increasing the data capacity without compromising size, weight, power, and cost (SWaP-C) constraints. Especially during the past decade, photonics integration and coherent detection technologies were booming tremendously. From this trend, we are expecting the co-packaged optics (CPO) to dominate the next-generation optical fiber communication systems. Hopefully, more and more novel technologies, such as SDM, can be commercialized in the near future, to enable new growth for the industry and serve the ever-growing data-traffic demand from society.

**Conflicts of Interest:** The authors declare no conflict of interest.

## References


1. Kao, K.C.; Hockham, G.A. Dielectric-fibre surface waveguides for optical frequencies. *Proc. Inst. Electr. Eng.* **1966**, *113*, 1151–1158. [CrossRef]
2. Bozinovic, N.; Yue, Y.; Ren, Y.; Tur, M.; Kristensen, P.; Huang, H.; Willner, A.E.; Ramachandran, S. Terabit-scale orbital angular momentum mode division multiplexing in fibers. *Science* **2013**, *340*, 1545–1548. [CrossRef] [PubMed]
3. Van Uden, R.G.H.; Correa, R.A.; Lopez, E.A.; Huijskens, F.M.; Xia, C.; Li, G.; Schülzgen, A.; De Waardt, H.; Koonen, A.M.J.; Okonkwo, C.M. Ultra-high-density spatial division multiplexing with a few-mode multicore fibre. *Nat. Photon.* **2014**, *8*, 865–870. [CrossRef]
4. Fang, Y.; Bao, C.; Wang, Z.; Liu, Y.; Zhang, L.; Huang, H.; Ren, Y.; Pan, Z.; Yue, Y. Polarization Beam Splitter Based on $Si_3N_4/SiO_2$ Horizontal Slot Waveguides for On-Chip High-Power Applications. *Sensors* **2020**, *20*, 2862. [CrossRef] [PubMed]
5. Rong, J.; Yang, H.; Xiao, Y. Accurately Shaping Supercontinuum Spectrum via Cascaded PCF. *Sensors* **2020**, *20*, 2478. [CrossRef]
6. Yang, J.; Yang, T.; Wang, Z.; Jia, D.; Ge, C. A Novel Method of Measuring Instantaneous Frequency of an Ultrafast Frequency Modulated Continuous-Wave Laser. *Sensors* **2020**, *20*, 3834. [CrossRef] [PubMed]
7. Tseng, S.-P.; Wijanto, E.; Lai, P.-H.; Cheng, H.-C. Bipolar Optical Code Division Multiple Access Techniques Using a Dual Electro-Optical Modulator Implemented in Free-Space Optics Communications. *Sensors* **2020**, *20*, 3583. [CrossRef]
8. Zhang, W.; Ge, L.; Zhang, Y.; Liang, C.; He, Z. Compressed Nonlinear Equalizers for 112-Gbps Optical Interconnects: Efficiency and Stability. *Sensors* **2020**, *20*, 4680. [CrossRef]
9. Ding, J.; Xu, T.; Jin, C.; Wang, Z.; Zhao, J.; Liu, T. Impact of Equalization-Enhanced Phase Noise on Digital Nonlinearity Compensation in High-Capacity Optical Communication Systems. *Sensors* **2020**, *20*, 4149. [CrossRef]
10. Wu, B.; Lu, J.; Gao, M.; Ren, H.; Le, Z.; Qin, Y.; Guo, S.; Hu, W. Time-Domain Blind ICI Compensation in Coherent Optical FBMC/OQAM System. *Sensors* **2020**, *20*, 6397. [CrossRef]
11. Wu, B.; Huang, Y.-K. Optical-Amplifier-Compatible Long-Distance Secure Key Generation Based on Random Phase Fluctuations for WDM Systems. *Sensors* **2020**, *20*, 6296. [CrossRef]
12. Holik, M.; Horvath, T.; Oujezsky, V.; Munster, P.; Tomasov, A.; Valach, S. MongoDB Database as Storage for GPON Frames. *Sensors* **2020**, *20*, 6208. [CrossRef]
13. He, C.; Wang, R. A QoE-Aware Energy Supply Scheme over a FiWi Access Network in the 5G Era. *Sensors* **2020**, *20*, 3794. [CrossRef] [PubMed]
14. He, S.; Qiu, Y.; Xu, J. Invalid-Resource-Aware Spectrum Assignment for Advanced-Reservation Traffic in Elastic Optical Network. *Sensors* **2020**, *20*, 4190. [CrossRef]
15. Rodrigues, E.; Cerqueira, E.; Rosário, D.; Oliveira, H. Hybrid Routing, Modulation, Spectrum and Core Allocation Based on Mapping Scheme. *Sensors* **2020**, *20*, 6393. [CrossRef] [PubMed]
16. Zong, Y.; Feng, C.; Guan, Y.; Liu, Y.; Guo, L. Virtual Network Embedding for Multi-Domain Heterogeneous Converged Optical Networks: Issues and Challenges. *Sensors* **2020**, *20*, 2655. [CrossRef]
17. Wu, Z.; Wu, P.; Kudinova, M.; Zhang, H.; Shum, P.P.; Shao, X.; Humbert, G.; Auguste, J.-L.; Dinh, X.Q.; Pu, J. Bragg Grating Assisted Sagnac Interferometer in $SiO_2$-$Al_2O_3$-$La_2O_3$ Polarization-Maintaining Fiber for Strain–Temperature Discrimination. *Sensors* **2020**, *20*, 4772. [CrossRef] [PubMed]
18. Xu, X.; Luo, M.; Liu, J.; Luan, N. Fluorinated Polyimide-Film Based Temperature and Humidity Sensor Utilizing Fiber Bragg Grating. *Sensors* **2020**, *20*, 5469. [CrossRef]
19. Han, H.; Hou, D.; Luan, N.; Bai, Z.; Song, L.; Liu, J.; Hu, Y. Surface Plasmon Resonance Sensor Based on Dual-Side Polished Microstructured Optical Fiber with Dual-Core. *Sensors* **2020**, *20*, 3911. [CrossRef]

20. Wu, Y.; Yang, C.; Pan, J.; Sui, Q.; Wang, D. Ultrafast Resolution-Enhanced Digital Optical Frequency Comb-Based BOTDA with Pump Pulse Array Coding. *Sensors* **2020**, *20*, 6411. [CrossRef]
21. Cheng, Y.; Luo, M.; Liu, J.; Luan, N. Numerical Analysis and Recursive Compensation of Position Deviation for a Sub-Millimeter Resolution OFDR. *Sensors* **2020**, *20*, 5540. [CrossRef] [PubMed]

*sensors*

MDPI

*Article*

# Polarization Beam Splitter Based on $Si_3N_4/SiO_2$ Horizontal Slot Waveguides for On-Chip High-Power Applications

**Yuxi Fang [1,†], Changjing Bao [2,†], Zhonghan Wang [1], Yange Liu [1], Lin Zhang [3], Hao Huang [2], Yongxiong Ren [2], Zhongqi Pan [4] and Yang Yue [1,*]**

1 Institute of Modern Optics, Nankai University, Tianjin 300350, China; 1120180104@mail.nankai.edu.cn (Y.F.); 1711143@mail.nankai.edu.cn (Z.W.); ygliu@nankai.edu.cn (Y.L.)
2 Department of Electrical Engineering, University of Southern California, Los Angeles, CA 90089, USA; changjib@usc.edu (C.B.); haoh@usc.edu (H.H.); yongxior@usc.edu (Y.R.)
3 School of Precision Instrument and Opto-Electronics Engineering, Tianjin University, Tianjin 300072, China; lin_zhang@tju.edu.cn
4 Department of Electrical and Computer Engineering, University of Louisiana at Lafayette, Lafayette, LA 70504, USA; zpan@louisiana.edu
* Correspondence: yueyang@nankai.edu.cn
† These authors contributed equally to this work.

Received: 1 April 2020; Accepted: 15 May 2020; Published: 18 May 2020

check for updates

**Abstract:** In this paper, we propose an $Si_3N_4/SiO_2$ horizontal-slot-waveguide-based polarization beam splitter (PBS) with low nonlinearity for on-chip high-power systems. The coupling length ratio between the quasi-TE and quasi-TM modes ($L_{TE}/L_{TM}$) was optimized to 2 for an efficient polarization splitting. For the single-slot design, the coupling length of the PBS was 281.5 µm, while the extinction ratios (ER) of the quasi-TM and quasi-TE modes were 23.9 dB and 20.8 dB, respectively. Compared to PBS based on the $Si_3N_4$ strip waveguide, the coupling length became 22.6% shorter. The proposed PBSs also had a relatively good fabrication tolerance for an ER of >20 dB. For the multi-slot design, the coupling length of the PBS was 290.3 µm, while the corresponding ER of the two polarizations were 24.0 dB and 21.0 dB, respectively. Furthermore, we investigated the tradeoff between the ER and coupling length for the optimized PBSs with single slot or multiple slots.

**Keywords:** optical waveguide; silicon photonics; silicon nitride; optical polarization

## 1. Introduction

Photonic integrated circuit (PIC) has attracted much interest recently due to its fundamental advantages in terms of low power consumption, compact size, and low cost [1,2]. Specifically, silicon nitride ($Si_3N_4$) is considered to be a promising platform for PIC, due to its CMOS compatibility [3,4]. Compared to the silicon (Si) waveguide, $Si_3N_4$ has the advantages of a smaller roughness-induced scattering loss due to the lower refractive index, larger fabrication tolerance due to the weaker light confinement [5–8], and lower nonlinear loss induced by two-photon absorption (TPA) in the telecommunication band [9–13]. Thus, the $Si_3N_4$ waveguide has been widely considered for the high-power applications, such as nonlinear optics (optical frequency comb generation [14], chip optical parametric oscillators, supercontinuum generation [10]), material processing, laser medicine [15], etc.

Polarization beam splitter (PBS) is widely used to mitigate the polarization randomness problem in photonics systems by controlling the polarization state, thus, ensuring proper functioning [16–20]. PBS has been extensively used in sensor applications, such as polarization navigation sensor [21], photoacoustic remote sensing microscopy [22], distributed vibration sensor [23], magnetic field sensor [24], and fiber-optic gyroscope [25]. Furthermore, integrated PBS is an indispensable component of the transceiver in coherent optical communication system. The function of PBS is to separate the transverse electric (TE) and transverse magnetic (TM) polarization beams into different paths, due to their different propagation constants. Previously, various PBSs with different structures have been demonstrated, such as directional couplers, photonic crystals, multimode interferometers (MMIs), Mach-Zehnder interferometers (MZIs), and subwavelength gratings [16,26–28]. Recently, slot waveguide has been implemented into directional couplers for improved polarization-splitting performance [17,18,29–35]. In the slot waveguide, a low refractive index layer is sandwiched by two high refractive index layers. Due to the discontinuity of the electric field at the high-refractive-index-contrast interfaces, the mode with a polarization normal to the interfaces is significantly enhanced within the slot region. Either vertical-slot or horizontal-slot waveguides have been used for on-chip PBSs. The fabrication process of the vertical-slot-waveguide-based PBS usually introduces large roughness in the vertical interfaces [36]. A horizontal-slot-waveguide-based PBS has smoother interfaces, as it only needs to deposit layers with different materials and then etch the waveguide vertically.

In this study, we investigated the integrated PBS using $Si_3N_4$-based, coupled, horizontal slot waveguides that could be used for kW-level peak power on-chip systems. The designed $Si_3N_4$-based PBS could have negligible linear and nonlinear loss, and its Kerr nonlinear coefficient was approximately two orders of magnitude less than that of PBS using silicon waveguides. By manipulating the coupling length ratio of the quasi-TE and quasi-TM modes ($L_{TE}/L_{TM}$) to be around 2, efficient polarization splitting can be realized. First, we proposed and designed the $Si_3N_4$-based single-slot waveguide PBS, which had a 281.5-μm coupling length. The corresponding extinction ratios (ER) were 23.9 dB and 20.8 dB at 1550 nm for the quasi-TM and quasi-TE modes, respectively. Compared with the PBS based on $Si_3N_4$ strip waveguide, the coupling length becomes 22.6% shorter. Our simulation also showed good fabrication tolerance, the ER maintained >20 dB for the quasi-TM mode with a waveguide width variation of ±20 nm. Additionally, we designed a $Si_3N_4$-based multi-slot waveguide PBS with a 290.3-μm coupling length. Its ERs of the quasi-TM and quasi-TE modes were 24.0 dB and 21.0 dB at 1550 nm, respectively. Moreover, the tradeoff between the ER and coupling length for both the single-slot and multi-slot waveguides was also investigated at the end.

## 2. Concept

Figure 1a shows two basic polarization splitting mechanisms. Type I illustrates that the coupling length of two orthogonal polarizations has a large difference, such as the coupling length ratio ($L_{TE}/L_{TM}$), which was up to 21 [18]. In this condition, most of the quasi-TE mode remained in the bar port, while a fraction of the quasi-TE mode was still coupled to the cross port. The quasi-TM mode was fully coupled to the cross port. The coupling length of the quasi-TE mode was designed to reach to the integer multiple of the one for the quasi-TM mode in Type II. In such a case, the quasi-TM mode was coupled out and then back to the bar port, while the quasi-TE mode was completely coupled to the cross port. As a result, an efficient polarization splitting was achieved. The schematic of the $Si_3N_4$ horizontal-single-slot-waveguide-based PBS is depicted in Figure 1b, which was based on Type II. The quasi-TE and quasi-TM modes had different coupling lengths when propagating along a waveguide. The blue and red lines represented the power transformation of the quasi-TE and quasi-TM mode, respectively.

The cross-section of the proposed $Si_3N_4$ horizontal-slot-waveguide-based PBS is also illustrated in Figure 1b. High index material $Si_3N_4$ was used for the top and bottom regions, while a low index silicon dioxide ($SiO_2$) was chosen for the middle slot region. The material refractive indices of $SiO_2$ and $Si_3N_4$ were obtained according to the Sellmeier equations in our model [37,38]. $H_s$ is the slot thickness, $H_u$ is the upper $Si_3N_4$ thickness, $H_l$ is the lower $Si_3N_4$ thickness, and $d$ is the spacing between the two slot waveguides. We keep the total height $H$ and the width $W$ of the slot waveguide to 800 nm and 1050 nm, respectively.

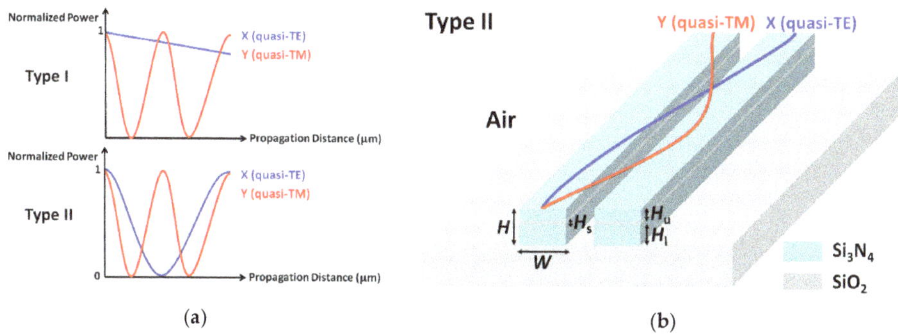


**Figure 1.** Schematic of two basic polarization splitting mechanisms (**a**) and the $Si_3N_4$ horizontal-single-slot-waveguide-based polarization beam splitter (PBS) (**b**).

Based on the mode-coupling theory, the propagating behavior in the designed PBS could be viewed as the summation of the even and odd modes along the device. The coupling length of two orthogonal polarizations can be expressed as [18]:

$$L_{TE} = \lambda/[2^*(n_{e,TE} - n_{o,TE})], \tag{1}$$

$$L_{TM} = \lambda/[2^*(n_{e,TM} - n_{o,TM})], \tag{2}$$

where $L_{TE}$ and $L_{TM}$ are the coupling lengths of the quasi-TE and quasi-TM modes, $n_{e,TE}$, $n_{o,TE}$, $n_{e,TM}$, and $n_{o,TM}$ are the effective refractive indices of the even and odd supermodes of the $x$ and $y$ polarizations, respectively. We calculated the effective indices of the supermodes for the coupled-waveguide structure, using a finite-element mode solver. The electric field distributions of the quasi-TE ($n_{e,TE}$ and $n_{o,TE}$) and quasi-TM ($n_{e,TM}$ and $n_{o,TM}$) modes at the wavelength of 1550 nm are shown in Figure 2a–d.

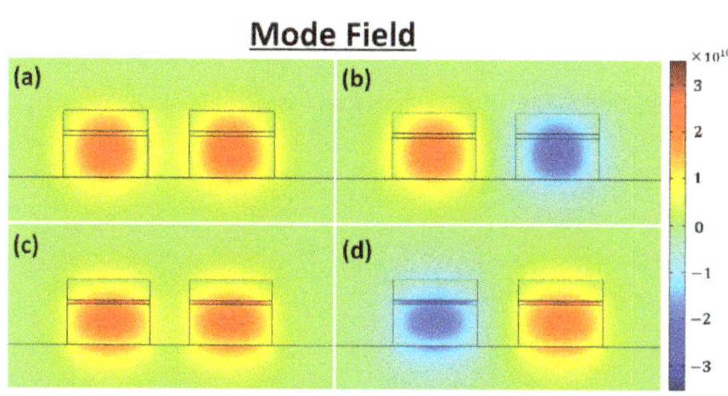


**Figure 2.** The electric field distributions of (**a**) the even quasi-transverse electric (TE) mode, (**b**) the odd quasi-TE mode, (**c**) the even quasi-TM mode, and (**d**) the odd quasi-TM mode.

## 3. Resultant Design of the Polarization Beam Splitter

### 3.1. Material Characteristics

We obtained the normalized power of the silicon slot waveguide [39] with different input peak power based on Equation (3)

$$dI/dz = -\alpha I - \beta_{TPA} I^2, \tag{3}$$

where $I$ is the input peak power, $z$ is the propagation distance, $\alpha$ is the linear loss, and $\beta_{TPA}$ is the TPA coefficient. As per the literature results, 6.3 dB/cm [36] and 0.01 dB/cm [40,41] were chosen for the linear losses of the Si and $Si_3N_4$ slot waveguides, the TPA coefficient $\beta_{TPA}$ of Si was set to $9.95 \times 10^{-12}$ m/W [41].

In Figure 3, the normalized power of the silicon slot waveguide decreased rapidly as the increase of the input power to the waveguide due to nonlinear loss process. For high power on-chip applications, kW-level power was used in some experiments [42]. The normalized power only remained 27.8% after 0.1-mm propagation, when the input peak power was up to 1 kW. The dashed black line represented the power evolution in the $Si_3N_4$-based slot waveguide. In contrast, the normalized power after the 1-mm $Si_3N_4$-based slot waveguide still approached 99.98%. Consequently, compared to silicon horizontal slot waveguide splitter [18], it was more suitable for high power integrated systems.

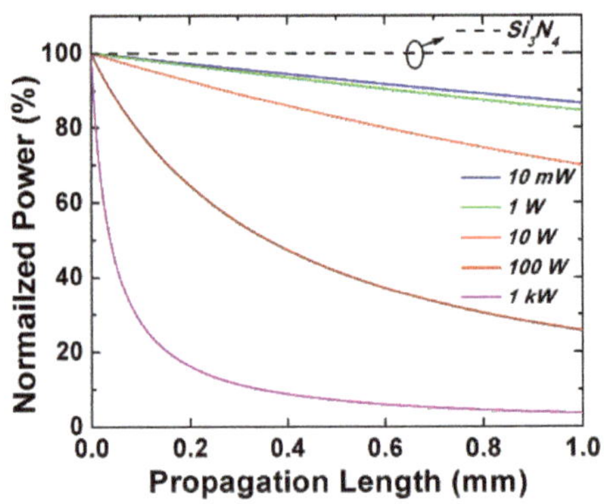


**Figure 3.** Normalized power of the inputs to the Si and $Si_3N_4$ horizontal slot waveguides.

We used a full-vector model [43] to guarantee an accurate result, which considered the modal distribution of different materials to the overall nonlinear coefficient. Equation (4) was used to characterize the nonlinear coefficient:

$$\gamma = 2\pi n_2/\lambda A_{eff}, \tag{4}$$

where $A_{eff}$ is the effective mode area, $n_2$ is the nonlinear refractive index averaged over an inhomogeneous cross-section weighted with respect to field distribution. At 1550 nm, the nonlinear refractive indices of Si, $Si_3N_4$, and $SiO_2$ were $4.5 \times 10^{-18}$ m$^2$/W, $2.4 \times 10^{-19}$ m$^2$/W, and $2.6 \times 10^{-20}$ m$^2$/W, respectively [3,44,45]. The Kerr nonlinear coefficient $\gamma$ of the silicon horizontal slot waveguide splitter [18] is shown in Figure 4a. The nonlinear coefficients of the quasi-TE modes were >68 /W/m, which were bigger than that of the quasi-TM modes. This was because the quasi-TE modes remained mainly in the silicon part, while a large portion of the optical power for the quasi-TM modes resided within the $SiO_2$ slot part. Figure 4b shows the nonlinear coefficient of the modes in the proposed $Si_3N_4$ slot-waveguide-based PBS. The corresponding geometric parameters were $H$ = 800 nm, $W$ = 1050 nm, $H_s$ = 50 nm, $H_u$ = 250 nm, and $d$ = 500 nm. We can see that the nonlinear coefficients of the quasi-TE modes were near 0.8 /W/m, the ones of the quasi-TM modes were near 0.6 /W/m. The Kerr nonlinear coefficients $\gamma$ of the proposed PBS was approximately one percent of the one of the silicon horizontal slot waveguide splitter for the quasi-TE modes. For the quasi-TM modes, $\gamma$ of the proposed PBS was near 5% of the one for the silicon horizontal slot waveguide splitter. Compared to the silicon horizontal slot waveguide splitter, the proposed $Si_3N_4$ slot-waveguide-based PBS had negligible nonlinearity under certain power level, and thus could efficiently separate the polarizations.

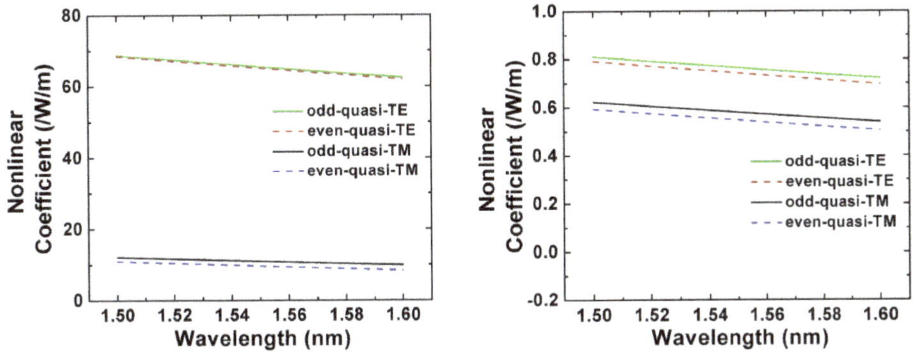


(a) (b)

**Figure 4.** Nonlinear coefficient of the Si (**a**) and $Si_3N_4$ (**b**) horizontal slot waveguide PBS.

### 3.2. Single-Slot Waveguide Polarization Beam Splitter

We first investigated the coupling length of the quasi-TM mode ($L_{TM}$) and the coupling length ratio of quasi-TE and TM mode ($L_{TE}/L_{TM}$) by varying the slot thickness in the single-slot-waveguide-based PBS, as shown in Figure 5a. We found that the coupling length of the quasi-TM mode decreased with the increasing of slot thickness, and the coupling length ratio ($L_{TE}/L_{TM}$) increased. Here, the waveguide spacing $d$ was 500 nm, and the slot thickness $H_s$ was 50 nm. In this condition, the coupling length of the TM mode was 140.5 μm, and the coupling length ratio of the two polarizations approached to 2, thus, the two orthogonal polarizations could be effectively separated at 1550 nm. Figure 5b depicts the effects of varying the waveguide spacing $d$. For a PBS with 50-nm slot thickness, $L_{TM}$ increases when $d$ increases from 100 to 900 nm. The coupling becomes weaker for larger waveguide spacing, and thus the coupling length increases. The waveguide spacing $d$ = 500 nm was chosen for the following study.

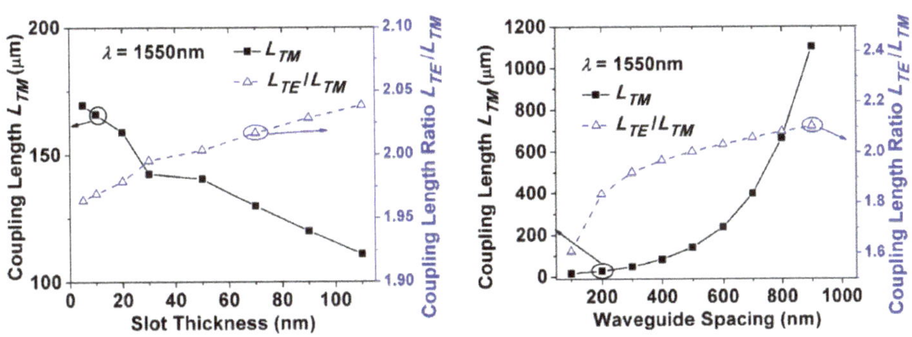


(a) (b)

**Figure 5.** Coupling length of the quasi-TM mode and the coupling length ratio as a function of (**a**) the slot thickness $H_s$ ($d$ = 500 nm); and (**b**) the waveguide spacing $d$ ($H_s$ = 50 nm).

As shown in Figure 6, the normalized power of the quasi-TE and quasi-TM modes in the bar port was a function of propagation distance. When the propagation distance was 281.5 μm, the quasi-TE mode was almost completely coupled to the cross port, the residual output in the bar port was only ~0.4%. Furthermore, when the quasi-TM mode was coupled out and brought back to the bar port, the output remained at 99.22%. The device length $L = L_{TE} = 2L_{TM}$ = 281.5 μm was chosen. Finally, an efficient PBS could be achieved by incorporating the coupled horizontal single-slot waveguides with the parameters of $H$ = 800 nm, $W$ = 1050 nm, $H_s$ = 50 nm, $H_u$ = 250 nm, $d$ = 500 nm, and $L$ = 281.5 μm.

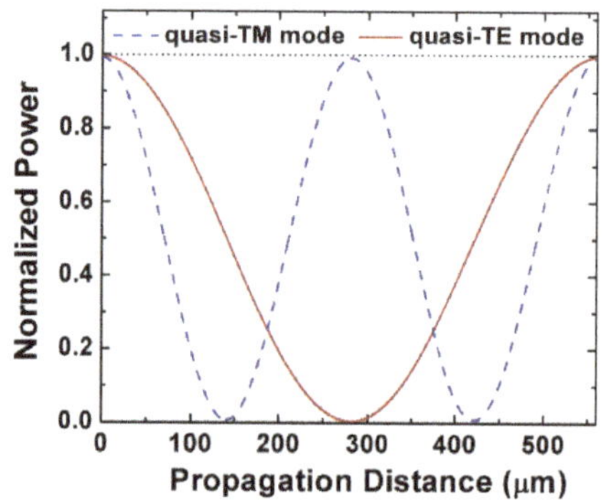


**Figure 6.** Normalized power exchange of the quasi-TM and quasi-TE modes at the bar port.

Equations (5) and (6) are the general expression of the ERs:

$$ER_{\text{bar}} = 10^{*}\text{Log}_{10}[\cos^2(\pi z/2L_{\text{TE}})/\cos^2(\pi z/2L_{\text{TM}})], \quad (5)$$

$$ER_{\text{cross}} = 10^{*}\text{Log}_{10}[\sin^2(\pi z/2L_{\text{TM}})/\sin^2(\pi z/2L_{\text{TE}})], \quad (6)$$

where $ER_{\text{bar}}$ and $ER_{\text{cross}}$ are the extinction ratios of the bar port and the cross port of the PBS, respectively. These formulas are suitable for both the Type I and Type II polarization splitting mechanisms. Considering $L = L_{\text{TE}} = 2L_{\text{TM}}$ in our Type II based design, the ERs in Equations (5) and (6) could be further expressed as:

$$ER_{\text{TM}} = 10^{*}\text{Log}_{10}[\cos^2(\pi z/L)/\cos^2(\pi z/2L)], \quad (7)$$

$$ER_{\text{TE}} = 10^{*}\text{Log}_{10}[\sin^2(\pi z/2L)/\sin^2(\pi z/L)], \quad (8)$$

where $L$ is the coupling length and $z$ is the propagation distance. In the critical scenario, i.e., the beam propagation distance $z = L = L_{\text{TE}} = 2L_{\text{TM}}$, the theoretical values of both $ER_{\text{TM}}$ and $ER_{\text{TE}}$ from the Equations (7) and (8) equaled to infinite. In practice, the fabrication imperfections from the design target, such as the length ratio of the two waveguides, the spacing between the two waveguides, and the width of the waveguides, resulted in incomplete power transfer and the degradation of the $ER_{\text{TM}}$ and $ER_{\text{TE}}$. We defined the bar port as the TM mode port, and the cross port as the quasi-TE mode port. Figure 7a shows the ERs of the proposed $Si_3N_4$ horizontal slot waveguide PBS for the quasi-TE and quasi-TM modes, which had a strong wavelength dependence. At the optimized wavelength of 1550 nm, the ERs of the quasi-TM and quasi-TE mode were 23.9 dB and 20.8 dB, respectively. The extinction ratio of the designed $Si_3N_4/SiO_2$ horizontal-slot-waveguide-based PBS could almost satisfy the practical demand of optical communication systems within the total C-band [46]. The ER of the quasi-TM mode was larger than the one of the quasi-TE mode, as the electric field of the quasi-TE mode was discontinuous at vertical interfaces, the overlap between two waveguides was stronger, and the crosstalk was higher.

Furthermore, we investigated the corresponding $Si_3N_4$ strip-waveguide-based PBS and simulated its properties. Based on the same mechanism, efficient polarization splitting could be realized using the coupled $Si_3N_4$ strip-waveguide-based PBS. The Kerr nonlinear coefficients $\gamma$ of the modes in the $Si_3N_4/SiO_2$ slot waveguide was smaller than the ones in the $Si_3N_4$ strip waveguide for both polarizations. The optimized geometric parameters of the $Si_3N_4$ strip waveguide was $H = 800$ nm, $W = 1070$ nm, and the coupling length $L = L_{\text{TE}}$ equaled 363.4 μm, when the corresponding ERs could achieve the values of the proposed PBS, i.e., 25.2 dB for the quasi-TM mode and 22.5 dB for the quasi-TE mode, respectively. Thus, the proposed $Si_3N_4/SiO_2$ slot-waveguide-based PBS was improved on the properties for both coupling length and nonlinearity. As shown in Figure 7b, the ERs for the $Si_3N_4$-strip-waveguide-based PBS basically equaled to the ERs for the slot one in the wavelength range of 1500–1600 nm, which also had a relatively strong wavelength dependence. In comparison, the coupling length of the proposed $Si_3N_4/SiO_2$ horizontal-slot-waveguide-based PBS was 22.6% shorter. Consequently, the $Si_3N_4/SiO_2$ horizontal slot waveguide structure could effectively reduce the length of the elements.

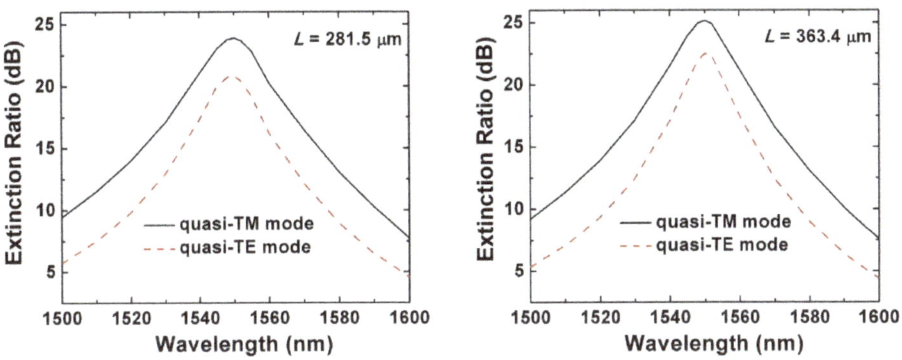


(a) (b)

**Figure 7.** Extinction ratio of the quasi-TM mode at the bar port and the quasi-TE mode at the cross port of the proposed $Si_3N_4$ horizontal slot (**a**) and strip (**b**) waveguide PBS.

We used the finite-difference time-domain method to simulate the modal evolution along the propagation of the proposed PBS. Figure 8 shows the power evolution along the propagation distance for the quasi-TE and quasi-TM modes at 1550 nm. We could observe that the optical power almost transferred from the bar port to the cross port at the coupling length (i.e., $L_{TE}$) for the quasi-TE mode. Meanwhile, the quasi-TM mode was first transferred to the cross port when the propagation distance equaled to $L_{TM}$ and then came back to the bar port. We could also observe that the coupling length of the quasi-TE polarization was twice as long as the one of the quasi-TM polarization.

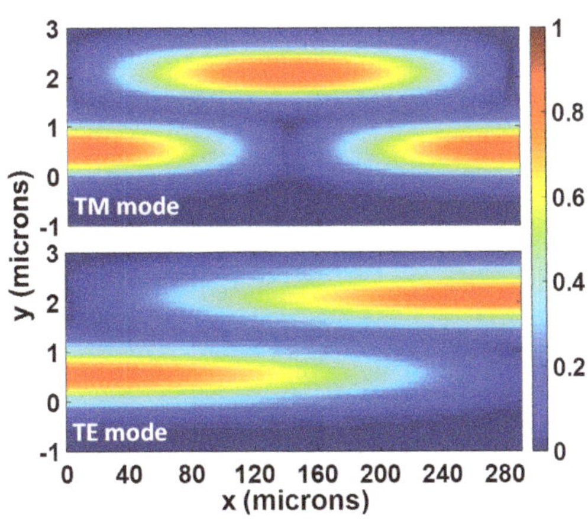


**Figure 8.** The power evolutions of the quasi-TM and quasi-TE modes in the designed PBS at 1550 nm.

We simulated the fabrication tolerance of the proposed $Si_3N_4$ horizontal slot-waveguide-based PBS. Taking into account the practical fabrication, the thickness of the slot and the strip thickness of the horizontal waveguide were defined by low-pressure chemical vapor deposition (LPCVD) or plasma-enhanced chemical vapor deposition (PECVD) [40]. These processes could be well controlled. The fabrication error was mainly from the etching process, i.e., the waveguide width was mainly the influence factor of the fabrication tolerance. Here, we defined the $\Delta w$ as the width deviation changing from −20 nm to +20 nm. When the waveguide width had an increase of $\Delta w$, the waveguide spacing would decrease by a value of $\Delta w$. As shown in Figure 9a, the ER maintained above 20 dB when $\Delta w$ varied from −20 to +20 nm for the quasi-TM mode at 1550 nm. For the quasi-TE mode, the fabrication tolerance was relatively low, and the ER dependence of the width deviation was more sensitive. The quasi-TE mode distribution was $x$-polarized, so the ER was more sensitive to the geometric parameters along the $x$-axis, such as the width of the waveguide $W$. In Figure 9b, we also simulated the maximum ER in the wavelength range of 1500–1600 nm, when the width deviation $\Delta w$ changed. The ER mainly fluctuated around the optimized value mentioned above, i.e., 23.9 dB for quasi-TM and 20.8 dB for quasi-TE. The center wavelength of the maximum ER drifted towards the shorter wavelength when the waveguide width increased by $\Delta w$. Conversely, the center wavelength moved to a longer one as the waveguide width was smaller than the design target. Accordingly, we could adjust the waveguide width slightly to realize that the applications of the other central wavelengths in the C-band, without changing any other geometric parameter.

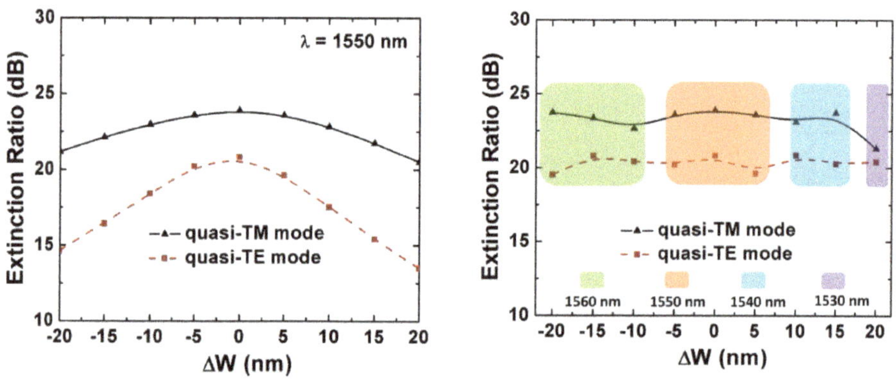


(a) (b)

**Figure 9.** (**a**) Extinction ratio at 1550 nm of the quasi-TM and quasi-TE mode as a function of the waveguide width deviation $\Delta w$; and (**b**) the maximum extinction ratio in the wavelength range of 1500–1600 nm of the quasi-TM and quasi-TE mode as a function of the waveguide width deviation $\Delta w$.

### *3.3. Multi-Slot Waveguide Polarization Beam Splitter*

We also investigated coupled multiple-slotted waveguides for the polarization beam splitting. The schematic of the proposed multi-slot-waveguide-based PBS is shown in Figure 10a. Based on the optimized results of the single-slot waveguide, the multi-slot waveguide maintained the same parameters $H$ = 800 nm, $W$ = 1050 nm, $H_u$ = 250 nm, and $d$ = 500 nm. The slot structure was composed of two $SiO_2$ layers with equal thicknesses, and the middle $Si_3N_4$ layer was embedded in between. Here, $H_m$ was the thickness of the middle $Si_3N_4$ layer. The electric field distributions of the quasi-TE and quasi-TM modes of the multi-slot $Si_3N_4$-based waveguide PBS at the wavelength of 1550 nm are shown in Figure 10b–e.

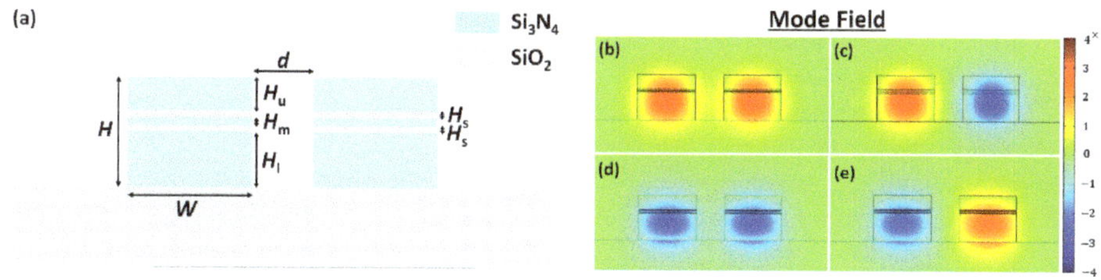


**Figure 10.** Schematic cross section (**a**) and the electric field distributions (**b–e**) of the multi-slot $Si_3N_4$-based waveguide PBS.

In such a multi-slot waveguide, the thickness of the $SiO_2$ and the inserted $Si_3N_4$ layers could determine the electromagnetic field distribution. The thicknesses of the two $SiO_2$ slot layers were chosen as $H_s$ = 20 nm, and the thickness of the middle $Si_3N_4$ layer was adjusted with a step of 15 nm. With the increase in the middle $Si_3N_4$ layer thickness, the power concentration was slightly decreased. The coupling between the two waveguides was, thus, increased. Consequently, the coupling length could be reduced as shown in Figure 11a. The middle $Si_3N_4$ layer thickness $H_m$ was chosen to be 25 nm to keep the coupling length ratio of the quasi-TE and quasi-TM modes ($L_{TE}/L_{TM}$) close to 2. The corresponding coupling length $L = L_{TE}$ equaled to 290.3 μm. Figure 11b shows the ERs of the quasi-TE and quasi-TM modes for the multi-slot-waveguide-based PBS from 1500 to 1600 nm. The ERs of the quasi-TM and quasi-TE modes were 24.0 dB and 21.0 dB at 1550 nm, respectively. We also investigated the fabrication tolerance of the multi-slot-waveguide-based PBS, the ERs were closed to the one based on the single-slot-waveguide for both modes at 1550 nm, as the width deviation $\Delta w$ changed from −20 nm to +20 nm. The wavelength dependence of the multi-slot PBS had a similar trend as the one of the single-slot PBS. From the previous research work, multiple slot waveguide could realize a higher confinement factor and power concentration [36,47], and further shorten the coupling length [48]. Compared with the single-slot-waveguide-based PBS, the properties of multi-slot-waveguide-based PBS did not show much improvement through our simulation, under the same overall size of the waveguides, materials, and simulation conditions, except for the number of slot layers.

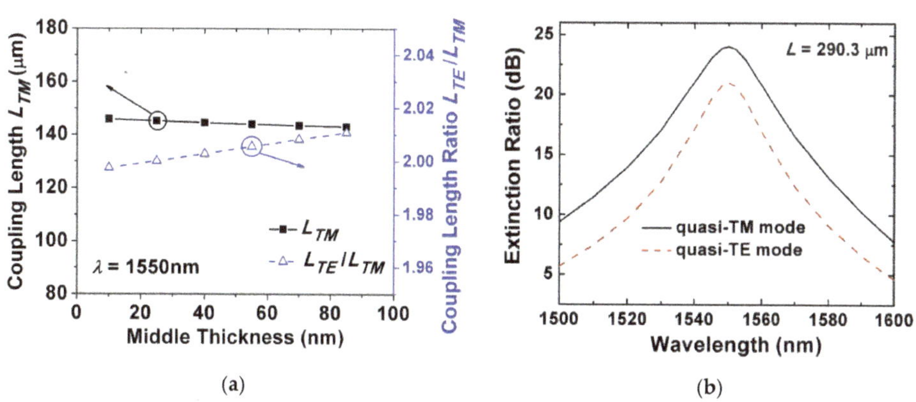


(a) (b)

**Figure 11.** (**a**) Coupling length of the quasi-TM mode and the coupling length ratio as a function of the middle $Si_3N_4$ layer thickness $H_m$ between the two $SiO_2$ slots; and (**b**) the extinction ratio of the quasi-TM mode at the bar port and the quasi-TE mode at the cross port for the multi-slot-waveguide-based PBS.

### 3.4. Tradeoff between the ER and Coupling Length

Figure 12 shows the relationship between the ER and coupling length for the designed PBSs based on the $Si_3N_4$ single- and multi-slot waveguide. The square and triangle symbols represent single- and multi-slot-waveguide-based PBSs, respectively. In Table 1, single-slot-waveguide-based PBS with different parameters were used to investigate the polarization splitting effect. The table shows the relatively lower ER in the cross port (quasi-TE mode) at the wavelength of 1550 nm. The PBSs with the structure parameters in the table provide ~2 coupling length ratio ($L_{TE}/L_{TM}$), and the offset was within 0.5%. In Table 2, we list the different parameters of the multi-slot-waveguide-based PBS, based on similar conditions. One can see that both the single- and multiple-waveguide-based PBS showed a similar trend, such that although a shorter coupling length can be achieved by using smaller waveguide space or adjusting other parameters, the corresponding ER would be worse. In sum, we could obtain a shorter coupling length at the sacrifice of the ERs for both types of PBSs.

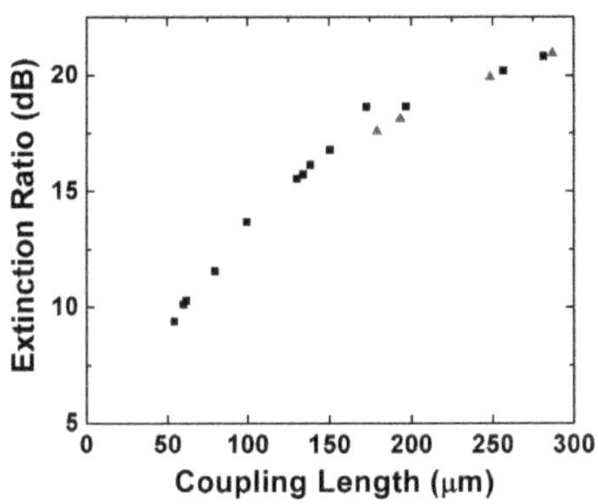


**Figure 12.** The tradeoff between the extinction ratios (ER) and coupling length of single-slot-waveguide-based PBS (square) and multi-slot-waveguide-based PBS (triangle).

**Table 1.** Optimized Parameters of the Single-Slot Waveguide-Based PBS.

| *H* (nm) | *W* (nm) | $H_s$ (nm) | $H_u$ (nm) | *d* (nm) | $L_{TE}/L_{TM}$ | $L_{TE}$ (μm) | *ER* (dB) |
|---|---|---|---|---|---|---|---|
| 620 | 1200 | 100 | 260 | 200 | 2.0015 | 54.20 | 9.38 |
| 780 | 1150 | 100 | 410 | 200 | 2.0022 | 59.74 | 10.12 |
| 800 | 1150 | 90 | 410 | 200 | 1.9985 | 61.69 | 10.29 |
| 1200 | 1200 | 300 | 260 | 200 | 2.0040 | 79.48 | 11.54 |
| 840 | 1100 | 90 | 300 | 300 | 1.9995 | 99.75 | 13.68 |
| 950 | 1100 | 150 | 300 | 350 | 2.0001 | 130.14 | 15.53 |
| 1000 | 1100 | 150 | 340 | 350 | 1.9991 | 134.13 | 15.72 |
| 950 | 1050 | 150 | 410 | 400 | 1.9974 | 138.59 | 16.12 |
| 950 | 1050 | 120 | 440 | 400 | 1.9995 | 150.57 | 16.78 |
| 720 | 1000 | 90 | 250 | 500 | 1.9963 | 172.86 | 18.62 |
| 800 | 1000 | 90 | 390 | 500 | 2.0001 | 196.96 | 18.63 |
| 830 | 1050 | 90 | 250 | 500 | 2.0027 | 256.65 | 20.19 |
| 800 | 1050 | 50 | 250 | 500 | 2.0035 | 281.46 | 20.82 |

**Table 2.** Optimized Parameters of the Multi-Slot Waveguide-Based PBS.

| $H$ (nm) | $W$ (nm) | $H_s$ (nm) | $H_u$ (nm) | $H_m$ (nm) | $d$ (nm) | $L_{TE}/L_{TM}$ | $L_{TE}$ (μm) | $ER$ (dB) |
|---|---|---|---|---|---|---|---|---|
| 840 | 1050 | 100 | 150 | 150 | 500 | 1.9990 | 179.29 | 17.58 |
| 840 | 1050 | 90 | 150 | 150 | 500 | 2.0041 | 193.41 | 18.12 |
| 800 | 1050 | 70 | 50 | 100 | 500 | 1.9966 | 248.44 | 19.93 |
| 830 | 1050 | 25 | 250 | 120 | 500 | 2.0017 | 286.89 | 20.95 |

## 4. Conclusions

In summary, we presented $Si_3N_4$ single- and multi-slot-waveguide-based PBSs. Compared with Si PBS, $Si_3N_4$-based devices were more suitable for high-power integrated photonic applications due to its much lower Kerr nonlinearity and negligible nonlinear loss. Furthermore, we analyzed the effect of the geometric parameters and propagation length on the coupling length and ER. The coupling length of the proposed $Si_3N_4/SiO_2$ horizontal-slot-waveguide-based PBS became 22.6% shorter than the one of coupled $Si_3N_4$ strip-waveguide-based PBS. The proposed structure could effectively reduce the length of the elements. The wavelength dependence of the quasi-TM and quasi-TE modes for these two designs were also investigated. The ERs for the quasi-TM and quasi-TE modes could both be >20 dB at 1550 nm for both the single-slot and multi-slot designs. The two designs showed good fabrication tolerance for up to ±20 nm width deviations. Finally, we studied the relationship between the ER and coupling length, and found that one needed to consider the tradeoff when optimizing the device performance or its size. To better serve the needs for different applications in on-chip high-power systems, one might introduce new design to further expand the operating bandwidth of the proposed PBS in the future.

**Author Contributions:** Conceptualization, Y.Y.; data curation, Y.F. and Z.W.; formal analysis, Y.F., C.B., and Y.Y.; funding acquisition, Y.Y.; investigation, Y.F., C.B., and Z.W., and Y.Y.; methodology, Y.F., C.B., and L.Z., and Y.Y.; project administration, Y.Y.; resources, Y.F. and Y.Y.; software, Y.F., C.B., and L.Z., and Y.Y.; supervision, C.B., Y.L., L.Z., Z.P., and Y.Y.; validation, Y.F. and Z.W.; visualization, Y.F.; writing—original draft, Y.F.; writing—review & editing, Y.F., C.B., Y.L., L.Z., H.H., Y.R., Z.P., and Y.Y. All authors have read and agreed to the published version of the manuscript.

**Funding:** This work was supported in part by the National Key Research and Development Program of China under Grant 2019YFB1803700, Grant 2018YFB0703500, and Grant 2018YFB0504400, in part by the National Natural Science Foundation of China (NSFC), under Grant 11774181 and Grant 61775107, and in part by the Fundamental Research Funds for the Central Universities, Nankai University under Grant 63191511.

**Conflicts of Interest:** The authors declare no conflicts of interest.

## References

1. Asghari, M.; Krishnamoorthy, A.V. Silicon photonics: Energy-efficient communication. *Nat. Photonics* **2011**, *5*, 268–270. [CrossRef]
2. Sun, C.; Wade, M.T.; Lee, Y.; Orcutt, J.S.; Alloatti, L.; Georgas, M.S.; Waterman, A.S.; Shainline, J.M.; Avizienis, R.R.; Lin, S.; et al. Single-chip microprocessor that communicates directly using light. *Nature* **2015**, *528*, 534–538. [CrossRef] [PubMed]
3. Ikeda, K.; Saperstein, R.E.; Alic, N.; Fainman, Y. Thermal and Kerr nonlinear properties of plasma-deposited silicon nitride/silicon dioxide waveguides. *Opt. Express* **2008**, *16*, 12987–12994. [CrossRef] [PubMed]
4. Tan, D.T.H.; Ikeda, K.; Sun, P.C.; Fainman, Y. Group velocity dispersion and self phase modulation in silicon nitride waveguides. *Appl. Phys. Lett.* **2010**, *96*, 061101. [CrossRef]
5. Gorin, A.; Jaouad, A.; Grondin, E.; Aimez, V.; Charette, P. Fabrication of silicon nitride waveguides for visible-light using PECVD: A study of the effect of plasma frequency on optical properties. *Opt. Express* **2008**, *16*, 13509–13516. [CrossRef]
6. Melchiorri, M.; Daldosso, N.; Sbrana, F.; Pavesi, L.; Pucker, G.; Kompocholis, C.; Bellutti, P.; Lui, A. Propagation losses of silicon nitride waveguides in the near-infrared range. *Appl. Phys. Lett.* **2005**, *86*, 121111. [CrossRef]

7. Bauters, J.F.; Heck, M.J.R.; John, D.; Dai, D.; Tien, M.-C.; Barton, J.S.; Leinse, A.; Heideman, R.G.; Blumenthal, D.J.; Bowers, J.E. Ultra-low-loss high-aspect-ratio $Si_3N_4$ waveguides. *Opt. Express* **2011**, *19*, 3163–3174. [CrossRef]
8. Li, Q.; Eftekhar, A.A.; Sodagar, M.; Xia, Z.; Atabaki, A.H.; Adib, A. Vertical integration of high-Q silicon nitride microresonators into silicon-on-insulator platform. *Opt. Express* **2013**, *21*, 18236–18248. [CrossRef]
9. Lacava, C.; Stankovic, S.; Khokhar, A.Z.; Bucio, T.D.; Gardes, F.Y.; Reed, G.T.; Richardson, D.J.; Petropoulos, P. Si-rich Silicon Nitride for Nonlinear Signal Processing Applications. *Sci. Rep.* **2017**, *7*, 22. [CrossRef]
10. Moss, D.J.; Morandotti, R.; Gaeta, L.A.; Lipson, M. New CMOS-compatible platforms based on silicon nitride and Hydex for nonlinear optics. *Nat. Photonics* **2013**, *7*, 597–607. [CrossRef]
11. Bristow, A.D.; Rotenberg, N.; Driel, H.M.V. Two-photon absorption and Kerr coefficients of silicon for 850–2200 nm. *Appl. Phys. Lett.* **2007**, *90*, 191104. [CrossRef]
12. Jalali, B. Nonlinear optics in the mid-infrared. *Nat. Photonics* **2010**, *4*, 506–508. [CrossRef]
13. Lin, Q.; Zhang, J.; Piredda, G.; Boyd, R.W.; Fauchet, P.M.; Agrawal, G.P. Dispersion of silicon nonlinearities in the near-infrared region. *Appl. Phys. Lett.* **2007**, *91*, 021111. [CrossRef]
14. Ferdous, F.; Miao, H.; Leaird, D.E.; Srinivasan, K.; Wan, J.; Chen, L.; Varghese, L.T.; Weiner, A.M. Spectral line-By-Line Pulse Shaping of On-Chip Microresonator Frequency Combs. *Nat. Photonics* **2011**, *5*, 1–2. [CrossRef]
15. Hirose, K.; Liang, Y.; Kurosaka, Y.; Watanabe, A.; Sugiyama, T.; Noda, S. Watt-class high-power, high-beam-quality photonic-crystal lasers. *Nat. Photonics* **2014**, *8*, 406–411. [CrossRef]
16. Sun, X.; Alam, M.Z.; Aitchison, J.S.; Mojahed, M. Compact and broadband polarization beam splitter based on a silicon nitride augmented low-index guiding structure. *Opt. Lett.* **2016**, *41*, 163–166. [CrossRef] [PubMed]
17. Wu, H.; Tan, Y.; Dai, D. Ultra-broadband high-performance polarizing beam splitter on silicon. *Opt. Express* **2017**, *25*, 6069–6075. [CrossRef]
18. Yue, Y.; Zhang, L.; Yang, J.-Y.; Beausoleil, R.G.; Willner, A.E. Silicon-on-insulator polarization splitter using two horizontally slotted waveguides. *Opt. Lett.* **2010**, *35*, 1364–1366. [CrossRef]
19. Ong, J.; Ang, T.Ş.; Sahin, E.; Pawlina, B.; Chen, G.F.R.; Tan, D.T.H.; Limb, S.T.; Png, C.E. Broadband silicon polarization beam splitter with a high extinction ratio using a triple-bent-waveguide directional coupler. *Opt. Lett.* **2017**, *42*, 4450–4453. [CrossRef]
20. Soudi, S.; Rahman, B.M.A. Compact Polarization Splitter by Using Identical Coupled Silicon Nanowires. *J. Lightwave Technol.* **2016**, *34*, 4168–4176. [CrossRef]
21. Yang, J.; Du, T.; Niu, B.; Li, C.; Qian, J.; Guo, L. A Bionic Polarization Navigation Sensor Based on Polarizing Beam Splitter. *IEEE Access* **2018**, *6*, 11472–11481. [CrossRef]
22. Hajireza, P.; Shi, W.; Bell, K.; Paproski, R.J.; Zemp, R.J. Non-interferometric photoacoustic remote sensing microscopy. *Light-Sci. Appl.* **2017**, *6*, e16278. [CrossRef] [PubMed]
23. Qin, Z.; Zhu, T.; Chen, L.; Bao, X. High Sensitivity Distributed Vibration Sensor Based on Polarization-Maintaining Configurations of Phase-OTDR. *IEEE Photonics Technol. Lett.* **2011**, *23*, 1091–1093. [CrossRef]
24. Li, X.; Ma, R.; Xia, Y. Magnetic Field Sensor Exploiting Light Polarization Modulation of Microfiber with Magnetic Fluid. *J. Lightwave Technol.* **2018**, *36*, 1620–1625. [CrossRef]
25. Zhou, K.; Pan, S.; Liu, S.; Hu, K. Fiber gyroscope with a double sensitivity employing a polarization splitter. *Opt. Lett.* **2013**, *38*, 1337–1339. [CrossRef] [PubMed]
26. Zhang, H.; Huang, Y.; Li, S.; Das., C.; Yu, M.; Lo, P.G.-Q.; Hong, M.; Thong, J. Polarization splitter using horizontal slot waveguide. *Opt. Express* **2013**, *21*, 3363–3369. [CrossRef]
27. Zhang, Y.; He, Y.; Wu, J.; Jiang, X.; Liu, R.; Qiu, C.; Jiang, X.; Yang, J.; Tremblay, C.; Su, Y. High-extinction-ratio silicon polarization beam splitter with tolerance to waveguide width and coupling length variations. *Opt. Express* **2016**, *24*, 6586–6593. [CrossRef]
28. Sun, X.; Aitchison, J.S.; Mojahedi, M. Realization of an ultra-compact polarization beam splitter using asymmetric MMI based on silicon nitride/silicon-on-insulator platform. *Opt. Express* **2017**, *25*, 8296–8305. [CrossRef]
29. Lin, S.; Hu, J.; Crozier, K.B. Ultracompact, broadband slot waveguide polarization splitter. *Appl. Phys. Lett.* **2011**, *98*, 151101. [CrossRef]
30. Cheng, N.-C.; Ma, Y.-F.; Fu, P.-H.; Chin, C.-C.; Huang, D.-W. Horizontal slot waveguides for polarization branching control. *Appl. Opt.* **2015**, *54*, 436–443. [CrossRef]

31. Yin, X.; Xiao, J. A Compact TE-Pass Polarizer for Silicon-Based Slot Waveguides. *IEEE Photonic Technol. Lett.* **2015**, *27*, 2071–2074.
32. Yin, X.; Xiao, J.; Sun, X. Design of a compact polarization demultiplexer for silicon-based slot waveguides. *Appl. Opt.* **2014**, *53*, 8305–8312.
33. Yin, X.; Xiao, J.; Sun, X. Compact Polarization Beam Splitter for Silicon-Based Slot Waveguides Using an Asymmetrical Multimode Waveguide. *J. Lightwave Technol.* **2014**, *32*, 4884–4890. [CrossRef]
34. Feng, J.; Akimoto, R.; Zeng, H. Asymmetric Silicon Slot-Waveguide-Assisted Polarizing Beam Splitter. *IEEE Photonic Technol. Lett.* **2016**, *28*, 1294–1297. [CrossRef]
35. Liu, Y.; Chang, L.; Li, Z.; Liu, L.; Guan, H.; Li, Z. Polarization beam splitter based on a silicon nitride–silica–silicon horizontal slot waveguide. *Opt. Lett.* **2019**, *44*, 1335–1338. [CrossRef]
36. Sun, R.; Dong, P.; Feng, N.; Hong, C.; Michel, J.; Lipson, M.; Kimerling, L. Horizontal single and multiple slot waveguides: Optical transmission at $\lambda = 1550$ nm. *Opt. Express* **2007**, *15*, 17968–17972. [CrossRef]
37. Palik, E.D. *Handbook of Optical Constants of Solids*; Academic Press: San Diego, CA, USA, 1998.
38. Baak, T. Silicon oxynitride; a material for GRIN optics. *Appl. Opt.* **1982**, *21*, 1069–1072. [CrossRef]
39. Liang, T.-K.; Tsang, H.-K. Nonlinear Absorption and Raman Scattering in Silicon-on-Insulator Optical Waveguides. *J. Sel. Top. Quantum Electron.* **2004**, *10*, 1149–1153. [CrossRef]
40. Roeloffzen, C.G.H.; Hoekman, M.; Klein, E.J.; Wevers, L.S.; Timens, R.B.; Marchenko, D.; Geskus, D.; Dekker, R.; Alippi, A.; Grootjans, R.; et al. Low-Loss $Si_3N_4$ TriPleX Optical Waveguides: Technology and Applications Overview. *J. Sel. Top. Quantum Electron.* **2018**, *24*, 1–21. [CrossRef]
41. Zhang, L.; Agarwal, A.M.; Kimerling, L.C.; Michel, J. Nonlinear Group IV photonics based on silicon and germanium: From near-infrared to mid-infrared. *Nanophotonics* **2014**, *3*, 247–268. [CrossRef]
42. Grassani, D.; Tagkoudi, E.; Guo, H.; Herkommer, C.; Yang, F.; Kippenberg, J.T.; Brès, C.-S. Mid infrared gas spectroscopy using efficient fiber laser driven photonic chip-based supercontinuum. *Nat. Commun.* **2019**, *10*, 1553. [CrossRef] [PubMed]
43. Afshar, S.; Monro, T.M. A full vectorial model for pulse propagation in emerging waveguides with subwavelength structures part I: Kerr nonlinearity. *Opt. Express* **2009**, *17*, 2298–2318. [CrossRef] [PubMed]
44. Kim, K.S.; Stolen, R.H.; Reed, W.A.; Quoi, K.W. Measurement of the nonlinear index of silica-core and dispersion shifted fibers. *Opt. Lett.* **1994**, *19*, 257–259. [CrossRef] [PubMed]
45. Dinu, M.; Quochi, F.; Garcia, H. Third-order nonlinearities in silicon at telecom wavelengths. *Appl. Phys. Lett.* **2003**, *82*, 2954–2956. [CrossRef]
46. Bozinovic, N.; Yue, Y.; Ren, Y.; Tur, M.; Kristensen, P.; Huang, H.; Willner, A.E.; Ramachandran, S. Terabit-Scale Orbital Angular Momentum Mode Division Multiplexing in Fibers. *Science* **2013**, *340*, 1545–1548. [CrossRef]
47. Feng, N.-N.; Michel, J.; Kimerling, L.C. Optical Field Concentration in Low-Index Waveguides. *IEEE J. Quantum Electron.* **2006**, *42*, 885–890. [CrossRef]
48. Jiang, W.; Sun, X.; Rahman, B.M.A. Compact and fabrication-tolerant polarization splitter based on horizontal triple-slot waveguide. *Appl. Opt.* **2017**, *56*, 2119–2126. [CrossRef]

sensors

MDPI

*Article*

# Accurately Shaping Supercontinuum Spectrum via Cascaded PCF

**Jifang Rong [1], Hua Yang [1,2,*] and Yuzhe Xiao [3]**

1 College of Computer Science and Electronic Engineering, Key laboratory for Micro/Nano Optoelectronic Devices of Ministry of Education, Hunan University, Changsha 410082, Hunan Province, China; rongjifang@hnu.edu.cn
2 Synergetic Innovation Center for Quantum Effects and Application, Hunan Normal University, Changsha 410082, Hunan Province, China
3 Department of Electrical Engineering, University of Wisconsin-Madison, Madison, WI 53715, USA; xiao5@wisc.edu
* Correspondence: huayang@hnu.edu.cn

Received: 11 March 2020; Accepted: 23 April 2020; Published: 27 April 2020

check for updates

**Abstract:** Shaping is very necessary in order to obtain a wide and flat supercontinuum (SC). Via numerical simulations, we accurately demonstrated shaping the SC using the fiber cascading method to significantly increase the width as well as the flatness of the spectrum in silica photonic crystal fiber (PCF). The cascaded PCF contains two segments, each of which has dual zero-dispersion frequencies (ZDFs). The spectral range of the SC can be expanded tremendously by tuning the spacing between the two ZDFs of the first segmented cascaded PCF. Increasing the pump power generates more solitons at the red edge, which accelerates solitons trapping and improves the spectral flatness of the blue edge. Furthermore, cascading the second segmented PCF by choosing appropriate fiber parameters ensures the flatness of the red end of SC. Therefore, a cost-effective alternative method for broad and flat supercontinuum generation in the near-infrared range is proposed here, which can be implemented easily in any photonics laboratory, where dual ZDFs PCFs are commonly found.

**Keywords:** spectral shaping; photonic crystal fiber cascade; zero-dispersion frequencies spacing; supercontinuum generation

## 1. Introduction

Since Alfano and Shapiro firstly reported supercontinuum (SC) in 1970 in bulk glass [1], it has been explosively used in various fields, including spectroscopy [2], optical coherent tomography [3], metrology [4], and so on. Up to now, both numerical simulations and experiments have shown that SC can be obtained by the extremely nonlinear broadening of narrow-band incident pulses, whose widths vary from femtosecond to picosecond [5]. As is well known, the distinguishing features of SC are related to spectral range and spectral flatness, both of which are essential to promoting the technical applications of SC. The femtosecond pulse pumped in the anomalous dispersion region can lead to abundant spectral components and relatively broadband spectrum, via soliton dynamics [5–7]. Therefore, most of the studies on SC have chosen femtosecond pulses as the pump sources. In this case, stimulated Raman scattering (SRS) and high-order dispersion (HOD) are crucial to the evolution of ideally periodic high-order soliton [8–11]. Accordingly, the high-order soliton splits and sheds away energy to resonant nonsolitonic radiation, depending on the phase-matching condition [5]. When the group velocity matches between solitons and dispersion waves (DWs), stable trapping dispersion waves (Trapping DWs) can be generated under the effect of four-wave mixing (FWM) [12]. The contribution of Raman solitons, DWs, and Trapping DWs to SC has been investigated thoroughly in photonic crystal fiber (PCF) [13–15].

Recently, studies on ultrafast dynamics in mode-locked lasers have broadened our horizons in complex nonlinear systems and contributed to laser design. It is convenient for us to obtain a femtosecond pulse with few-cycle and high peak power as a pump source for supercontinuum generation (SCG). That further revives interests in broad and flat SCG [16,17]. Numerous studies have shown that red-shifted dispersion wave (R-DW) and blue-shifted dispersion wave (B-DW) can be generated simultaneously in photonic crystal fibers (PCFs) with two or more zero-dispersion frequencies (ZDFs) to improve spectral width. For example, the authors of [18–25] show that the DWs and Trapping DWs can contribute to the generation of wider or smoother SC. Poletti et al. have mathematically explained the soliton spectral tunneling effect in an index-guiding holey fiber with an adjustable group velocity dispersion (GVD) barrier over a large frequency range, which can be used to adjust the spectral width [26]. All these results show that increasing the number of ZDFs will increase spectral width, while spectral flatness will decrease. Subsequently, many works have shown that the cascaded method can improve the spectral flatness or spectral width of SC [27–29]. For example, Guo et al. have shown that soliton spectral tunneling effect in a segmented cascaded PCF with engineered dispersion can prompt the soliton pulse to travel over a wide spectrum range [27]. Chunyu Guo et al. have experimentally demonstrated a flat SCG in cascaded fibers pumped by a continuous wave laser, but the spectrum width is just 420 nm at −10 dB level [28]. Saili Zhao et al. have reported harnessing rogue waves in cascaded PCFs for SCG [29]. Haihuan Chen et al. have reported the fabrication of cascaded PCF tapers in monolithic design, and demonstrated flat broadband SCG in cascaded PCF tapers [30]. All of these cascaded methods have improved spectral span or spectral flatness to some extent, but few works have acquired large bandwidth and high flatness simultaneously. As a result, much research about SCG has focused on chalcogenide PCFs whose bandwidth can reach microscale and flatness is fine too, due to its wider transmission window into long-wavelength range and higher nonlinearity [31–34]. However, the poor thermal stability of this material is not conducive to the widespread industrial application of SC, not replacing the position of silica PCF in SCG. More recently, the research on SCG has grown more and more extensive, even extending to the generation of SCG by dark solitons [35–37]. Therefore, it is meaningful to accurately shape the SC to obtain a wide and flat supercontinuum spectrum in silica PCF.

In this paper, we propose a method of accurately shaping SC via cascaded PCF. Here, each PCF needs to be tapered and spliced to the next stage; the splicing loss is directly ignored to get a better physical understanding of various effects. In the first part of our simulation, several PCFs with different ZDF spacing are used as the propagation medium of the first segmented cascaded PCF. We found that increasing the ZDF spacing is beneficial for extending the SC by R-DW and trapping blue-shifted dispersion wave (Trapping B-DW). Although increasing pump power is adequate to generate more solitons, accelerating solitons capture will improve the spectral flatness of the blue edge. Then, in the second part of our simulation, we analyze the output spectrum of the first segment and appropriately choose the fiber parameters of the second segmented cascaded PCF, realizing the accurate shaping of the red edge spectrum.

## 2. Propagation Model

Femtosecond pulse propagation in cascaded single-mode PCF can be described by the generalized nonlinear Schrödinger equation(G-NLSE) [5,18]:

$$\begin{aligned}\frac{\partial A(z,T)}{\partial z} &= \sum_{k\geq 2} \frac{i^{k+1}\beta_k}{k!}\frac{\partial^k A}{\partial T^k} \\ &+ i\gamma\left(1+\frac{i}{\omega_0}\frac{\partial}{\partial T}\right) \times \left([A(z,T)\int_{-\infty}^{+\infty} R(T')|A(z,T-T')|^2 dT' + i\Gamma_R(z,T)]\right),\end{aligned} \tag{1}$$

where is $A(z,T)$ complex amplitude of time-domain pulse envelope; $T$ and $z$ are, respectively, the time parameter related to group velocity at the central frequency $\omega_0$ and propagation distance of optical pulse; $\gamma$ is the nonlinear coefficient of the optical fiber; $\beta_k$ is the kth-order dispersion coefficient at the central frequency $\omega_0$; $\Gamma_R$ describes thermally-driven spontaneous Raman scattering [5]; and $R(T)$ is the nonlinear response function:

$$R(T) = (1 - f_R)\delta(T) + f_R h_R(T), \tag{2}$$

which includes Raman contribution and Kerr components. We use $f_R$ = 0.18 and $h_R(T)$ determined from the fused silica Raman cross section [5,10]. Dispersion effects relate to the first term on the right side hand of Equation (1) as well as nonlinear optical effects correspond to the second one including self-phase modulation (SPM), SRS, and self-steepening (SS). The linear loss of fiber is neglected because only a short transmission distance is considered. We adopted the non-chirped Gaussian pump pulse propagating in cascaded PCF: $A(0,T) = \sqrt{P_0}\exp(T^2/2T_0^2)$.

## 3. Numerical Simulation and Discussion

We adopt the split-step Fourier method with adaptive step size control to solve the G-NLSE [10,38–41]. Based on the G-NLSE, we will discuss the evolution process of the femtosecond pulse pumped in the anomalous dispersion region of the cascaded PCF. In the first section, PCFs, specifically PCF1, PCF2, and PCF3, which can be obtained by properly setting the geometry of the air–silica holey fiber according to the method proposed in [42,43], are candidates for the first segmented cascaded PCF to demonstrate how the spacing of ZDFs influences the output spectrum. The dispersion profile and relative group delay of PCF1, PCF2, and PCF3 as a function of the frequency are depicted in Figure 1a,b and the dispersion parameters of them are shown in Table 1. In Figure 1a, we can see that the two ZDFs divide the whole dispersion profile into three regions: the left and right regions are normal dispersion regions, and the middle part is the anomalous dispersion region. Meanwhile, the first (high-frequency) and the second (low-frequency) ZDFs provide positive and negative dispersion slopes, respectively. Here, $d_{PCF1}$ = 125 THz, $d_{PCF2}$ = 127 THz, and $d_{PCF3}$ = 178 THz represent the frequency spacing between the two ZDFs of PCF1, PCF2, and PCF3, respectively. The parameters of input pulse are chosen as follows; peak power is $P_0$ = 1660 W, initial pulse width is $T_0$ = 50 fs, which has been proved to be an appropriate temporal width to maintain the coherence of the input pulse [5]. Pump wavelengths are all at $\lambda$ = 795 nm (375 THz). Assuming that the effective mode areas of the three different fibers are the same, the nonlinear coefficient of the PCFs is $\gamma = 0.124\ \mathrm{W^{-1}m^{-1}}$. It can be seen from Figure 1a that second-order dispersion parameters at pump central frequency of the three PCFs are equal ($\beta_2 = -0.80352\ \mathrm{ps^2/km}$) to ensure the same solitons order $N$ under the same incident conditions. The soliton order $N$ of the input pulse is determined by both pulse and fiber parameters through $N^2 = L_D/L_{NL}$. Here, $L_D = T_0^2/|\beta_2|$ and $L_{NL} = 1/\gamma P_0$ are the characteristic dispersive and nonlinear length, respectively.

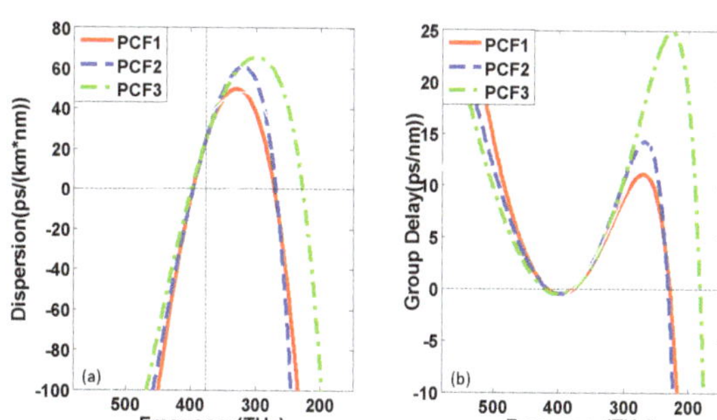


**Figure 1.** (**a**) The dispersion curve and (**b**) relative group delay curve of the adopted PCFs. Among them, red solid line, blue dashed line, and green dash-dot line represent PCF1, PCF2, and PCF3 respectively. The vertical dashed line in panel (**a**) marks the pump wavelength $\lambda$ = 795 nm (375 THz).

**Table 1.** The dispersion parameters of PCF1, PCF2, and PCF3.

| $\beta_k(ps^k)$/km | PCF1 | PCF2 | PCF3 |
|---|---|---|---|
| $\beta_2$ | $-8.0352$ | $-8.0352$ | $-8.0352$ |
| $\beta_3$ | $6.2527 \times 10^{-2}$ | 0.06662 | 0.060473 |
| $\beta_4$ | $6.5314 \times 10^{-5}$ | $0.9517 \times 10^{-5}$ | $-7.7942 \times 10^{-6}$ |
| $\beta_5$ | $-3.3086 \times 10^{-7}$ | $-2.2194 \times 10^{-7}$ | $-1.6566 \times 10^{-7}$ |
| $\beta_6$ | $1.2379 \times 10^{-9}$ | $7.7245 \times 10^{-10}$ | $1.9090 \times 10^{-10}$ |
| $\beta_7$ | $-4.3681 \times 10^{-12}$ | $-1.0368 \times 10^{-11}$ | $2.1444 \times 10^{-12}$ |
| $\beta_8$ | $1.7687 \times 10^{-14}$ | $9.6356 \times 10^{-14}$ | $1.0424 \times 10^{-14}$ |
| $\beta_9$ | $-8.9723 \times 10^{-17}$ | $-4.3407 \times 10^{-16}$ | $-2.1255 \times 10^{-16}$ |
| $\beta_{10}$ | $4.9736 \times 10^{-19}$ | $5.0646 \times 10^{-19}$ | $1.9090 \times 10^{-19}$ |
| $\beta_{11}$ | $-2.5042 \times 10^{-21}$ | $4.8947 \times 10^{-21}$ | $-5.4262 \times 10^{-22}$ |
| $\beta_{12}$ | $5.7033 \times 10^{-24}$ | $-2.9292 \times 10^{-23}$ | $-1.2854 \times 10^{-23}$ |

### 3.1. Pulse Evolution in PCFs with Different ZDF Spacing

To deeply discuss the influence of ZDF spacing in the first segmented cascaded PCF on SC, in Figure 2, we compare the evolution process of pulse spectrum with different ZDF spacing in 40 cm PCFs as well as the corresponding numerical cross-correlation frequency-resolved optical gating (X-FROG) trace at the output of PCFs. As shown in Figure 2a, during the initial stage of pulse propagation, the symmetrical spectrum broadening by SPM is observed. Subsequently, the higher-order soliton splits into fundamental solitons with different peak powers and almost simultaneously transfers energy to phase-matched B-DW. After soliton fission, the ejected fundamental solitons continuously red-shift to the low-frequency region under the effect of Raman induced soliton self-frequency shifting (SSFS). The first ejected soliton has a higher intensity and faster red-shifted speed than other solitons and experiences greater frequency downshift relative to the central frequency. This can be verified by spectrum evolution in Figure 2a. Besides, there is a distinct phenomenon by comparing Figure 2(a1–a3), with the increasing of ZDF spacing, the Trapping B-DW has much richer spectral components to broaden the spectrum. The ZDF spacing of PCF3 is larger than that of PCF1 and PCF2, so the first ejected red-shifted soliton in PCF3 has a larger red-shift distance to arrive at the second ZDF. From Figure 1b, we know that the time delay increases and the group velocity decreases with soliton red-shift. Therefore, the first ejected red-shifted soliton can gradually mix with B-DW from the low-frequency to the high-frequency in PCF3, leading to form stable Trapping B-DW at the trailing edge of B-DW. This process completely broadens the spectrum of the blue edge, as can be seen in Figure 2(b3). In contrast with that in PCF3, in PCF1, the Trapping B-DW generated by the interaction between the low-frequency part of B-DW and the first ejected soliton superposes with B-DW in frequency and does not broaden the spectrum, as is shown in Figure 2(b1), while the Trapping B-DW of PCF2 in Figure 2(b2) just broadens the spectrum a little. Besides, in all these three cases, an obvious gap (marked by red arrow) appears in the spectrum region of the B-DW as a result of the FWM effect transferring energy from B-DW to the Trapping B-DW, as shown in Figure 2a. However, trapping red-shifted dispersion waves (Trapping R-DWs) has no obvious effect to SC spectrum broadening. In our simulations, we find that when the input condition remains unchanged, that is, when the soliton order $N = 8$, Trapping B-DW can broaden the spectrum only if $d_{PCF1} > 125$ THz is satisfied. Therefore, we can conclude that the Trapping B-DW can broaden the spectrum only when the ZDF spacing reaches a certain degree.

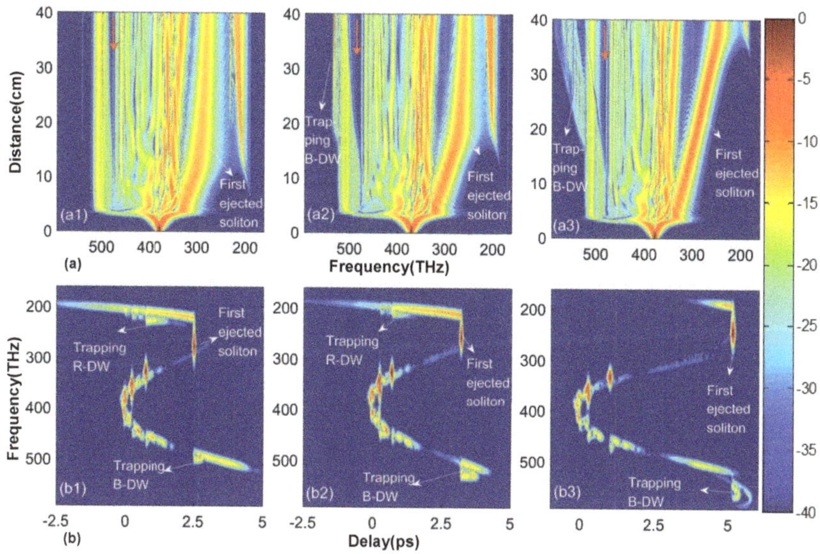


**Figure 2.** Evolution of the spectral characteristic of the incident optical pulse as a function of fiber length in three PCFs with different zero-dispersion frequency (ZDF) spacing and the corresponding spectrograms at 40 cm. Panels (**a1**,**b1**) correspond to $d_{PCF1}$ = 125 THz, Panels (**a2**,**b2**) correspond to $d_{PCF2}$ = 127 THz, Panels (**a3**,**b3**) correspond to $d_{PCF3}$ = 178 THz. Here, $d_{PCF1}$, $d_{PCF2}$ and $d_{PCF3}$ stand for the frequency spacing between two ZDFs of PCF1, PCF2, and PCF3, respectively. Red arrows in Panels (**a1**–**a3**) represent spectrum gaps in B-DW.

Until the central frequency of the first ejected soliton approaches to the second ZDF of PCFs with a negative dispersion slope, Trapping B-DW no longer keeps broadening the spectrum in the blue region, as shown in Figure 2(a2,a3). With the increasing of ZDF spacing, red-shifted soliton that greatly extends the red edge propagates over a wider spectral range to reach the second ZDF where R-DW radiates out according to the phase-matching condition. Thus, the larger the ZDF spacing is, the greater the translation distance and frequency downshift are needed to radiate R-DW, which is consistent with the results shown in Figure 2a. Therefore, the output spectral width of SC is strongly influenced by the ZDF spacing, making it possible to manipulate the SCG process by adjusting the ZDF spacing.

From the above analysis, we find that a large ZDF spacing in PCFs can improve the width of the output spectrum by both R-DW and Trapping B-DW. This, however, leads to a poor flatness of the SC spectrum. The authors of [18] indicate that increasing the pump power is adequate to generate more solitons, which can accelerate soliton trapping to improve spectral flatness. For this purpose, Figure 3 shows the evolution of the optical pulse in PCF3 when the pump power is increased to 2600 W. An important phenomenon observed in Figure 3c is that the second ejected red-shifted soliton overlaps with B-DW and R-DW in the time domain to generate Trapping B-DW and Trapping R-DW by FWM effects. Subsequently, the four pulses co-propagate in the PCF and the generated Trapping B-DW well fills the spectrum gap in B-DW (see the red rectangle region in Figure 3a). That is to say, by reasonable selection of the spacing between ZDFs, a pump power not only can broaden the width of the SC but also can ensure the flatness of the blue edge.

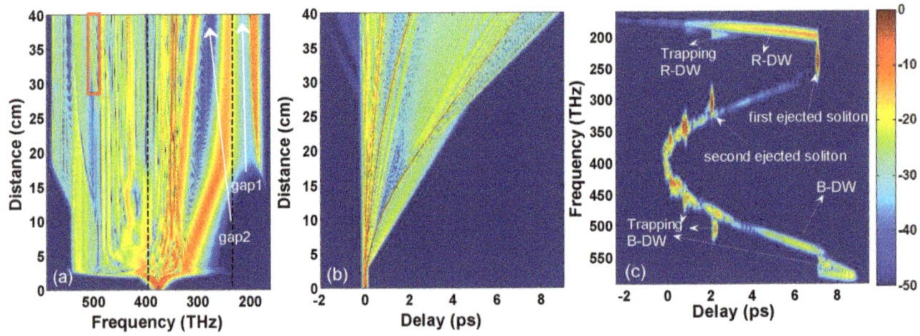


**Figure 3.** The evolution of the (**a**) spectral and (**b**) temporal signal as a function of the fiber length, and (**c**) the spectrogram at the output of PCF3 when the pump power is increased up to 2600 W. The black vertical dashed lines in panel (**a**) mark the two ZDFs of PCF3. The red rectangle area represents the place Trapping B-DW well fills the spectrum gap in B-DW.

*3.2. Pulse Evolution in Cascaded PCF*

From the output spectrum of PCF3 in Figure 3, we find that the flatness of the SC is better in the blue region than the red region where the energy is mainly distributed in a few solitons and DWs. There are two reasons for this: First, the Trapping R-DW superimposes with R-DW in the low-frequency region so that the spectrum gap (named gap1, see gap1 in Figure 3a) between R-DW and the first ejected soliton is not filled. Second, the large group velocity difference between the two fundamental solitons leads to a large spectrum gap (named gap2, see gap2 in Figure 3a) between them. The cascaded scheme is given as follows. When the pump pulse is transferred from the first segmented fiber (PCF3) into the second segmented fiber, the second ejected fundamental soliton in PCF3 turns into a higher-order soliton in the second segmented fiber. Then, the higher-order soliton will be split into fundamental soliton, so the red-shifted fundamental soliton and the corresponding R-DW fill the spectrum gap2 and the spectrum gap1, respectively. According to the definition of soliton order $N^2 = \gamma P_0 T_0^2/|\beta_2|$ , in order to realize the change from fundamental soliton to higher-order soliton of the second ejected soliton in the second segmented PCF (specifically, a larger value of $N$). The GVD at central frequency $\omega_0$ should be less than that of PCF3, or the nonlinear coefficient should be greater than that of PCF3 in the second segmented cascaded PCF. Moreover, the second ZDF of the second segmented fiber needs to be located in the spectrum range of the first ejected soliton at the output end of PCF3, so that the first ejected soliton cannot maintain the soliton shape and is dispersed, which further smoothens the SC spectrum. Based on the above requirements, we choose the appropriate PCF4, whose group delay and group dispersion curves are shown in Figure 4a and the dispersion parameter at the center frequency $\omega_0$ of the pump is $\beta_2 = -11.2346\, ps^2/\text{km}$, $\beta_3 = -0.0348\, ps^3/\text{km}$, $\beta_4 = 2.7608 \times 10^{-6}\, ps^4/\text{km}$, $\beta_5 = 6.9342 \times 10^{-8}\, ps^5/\text{km}$, $\beta_6 = 5.4154 \times 10^{-10}\, ps^6/\text{km}$, $\beta_7 = -4.2448 \times 10^{-12}\, ps^7/\text{km}$, $\beta_8 = 2.3240 \times 10^{-14}\, ps^8/\text{km}$, $\beta_9 = -8.3230 \times 10^{-17}\, ps^9/\text{km}$, $\beta_{10} = 1.9096 \times 10^{-19}\, ps^{10}/\text{km}$, $\beta_{11} = -2.5893 \times 10^{-22}\, ps^{11}/\text{km}$, $\beta_{12} = 1.5974 \times 10^{-25}\, ps^{12}/\text{km}$. The two ZDFs are 429 THz (699 nm) and 251 THz (1195 nm), respectively, and the nonlinear coefficient is $\gamma = 0.193\ \text{W}^{-1}\text{m}^{-1}$.

The total length of the cascaded PCF constructed by PCF3 and PCF4 is 60 cm, in which the first 40 cm is PCF3 and the remaining 20 cm is PCF4. The frequency domain and temporal domain evolution of the pump pulse in this cascaded PCF are shown in Figure 4b,c. Here, the pumping conditions are consistent with those in PCF3 in Figure 3. Thus, the evolution of the first 40 cm is the same as that in Figure 3. The changes in dispersion and relative group delay are presented as a bend of the solitons in the temporal domain at a distance of 40 cm, which is shown in Figure 4c, different from Figure 3b. In the frequency domain, only a small part of the energy of the first ejected soliton in PCF3 is still in the anomalous dispersion region. In contrast, most of the energy is in the normal dispersion region the balance between GVD and SPM is broken in PCF4. The spectral broadening of the partial energy in the anomalous dispersion region is much wider than it is in the normal dispersion region. This is due to the chirp generated by the SPM in the anomalous dispersion region, which causes the pulse to compress in the time domain and broaden in the spectral domain. In the normal dispersion region, the process is just the opposite: the partial energy just slightly broadens in the frequency domain due to SPM and diverges rapidly in the temporal domain due to different GVD. In a word, this process enriches the spectrum components which improves the flatness of the red region to a certain extent.

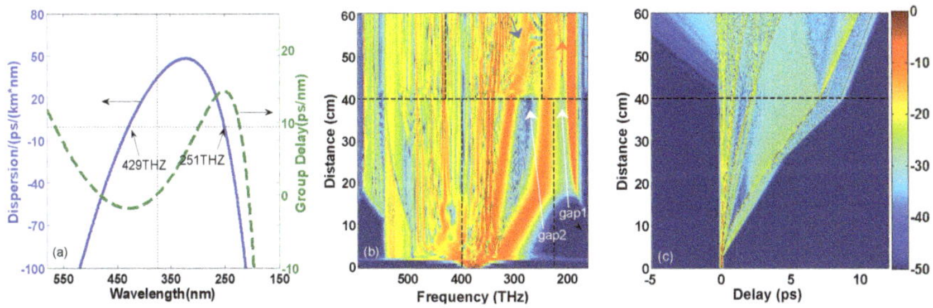


**Figure 4.** (**a**) The dispersion curve and relative group delay curve of PCF4. Evolution of the (**b**) spectral and (**c**) temporal signal of the input pulse as a function of propagation distance. The black vertical dashed lines in panels (**a**,**b**) mark the pump position of 375 THz and the two ZDFs of the first and second segmented cascaded PCF, respectively. The black horizontal dashed lines in panels (**b**,**c**) mark where PCF3 and PCF4 are connected. The red arrow and the blue arrow point to two filled spectrum gaps respectively.

However, the second ejected soliton in PCF3 plays a major role in shaping the SC, which turns into a higher-order soliton, but its central frequency is far from the two ZDFs of PCF4. Therefore, only SRS acts as the perturbation on its transmission and the HOD effect on the spectrum is negligible. Accordingly, only one fundamental soliton (marked by the blue arrow in Figure 4b) is split by the secondly ejected soliton, owing to its weak energy, as shown in Figure 4b. Under the effect of cascaded SRS, SSFS in the subsequent propagation process efficiently fills the spectrum gap2. As the SSFS approaches the second ZDF of PCF4, phase-matched R-DW (marked by the red arrow in Figure 4b) is gradually scattered out, filling gap1. With the increase in propagation distance, R-DW keeps tilting towards the higher frequency and its energy increases continuously in PCF4. When the spectrum recoil effect is balanced with SSFS, SSFS disappears. From the spectrum in Figure 4b, we can clearly see the shaping effect in the low-frequency region.

In Figure 5, from bottom to top, the three curves corresponding to the output spectrum in Figures 2(a3), 3a, and 4b, represent without increasing the pump power ($P_0$ = 1660 W, $N$ = 8), increasing the pump power ($P_0$ = 2600 W, $N$ = 10), and after cascade ($P_0$ = 2600 W, $N$ = 10), respectively. We use the standard deviation $\sigma = \sqrt{\frac{1}{N}\sum_{n=1}^{N}(x_i - \mu)^2}$ to describe the flatness of the output spectrum, where $x_i$ is the logarithmic representation of the output power at each frequency sampling point, $\mu$ is the average of $x_i$. The calculated standard deviations of the three mentioned cases above are 11.5, 9.8, and 7.6, respectively, which shows that the flatness from bottom to top is becoming increasingly optimal. This result is consistent with the conclusion in Figure 5. Compared with the black pentagram-symbol line, the flatness of the high-frequency part of the blue square-symbol line improves considerably; the two spectrum gaps on the black pentagram-symbol line indicated by the two green arrows are filled by the Trapping DWs after increasing the pump power. Relative to the blue square-symbol line, there is further improvement in the flatness of the low-frequency part of the red circle-symbol line. The un-flat areas indicated by the orange and the black arrow on the blue square-symbol line are improved by the red-shifted fundamental soliton and the corresponding R-DW in the second segmented cascaded fiber, respectively.

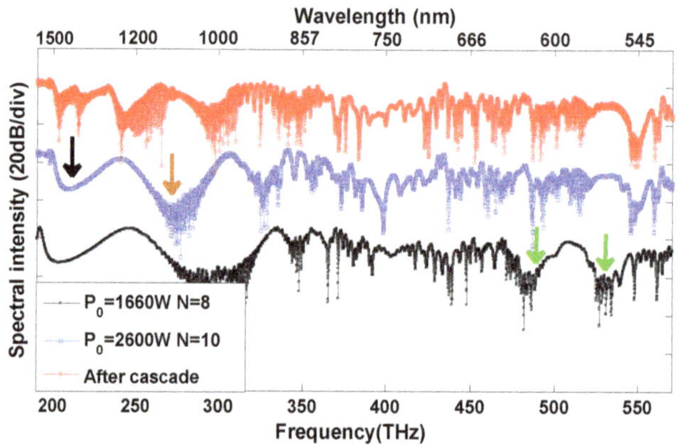


**Figure 5.** The spectral intensity of three different cases. The black pentagram-symbol line, blue square-symbol line, and red circle-symbol line represent the output spectrum shown in Figures 2(a3), 3a, and 4b, representing without increasing the pump power ($P_0$ = 1660 W, $N$ = 8), increasing the pump power ($P_0$ = 2600 W, $N$ = 10), and after cascade ($P_0$ = 2600 W, $N$ = 10), respectively. The three curves are staggered on the vertical axis only for comparison rather than actual intensity.

In order to clearly demonstrate the evolution of the pump pulse in the cascaded PCF, we used the X-FROG trace to record the time-frequency diagram of the pulse in nine selected locations. For a short propagation distance of 2 cm, the pump pulse rapidly symmetrical stretches in frequency by SPM (Figure 6a). Propagating from 2 to 5 cm, the pump pulse breaks due to the effects of HOD and SRS, transferring energy to B-DW in the blue region, as shown in Figure 6b. Until reaching 12 cm, the bandwidth of B-DW remains unchanged and its tail starts to tilt back because the high-frequency components of B-DW propagate slower than the low-frequency components in the normal dispersion region. The head edge of B-DW overlaps with the first ejected red-shifted soliton, as shown in Figure 6c. As the distance increases, more high-frequency components of B-DW interact with the slowing down fundamental soliton (see Figure 6d). It is apparent that the Trapping DW moves continuously to the backward of B-DW to maximize the spectrum of the blue region at 18 cm, where the soliton closes to the second ZDF and begins to emit R-DW. When propagating to 28 cm, the second fundamental soliton in front of the first ejected high-intensity soliton effectively mixes with B-DW, which generates the Trapping B-DW and fills in the spectral gap in B-DW frequency region to flatten the blue edge of SC, as seen in Figure 6e. When the pulse propagates for 45 cm in PCF4 (see Figure 6f), the first ejected soliton in PCF3 is dispersed and curved in shape as a result of the change in GVD. Meanwhile, the second ejected soliton in PCF3 splits into a red-shifted fundamental soliton and emits an extra-low energy R-DW simultaneously, which is generated due to the poorly met phase-matching condition. Between 45 and 50 cm, the newly generated fundamental soliton moves close to the second ZDF of PCF4, then R-DW is amplified (Figure 6g). In the last two figures, Figure 6h,i, the frequency shift of the fundamental soliton stops as the energy of the R-DW keeps increasing. The inclination of DWs increases as the pulse propagates further and the flatness of the red region is improved.

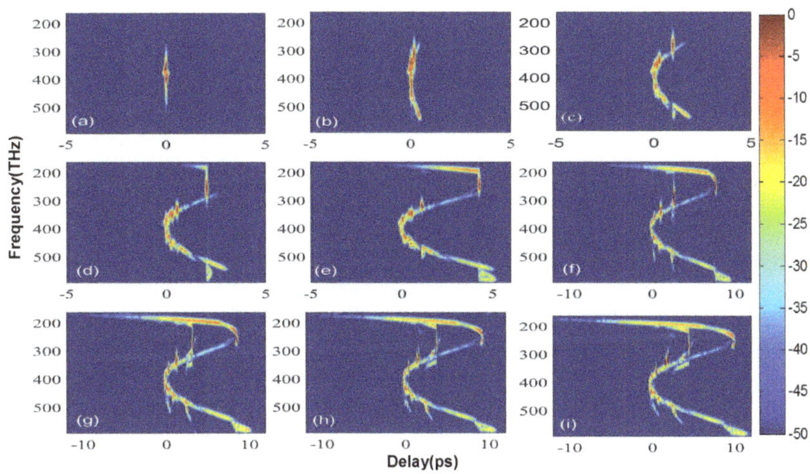

**Figure 6.** The spectrograms at different propagation distances z: (**a**) z = 2 cm, (**b**) z = 5 cm, (**c**) z = 12 cm, (**d**) z = 18 cm, (**e**) z = 28 cm, (**f**) z = 45 cm, (**g**) z = 50 cm, (**h**) z = 55 cm, (**i**) z = 60 cm.

## 4. Statistics with Spectral Flatness

This section mainly discusses the influence of several main factors on the output spectrum flatness of cascaded fiber. In Figure 7, we show average flatness of 100 individual simulations when changing the position of the second ZDF in the second segmented PCF, fiber length, and pump power, and the calculated results are, respectively, demonstrated in Figure 7a–c. The average flatness is defined as $\sigma_{mean} = \frac{\sum_{n=1}^{N} \sigma_i}{N}$, where $\sigma_i$ is the standard deviation of single simulation. The lower the value of $\sigma_{mean}$, the better the flatness.

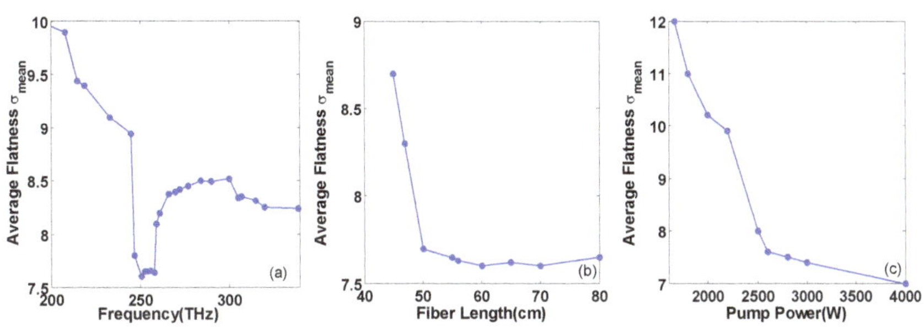

**Figure 7.** The evolution of the average flatness of 100 individual simulations ($\sigma_{mean}$) as a function of (**a**) the second ZDF in the second segmented PCF, (**b**) fiber length, and (**c**) pump power.

From Figure 4b, we can learn that the position of the second ZDF in the second segmented PCF is crucial to the behavior of spectral flatness. When the second ZDF is around 251 THz, the evolution processes of the spectrums are similar to that in Figure 4b. In this case, the newly generated red-shifted soliton in the second segmented fiber and the emitted DWs well fill the spectrum gaps. Therefore, it has a good flatness, which is named as flat area representing a low value of $\sigma_{mean}$ ($\approx$7.6). When the second ZDF moves from the flat area to the lower frequency region, the flatness deteriorates obviously and the value of $\sigma_{mean}$ quickly becomes higher, which arises from the ZDF in lower frequency region leading to a continued red-shift of the two fundamental solitons in the anomalous dispersion region so that the spectrum gaps are almost not filled. On the contrary, when the ZDF moves from the flat area toward the higher frequency region, the flatness slowly declines as the red-shift frequency range of the newly generated soliton is limited and the emitted DW partially overlaps the spectral components of the first ejected soliton. Therefore, spectrum gaps cannot be filled well and the value of $\sigma_{mean}$ shows a slow change. When the second ZDF reaches above 315 THz, where the two fundamental solitons generated in the first segmented fiber are directly translated into DWs in the normal dispersion region, the SC gets smoother and the flatness of the output spectrum is improved to some extent. With the continuous increase of the second ZDF, the average flatness will not change greatly, that is, $\sigma_{mean}$ will stay at a stable value (≈8.2). In a word, the output spectral flatness can be adjusted by the sencond ZDF in the second segmented cascaded PCF.

In Figure 7b, the length of the first segmented fiber is fixed to 40 cm, and when the total length of the fiber is less than 60 cm, the value of $\sigma_{mean}$ decreases with the increase in the fiber length. This is because the red-shift of the newly generated soliton in the second segmented fiber and the generation of DW are limited by the transmission distance. Until the fiber length reaches 60 cm, the best flatness of the spectrum is achieved where the spectrum recoil effect is balanced with SSFS. Continuing to increase the length of the fiber, the flatness of the fiber will not change obviously. As fiber length increases in actual applications, the loss will also increase. In this paper, 60 cm is selected as the optimal length.

In Figure 7c, increasing the pump power generates more solitons at the red edge, which accelerates solitons trapping and improves the spectral flatness, so the value of $\sigma_{mean}$ decreases gradually. When the pump power is less than 2600 W, the flatness is relatively poor because the spectral gap in B-DW is not filled by Trapping B-DW, which is similar to the situation in Figure 2(a3). Until the pump power is increased to 2600 W, the generated Trapping B-DW well fills the spectrum gap in B-DW, as shown in Figure 3a. When the pump power continues to increase, the flatness will gradually improve, but the improved speed is much lower than below 2600 W. Therefore, it is inefficient to continuously increase the pump power to more than 2600 W in order to improve the spectral flatness. This manuscript chooses exactly 2600 W as the optimum pump power.

We can conclude that, specifically in this manuscript, the optimum conditions for large spectral width and high spectral flatness (~1 um at −15 dB) are as follows; pumped at the frequency of 375 THz in the anomalous dispersion region in 60 cm fiber, incident power of 2600 W, the ZDF spacing of the first segmented fiber is 178 THz, and the second ZDF of the second segmented fiber is 251 THz. However, there are potential challenges and limitations in the practical applications of this method, where we need to consider access loss, transmission loss, and splicing loss. It is necessary to increase the pump power appropriately to compensate for actual total loss to ensure the same results as the theory analysis. Besides, when designing ZDF spacing, we need to consider the power endurability of the PCFs to ensure the spectral width and spectral flatness. What is more, it is difficult to achieve accurate control of the dispersion for practical devices, so detailed error analysis mechanism is very important in practical application.

## 5. Conclusions

In this work, we numerically demonstrate accurately shaping the SC to acquire wider and flatter SC via cascaded PCF, which consists of two silica PCFs with dual ZDFs. These results suggest that choosing a larger spacing between the two ZDFs of the first segmented cascaded PCF can not only redistribute energy between DWs and Trapping DWs, but also raise the red-shift of R-DW, which is beneficial to broaden the spectrum. Moreover, increasing the pump power can tremendously increase soliton trapping to improve the flatness of the blue edge of the spectrum. More importantly, we analyzed the output of the first segmented fiber and reasonably selected the second segmented PCF parameters, especially the second ZDF position, to reshape the SC to attain a flatter SC over the entire spectrum. We believe the result of this work can serve as a guideline in SCG, which can promote the technical applications of SC. In a next step, we will study the device with more cascaded PCFs to improve spectral width and spectral flatness more efficiently, which would be an interesting and important subject.

**Author Contributions:** H.Y. and J.R. designs simulation experiments, J.R. analyzes simulation results and writes manuscript, H.Y. and Y.X. revises manuscript. All authors have read and agreed to the published version of the manuscript

**Funding:** This work was supported by the National Natural Science Foundation of China (Grant No. 61275137, 61571186) and the Key Research and Development Program of Changsha City (kq1901027).

**Acknowledgments:** Thanks to my tutor and classmates for their guidance and help in my simulation experiments, and to the college for providing a good learning environment.

**Conflicts of Interest:** The authors declare no conflicts of interest.

## References

1. Alfano, R.; Shapiro, S.L. Emission in the region 4000 to 7000 Å via four-photon coupling in glass. *Phys. Rev. Lett.* **1970**, *24*, 584–587.
2. Holzwarth, R.R.; Udem, T.; Hänsch, T.W.; Knight, J.C.; Wadsworth, W.J.; Rusell, P.J. Optical frequency synthesizer for precision spectroscopy. *Phys. Rev. Lett.* **2000**, *85*, 2264–2267.
3. Marks, D.L.; Oldenburg, A. L.; Reynolds, J.J.; Boppart, S.A. Study of an ultrahigh-numerical-aperture fiber continuum generation source for optical coherence tomography. *Opt. Lett.* **2002**, *27*, 2010–2012.
4. Schliesser, G.A.; Picqué, N.; Hänsch, T.W. Mid-infrared frequency combs. *Nat. Photonics* **2012**, *6*, 440–449.
5. Dudley, J.M.; Genty, G.; Coen, S.; Supercontinuum generation in photonic crystal fiber. *Rev. Mod. Phys.* **2006**, *78*, 1135–1184.
6. Skryabin D.V.; Gorbach, A.V. Theory of supercontinuum and interaction of solitons with dispersive waves. In *Supercontinuum Generation in Optical Fibers;* Dudley, J.M., Taylor, J.R., Eds.; Cambridge University: Cambridge, UK, 2010; pp. 178–198.
7. Xiao, G.; Yang, H.; Zhao, S. Impact of the self-steepening effect on soliton spectral tunneling in PCF with three zero dispersion wavelengths. *Chin. Opt. Lett.* **2018**, *16*, 070601.
8. Klimczak, M.; Siwicki, B.; Heidt, A.; Buczyński, R. Coherent supercontinuum generation in soft glass photonic crystal fibers. *Photonics Res.* **2017**, *5*, 710–724.
9. Roy, S.; Ghosh, D.; Bhadra, S.K.; Agrawal, G.P. Role of dispersion profile in controlling the emission of dispersive waves by solitons in supercontinuum generation. *Opt. Commun.* **2010**, *283*, 3081–3088.
10. Foster, M.A.; Turne, A.C.; Lipson, M.; Gaeta, A. L. Nonlinear optics in photonic nanowires. *Opt. Express* **2008**, *16*, 1300–1320.
11. Latas, S.C.V.; Ferreira, M.F.S. Soliton propagation in the presence of intrapulse Raman scattering and nonlinear gain. *Opt. Commun.* **2005**, *251*, 415–422.
12. Skryabin, D.V.; Yulin, A.V. Theory of generation of new frequencies by mixing of solitons and dispersive waves in optical fibers. *Phys. Rev. E* **2005**, *72*, 016619.
13. Judge, A.C.; Bang, O.; Martijn de Sterke, C. Theory of Dispersive wave frequency shift via trapping by a soliton in an axially nonuniform optical fiber. *J. Opt. Soc. Am.* **2010**, *27*, 2195–2202.
14. Gorbach, A.V.; Skryabin, D.V. Light trapping in gravity-like potentials and expansion of supercontinuum spectra in photonic crystal fibers. *Nat. Photonics* **2007**, *1*, 653–657.
15. Gorbach, A.V.; Skryabin, D.V. Theory of radiation trapping by the accelerating solitons in optical fibers. *Phys. Rev. A* **2007**, *76*, 1–10.
16. Liu X.; Pang, M. Revealing the Buildup Dynamics of Harmonic Mode-Locking States in Ultrafast Lasers. *Laser Photonics Rev.* **2019**, *13*, 1800333.
17. Liu X.; Yao, X.; Cui, Y. Real-Time Observation of the Buildup of Soliton Molecules. *Phys. Rev. Lett.* **2019**, *121*, 023905.
18. Wang, W.; Yang, H.; Tang, P.; Zhao, C.; Gao, J. Soliton trapping of dispersive waves in photonic crystal fiber with two zero dispersive wavelength. *Opt. Express* **2013**, *21*, 11215–11226.
19. Frosz, M. H.; Falk, P.; Bang, O. The role of the second zero-dispersion wavelength in generation of supercontinua and bright-bright soliton-pairs across the zero-dispersion wavelength. *Opt. Express* **2005**, *13*, 6181–6192.
20. Mussot, A.; Beaugeois, M.; Bouazaoui, M.; Sylvestre, T. Tailoring CW supercontinuum generation in microstructured fibers with two-zero-dispersion frequencies. *Opt. Express* **2007**, *15*, 11553–11563.
21. Austin, D.R.; Desterke, C. M.; Eggleton, B. J.; Brown, T. G. Dispersive wave blue-shift in supercontinuum generation. *Opt. Express* **2006**, *14*, 11997–12007.

22. Li J.F.; Zhou, G. Y.; Hou, L. T. Numerical simulation of soliton trapping of the supercontinuum in photonic crystal fiber. *Acta Phys. Sin.* **2012**, *61*, 124203.
23. Hilligsøe, K.M.; Andersen, T.V.; Paulsen, H.N.; Nielsen, C. K. Supercontinuum generation in a photonic crystal fiber with two zero-dispersion frequencies. *Opt. Express* **2004**, *12*, 1045–1054.
24. Schreiber, T.; Andersen, T.V.; Schimpf, D.; Limpert, J.; Tiinnermann, A. Supercontinuum generation by femtosecond single and dual wavelength pumping in photonic crystal fibers with two zero-dispersion frequencies. *Opt. Express* **2005**, *13*, 9556–9569.
25. Luo, Y.P; Hua, Y.; Zhao, S.; Hu, H. Impact of potential barriers on soliton spectral tunneling in photonic crystal fiber. *Opt. Commun.* **2020**, *454*, 124330.
26. Poletti, F.; Horak, P.; Richardson, D.J. Soliton spectral tunneling in dispersion-controlled holey fibers. *IEEE Photonics Technol. Lett.* **2008**, *20*, 1414–1416.
27. Guo, H.; Wang, S.; Zeng, X.; Bache, M. Analysis of Cascaded Soliton Spectral Tunneling Effect in Segmented Fibers With Engineered Dispersion. *IEEE Photonics Technol. Lett.* **2013**, *5*, 1928.
28. Guo, C.; Ruan, S.; Yan, P.; Pan, E.; Wei, H. Flat supercontinuum generation in cascaded fibers pumped by a continuous wave laser. *Opt. Express* **2010**, *18*, 11046–11051.
29. Zhao, S.; Yang, H.; Chen, N.; Zhao, C. Harnessing rogue wave for supercontinuum generation in cascaded photonic crystal fiber. *Opt. Express* **2017**, *25*, 7192–7202.
30. Chen, H.; Chen, Z.; Zhou, X.; Hou, J. Cascaded PCF tapers for broadband supercontinuum generation. *Chin. Opt. Lett.* **2012**, *10*, 120603–120603.
31. Petersen, C.R.; Møller, U.; Kubat, I.; Zhou, B.; Dupont, S.; Ramsay, J.; Benson, T.; Sujecki, S.; Abdel-Moneim, N.; Tang, Z.; et al. Mid-infrared supercontinuum covering the 1.4–13.3 μm molecular fingerprint region using ultra-high NA chalcogenide step-index fiber. *Nat. Photonics* **2014**, *8*, 830–834.
32. Ali, R.A.H.; Hameed, M.F.O.; Obayya, S.S.A. Ultrabroadband Supercontinuum Generation Through Photonic Crystal Fiber With As2S3 Chalcogenide Core. *J. Light. Technol.* **2016**, *34*, 5423–5430.
33. Ma, P.; Choi, D.Y.; Yu, Y.; Gai, X.; Yang, Z.; Debbarma, S.; Madden, S.; Luther-Davies, B. Low-loss chalcogenide waveguides for chemical sensing in the mid-infrared. *Opt. Express* **2013**, *21*, 29927–29937.
34. Liu, L.; Cheng, T.; Nagasaka, K.; Tong, H.; Qin, G.; Suzuki, T.; Ohishi, Y. Coherent midinfrared supercontinuum generation in all-solid chalcogenide microstructured fibers with all-normal dispersion. *Opt. Lett.* **2016**, *41*, 392–395.
35. Milián, C.; Marest, T.; Kudlinski, A.; Skryabin, D.V. Spectral wings of the fiber supercontinuum and the dark-bright soliton interaction. *Opt. Express* **2017**, *25*, 10494–10499.
36. Zhao, S.; Yang, H.; Xiao, Y. Hu, H. Generation of tunable ultra-short pulse sequences in a quasi-discrete spectral supercontinuum by dark solitons. *Opt. Express* **2019**, *27*, 23539–23548.
37. Zhao, S.; Yang, H. Xiao, Y.; Hu, H. Generation of spectral wings and temporal ultra-short pulse sequence. *EPL Europhys. Lett.* **2019**, *126*, 24002.
38. Sahin, A.; Kutay M.A. Haldun, M.O. Nonseparable two-dimensional fractional Fourier transform. *Appl. Opt.* **1998**, *37*, 5444–5453.
39. Takeda, M.; Ina, H. Kobayashi, S. Fourier-transform method of fringe-pattern analysis for computer-based topography and interferometry. *J. Opt. Soc. Am.* **1982**, *72*, 156–160.
40. Zheng, W.; Li, K.; Li, K.Q. So, H.C. Fourier-transform method of fringe-pattern analysis for computer-based topography and interferometry. *IEEE. Acm Trans. Comput.* **2017**, *14*, 634–645.
41. Zheng, W.; Li, K.; Li, K.Q. Datapath-regular implementation and scaled technique for $N = 3 \times 2$ m DFTs. *Signal Process* **2015**, *11*, 68–79.
42. Saitoh, K.; Koshiba, M. Empirical relations for simple design of photonic crystal fibers. *Opt. Express* **2005**, *13*, 267–274.
43. Hong, J.; Hou, L. Effect of structure parameters changes on the properties of photonic crystal fiber with two zero-dispersion wavelengths. *Acta Phys. Sin.* **2010**, *59*, 1095–1100.

*sensors*

Article

# A Novel Method of Measuring Instantaneous Frequency of an Ultrafast Frequency Modulated Continuous-Wave Laser

**Jiewei Yang, Tianxin Yang *, Zhaoying Wang, Dongfang Jia and Chunfeng Ge**

Key Laboratory of the Ministry of Education on Optoelectronic Information Technology, School of Precision Instrument and Optoelectronics Engineering, Tianjin University, Tianjin 300072, China; yang_jw@tju.edu.cn (J.Y.); wangzy@tju.edu.cn (Z.W.); jiadf@tju.edu.cn (D.J.); gechunfeng@tju.edu.cn (C.G.)

* Correspondence: tyang@tju.edu.cn

Received: 20 May 2020; Accepted: 7 July 2020; Published: 9 July 2020

**Abstract:** Ultrafast linear frequency modulated continuous-wave (FMCW) lasers are a special category of CW lasers. The linear FMCW laser is the light source for many sensing applications, especially for light detection and ranging (LiDAR). However, systems for the generation of high quality linear FMCW light are limited and diverse in terms of technical approaches and mechanisms. Due to a lack of characterization methods for linear FMCW lasers, it is difficult to compare and judge the generation systems in the same category. We propose a novel scheme for measuring the mapping relationship between instantaneous frequency and time of a FMCW laser based on a modified coherent optical spectrum analyzer (COSA) and digital signal processing (DSP) method. Our method has the potential to measure the instantaneous frequency of a FMCW laser at an unlimited sweep rate. In this paper, we demonstrate how to use this new method to precisely measure a FMCW laser at a large fast sweep rate of 5000 THz/s by both simulation and experiments. We find experimentally that the uncertainty of this method is less than 100 kHz and can be improved further if a frequency feedback servo system is introduced to stabilize the local CW laser.

**Keywords:** frequency modulated continuous-wave (FMCW); light detection and ranging (LiDAR); instantaneous frequency; coherent optical spectrum analyzer (COSA); time-frequency curve

## 1. Introduction

An ultrafast linear frequency modulated continuous-wave (FMCW) laser in which the frequency changes continuously with time in a periodic fashion while its intensity is kept constant is extensively required in various applications, such as measurements for absolute distance, speed, and vibration [1–3], light detection and ranging (LiDAR) [4,5], coherent optical spectrum analyzer (COSA) [6–8], optical coherence tomography (OCT) [9–11], micro-cavity dispersion measurement [12], satellite formation flying [13], precision manufacturing [14], and so on. Many applications ideally need a linear FMCW light source with broad frequency sweep excursion, narrow instantaneous linewidth, fast frequency sweep rate, and excellent frequency modulation linearity (without mode-hopping) [15–17], so that a FMCW LiDAR could detect remote targets at ultrafast scanning rates with ultrahigh range resolution compared to its counterpart, a time-of-flight (ToF) LiDAR.

Among these characteristics, the frequency sweep rate is a key parameter for the application of a real-time optical sensing system in which a FMCW laser is embedded. The responsivity of the whole sensing system is related to the frequency sweep rate of the FMCW laser, while other parameters, including the linewidth and sweep excursion, are also important. For example, in manufacturing applications the inspection update rate is often more important than accuracy or precision.

Although it is challenging to simultaneously realize submicrometer accuracy (and/or precision) and high update rates in a remote sensing system, as in a linear FMCW LiDAR, both are required in the application of tight-formation flying of satellites, for which the ranging accuracy is required to reach the nanometer level [13]. Based on the principle of linear FMCW LiDAR, the detection range of $z$ is governed by Equation (1) [18].

$$z = \left(\frac{c}{2\gamma}\right) \cdot \Delta f_{FMCW} \tag{1}$$

where $c$ is the speed of light, $\gamma$ is the frequency sweep rate, and $\Delta f_{FMCW}$ is the frequency difference between the returning light from the remote target and a local reference light which is just emitting out from the linear FMCW laser. From Equation (1), the uncertainty of the detection range $\delta z$ obeys Equation (2).

$$\delta z = \overline{z}\left(\frac{\delta\gamma}{\overline{\gamma}} + \frac{\delta\Delta f_{FMCW}}{\overline{\Delta f}_{FMCW}}\right) \tag{2}$$

In Equation (2), we can see that, for detecting a long distance ($\overline{z}$ is large) accurately, the two terms in the parenthesis should be as small as possible. So, it is obvious that a faster frequency sweep rate $\overline{\gamma}$ is preferred for the sake of higher range accuracy. Normally, the frequency sweep rate $\overline{\gamma}$ is above THz/s over a broad optical bandwidth of several THz so that no general optical spectrum analyzer or electronic spectrometer could be able to catch up with the sweep CW light. Neither instrument could obtain the instantaneous frequency over the broad sweep excursion of several THz. A dedicated method which was able to measure the instantaneous frequency of a linear FMCW laser over a sweep excursion of 5 THz was proposed and completed by a group at the National Institute of Standards and Technology (NIST) in the United States [19]. They successfully retrieved the phase curve versus time from their experimental data when an under-test linear FMCW laser was swept against a one femtosecond mode-locked frequency comb in the well-known NIST dual-comb system [20]. Then, the time-frequency curve of the FMCW laser, i.e., the instantaneous frequencies versus time during the whole sweep period, was obtained by simply calculating a numerical first-order derivative with time of the phase curve they just retrieved. However, they encountered a technical problem in which the phase was ambiguous at the half way point between two adjacent comb teeth, because the in-phase/quadrature (IQ) demodulator that they used in their measurement system could not identify two different situations at the middle point of two comb teeth, i.e., the IQ detectors cannot judge whether the sweep is approaching to an individual tooth or moving away from the tooth. Therefore, the time-frequency curve became 'corrupted' at the middle of two adjacent comb teeth. That meant that the phase at that moment could not be retrieved from the data which were collected by the IQ demodulator. One year later, they overcame the problem technically by removing the 'corrupted' parts in two measurements with frequency-shifted combs and stitching the rest of the 'good parts' together to obtain a smooth phase curve versus time. So, they claimed that the maximum measurable frequency sweep rate was up to 1500 THz/s using the comb system for which the frequency spacing was about 100 MHz [21]. Very soon after this, the record was updated to 3400 THz/s by doubling the comb spacing to 200 MHz [2].

## 2. Principles for Measuring Instantaneous Frequency of an Under-Test FMCW Laser at Fast Sweep Rate

NIST's dual-comb system might be the most elaborate system for optical frequency measurement in the world [2,20–24], however in principle, it might not be the best choice for directly measuring the instantaneous frequency of a FMCW light, especially when the frequency sweeps at a fast rate. Because, according to the definition of frequency, a measuring time window has to be specified before the frequency measurement is initiated, then how many periods in this measuring time window is counted by the fact that, in one period, the optical phase changes by $2\pi$. For a CW laser with a constant frequency or wavelength, there is a rule that the longer the time window is, the smaller the frequency uncertainty that can be observed. However, for a FMCW laser with a high sweep rate, the rule does not hold anymore. Because the frequency of the FMCW laser changes dramatically during a long-time window, the instantaneous frequency cannot be defined well. Theoretically, according to the exact definition of instantaneous frequency for a frequency sweep laser, the measuring time window should be zero. Obviously, this is impossible to realize in NIST's system. So, the instantaneous frequency was measured in a very short time window with a tolerance uncertainty. It was found experimentally in NIST's paper that the accuracy was improved only by one order of magnitude while the length of the measuring time window had to be increased by two orders of magnitude. So, they claimed that their system was suitable for the FMCW laser where the sweep rate was less than 3400 THz/s [2].

In this paper, we avoid the issue of the measuring time window and propose a new scheme for measuring the instantaneous frequency of a FMCW laser at a fast sweep rate against a CW optical frequency comb rather than a traditional mode-locked femtosecond pulsed optical frequency comb. The CW optical frequency comb here actually is a single-longitudinal-mode multi-wavelength laser. It could be an electro-optic optical frequency comb generated by a recirculating frequency-shifted optical fiber loop based on a technique that uses an external modulation of a seed CW laser [17,25], or it could be generated by an optical fiber CW laser that has a Fabry–Perot comb filter inserted in the ring cavity [26]. In short, there is no fixed phase relationship among these individual comb teeth so that each comb tooth can be seen as a CW light wave with a single frequency component. Therefore, before the CW comb is used to measure the instantaneous frequency of a FMCW laser, the frequency or wavelength of each comb tooth can be known by measuring them precisely in a long time window using a general advanced optical spectrum analyzer which is available in the marketplace.

Assuming that the known absolute optical frequencies of the CW comb are labeled as a set of numbers in Hertz, noted as $\{\nu_1, \nu_2, \nu_3, \ldots, \nu_n\}$, where $\nu_n$ is the known frequency of the $n$th tooth and $\nu_n = \nu_1 + (n-1) \cdot \Delta\nu$, where $\Delta\nu$ is the frequency spacing of the CW comb, and the span of the CW comb $(\nu_n - \nu_1)$ should cover the full sweep excursion of the under-test FMCW laser. Now, let another set $\{t_1, t_2, t_3, \ldots, t_n\}$ represent the times of the moments when the under-test FMCW laser sweeps across each CW comb tooth which frequencies are known as $\{\nu_1, \nu_2, \nu_3, \ldots, \nu_n\}$ in the frequency domain. It is obvious that there is a one-to-one map between the two sets. Each element in the time set $\{t_1, t_2, t_3, \ldots, t_n\}$ can be determined by the method we present in this paper. By our method, $n$ number of instantaneous frequencies of the under-test FMCW laser can be obtained in the form of $\{\nu_1(t_1), \nu_2(t_2), \nu_3(t_3), \ldots, \nu_n(t_n)\}$, then the time-frequency curve of the FMCW light over the full sweep excursion can be depicted. This is the principle we propose for measuring the instantaneous frequency of a FMCW light. Based on this principle, we propose a new scheme for measuring the instantaneous frequency of a FMCW light wave using a CW frequency comb by building a one-to-one mapping between the known frequencies of the CW comb tooth and points in time of a FMCW light wave sweeping across these CW comb tooth.

In this paper, a basic technique for determining the point in time of a FMCW light wave sweeping across one tooth of the CW comb is presented by both simulation and experiments. The heterodyne signal between the under-test FMCW light source and a local stable CW laser with an accurately known frequency or wavelength is recorded by a pair of balanced photodetectors connected to an advanced broadband real-time oscilloscope (OSC). In order to precisely identify the time of the moment when the under-test FMCW light just swept across the local stable CW laser in the frequency domain, the power of the local stable CW laser is amplified so that the central part of heterodyne signal can be chopped off slightly due to the saturation effect of the balanced photodetectors. However, it would not reduce the accuracy for us to determine the point in time because we developed a new technique to deal with the problem. By taking a digital signal processing (DSP) of a numerical low-pass filtering of the recorded heterodyne signal, the well-defined profile of the interference pattern can be retrieved clearly. This profile should be centered at the time of the moment when the instantaneous frequency of the FMCW laser is equal to the frequency of the local CW laser which is already accurately known. The time at the center of the interference pattern can be determined precisely by fitting the profile numerically. Therefore, based on the new technique, the one-to-one mapping between the time and the instantaneous frequency of the FMCW laser could be achieved in this way if the CW laser was replaced by the CW frequency comb that we just described in former paragraph.

In this paper, a linear FMCW light is generated with a sweep excursion of 20 GHz in a period of 4 μs, so the sweep rate can reach to 5000 THz/s. Using the new technique, we developed, one instantaneous frequency of the fast sweep FMCW light can be accurately measured at one instant of time. It is shown by our experiment that the uncertainty of the measured instantaneous frequency is less than 100 kHz.

## 3. Generation of the Under-Test FMCW Light and Its Time-Frequency Curve Measured by an OSC

The under-test FMCW light wave is generated by an ultra-fast swept laser system which is shown in Figure 1. The seed laser is a distributed feedback (DFB) semiconductor laser emitting out a CW light with narrow linewidth. The CW light is modulated by an in-phase/quadrature (I/Q) modulator which is driven by an arbitrary waveform generator (AWG). The I/Q modulator is a double parallel Mach–Zehnder interference modulator (DPMZIM) operating in single-sideband (SSB) mode. The AWG generates a linearly frequency swept signal with a period of $T$ and it is divided into two branches, named as in-phase (I) branch and quadrature (Q) branch, feeding into the I input and Q input of the optical I/Q modulator, respectively. Based on the principle described in our previous paper [17,27], the linearly frequency swept CW light, which is actually the component of the +1 order side band, is generated from the I/Q modulator. The power of the linear FMCW light is then amplified by an erbium-doped fiber amplifier (EDFA), followed by a tunable optical filter to suppress the unnecessary residual high order sidebands such as the minus 3rd order side band. Finally, the linear FMCW light is generated and tested later in order to show how our measurement system works.

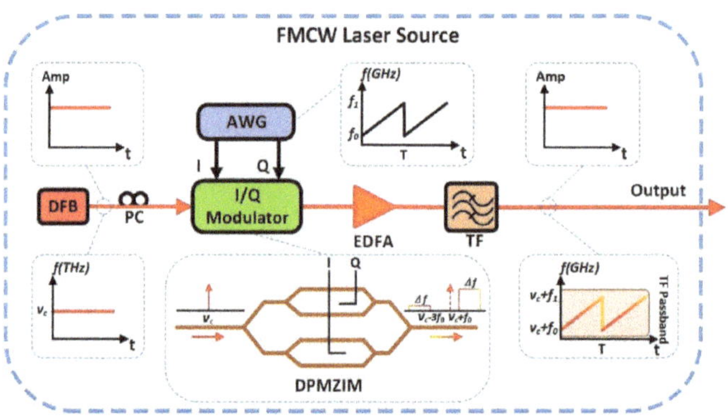


**Figure 1.** The generation system of the under-test FMCW light wave.

The under-test FMCW light could be generated from the system shown in Figure 1. The frequency of the FMCW light linearly sweeps over an excursion of $(f_1 - f_0)$ during a period of $T$ from $(\nu_c + f_0)$ to $(\nu_c + f_1)$, where $\nu_c$ is the optical frequency of the seed DFB laser. To check the frequency sweep linearity of the FMCW light, we use a heterodyne detection system shown in Figure 2 to record a heterodyne signal that generates at an optical coupler (OC) by photonic mixing the linear FMCW light with a stable CW laser which is a tunable external cavity laser diode (TECLD). The optical frequency $\nu_L$ of the TECLD is tuned to be close to the sweep excursion of the FMCW light in the frequency domain, but fixed at a point on the left side of the sweep excursion in the frequency domain, i.e., $\nu_L < (\nu_c + f_0)$ (see the spectrum in the inset of Figure 2). Then, the time-frequency curve of the FMCW light generated in the system shown in Figure 1 can be obtained by an algorithm of short time Fourier transform (STFT) for example, using a personal computer (PC), based on the data collected by a pair of balanced photodetectors and a real-time OSC in the system shown in Figure 2.

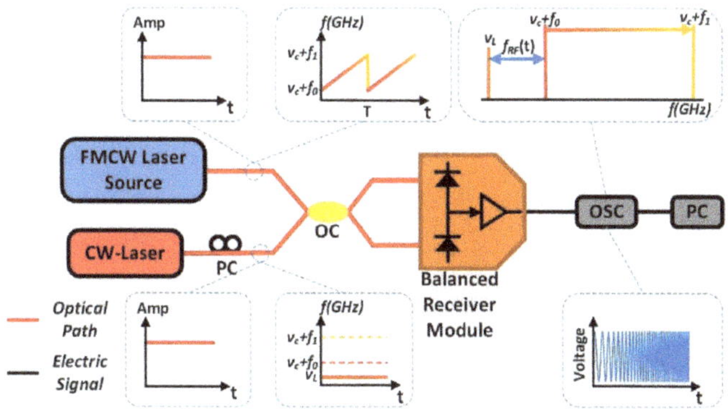


**Figure 2.** The heterodyne detection system of the under-test FMCW light wave.

The devices and equipment we used in our generation system (see Figure 1) and detection system (see Figure 2) are listed as follows: In Figure 1, the seed laser of the under-test FMCW light source is a DFB laser with a linewidth of about 20 MHz at the wavelength of 1550 nm. The electronic linear FMCW signal is generated from the AWG (Keysight, M8195A, sampling rate of 65 GSa/s) with peak-to-peak amplitude $V_{PP}$ = 1 V, a sweep excursion from $f_0$ = 5 GHz to $f_1$ = 25 GHz in a period $T$ = 4 µs. So, the sweep excursion $\Delta f$ is equal to 20 GHz and the sweep rate $\gamma$ is equal to 5000 THz/s. In Figure 2 the CW laser (TECLD, Keysight, 81606A) with a linewidth of less than 100 Hz, output power of 12 dBm, is tuned carefully to let $(\nu_c - \nu_L)$ be equal to 5.0 GHz. The tunable optical filter (Santec-OTF-350, minimum bandpass of 0.1 nm, 12.5 GHz @1550 nm) is set to be centered at 1550 nm with an appropriate bandwidth. The pair of balanced photodetectors is centered at 1550 nm with a 3dB bandwidth of 40 GHz. A real-time OSC (Keysight, MSOV334A, bandwidth of 33 GHz) with a sampling rate of 80 GSa/s records and stores the data of the heterodyne signal. The STFT is applied to those data with a frequency uncertainty of about 50 MHz due to the limited size of the time window we selected in STFT calculation. Large frequency uncertainty is the weakness of the traditional method based on STFT calculations. We will compare it with our new method presented in this paper.

The measurement results are shown in Figure 3. The heterodyne signal between the FMCW light generated by the system shown in Figure 1 and a CW laser (TECLD) is recorded by the OSC (see Figure 2), and then plotted in Figure 3a. There are some intensity ripples over the 5 periods. These ripples may be caused by the imperfect responsivity over the broadband of 20 GHz of the balanced photodetectors. However, what we are concerned with is the frequency characteristics, such as the sweep linearity $R^2$ and sweep rate $\gamma$. The STFT curve of the heterodyne signal is shown as a time-frequency curve of the FMCW light in Figure 3b. By zooming-in the time-frequency curve in one period of 4 µs, it can be seen in Figure 3c that the sweep linearity $R^2$ = 0.999849 and sweep rate $\gamma$ = 4.9812 GHz/µs = 4981.2 THz/s ≈ 5000 THz/s. It can be seen from the curve that the uncertainty of the instantaneous frequency of the FMCW light is about 50 MHz. This is because of the limited sampling rate of the OSC (although it is 80 GSa/s) and the limited time window we used in performing the STFT, which is mainly limited by time consuming computer calculations. However, the time window for performing the STFT cannot be significantly increased to reduce the uncertainty because we are dealing with a fast swept FMCW light. The longer the time window is, the worse the accuracy would be. We have to do a tradeoff between the sampling rate of the OSC and the window size of the time consuming STFT. So, an uncertainty of 50 MHz is the best result we could have at present for an ultra-fast FMCW light of 5000 THz/s by using the advanced real-time OSC with an ultrahigh sampling rate of 80 GSa/s.

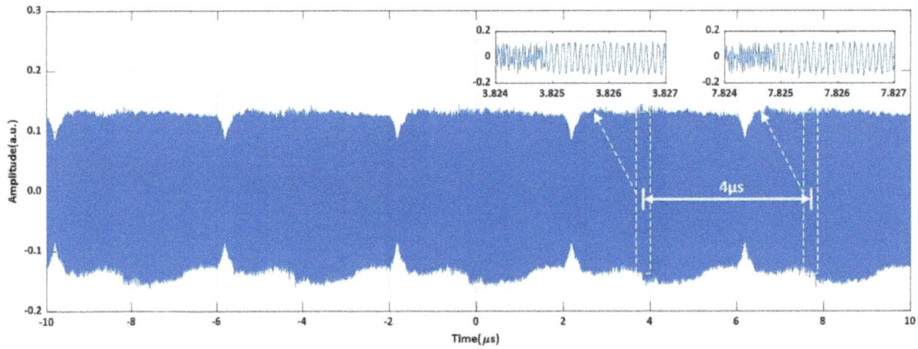


(**a**)

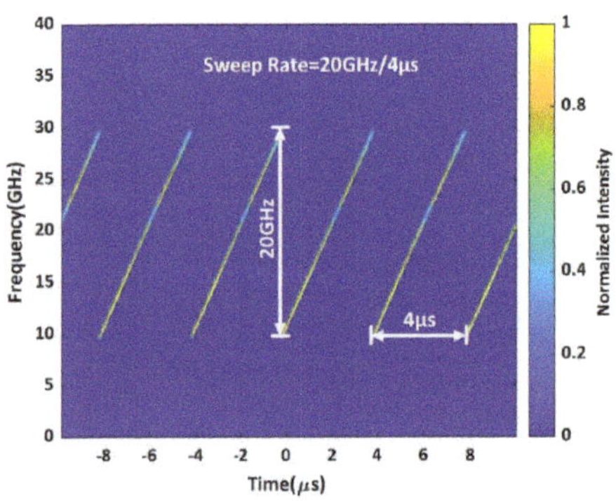


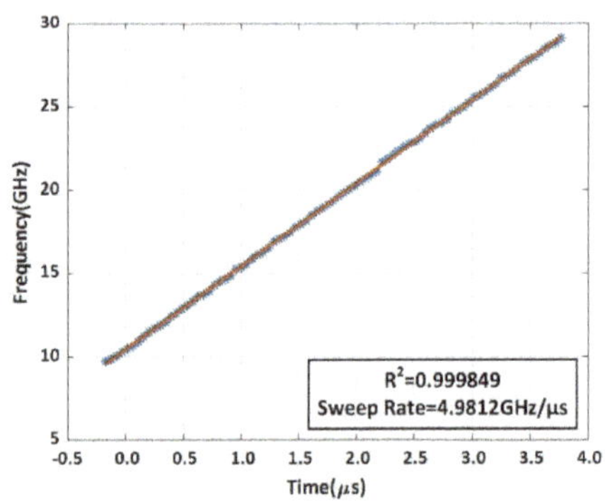


(**b**) (**c**)

**Figure 3.** The under-test FMCW light wave. (**a**) The heterodyne signal, (**b**) multi-period time-frequency curve, (**c**) zoomed-in time-frequency curve in one period.

## 4. Simulations of New Technique for Measuring the Instantaneous Frequency of Linear FMCW Light

The system for measuring instantaneous frequency of a FMCW light is shown in Figure 4. It may look like and resemble the configuration of a coherent optical spectrum analyzer (COSA) [6], however it is not. There are three differences from a COSA. First, the local laser here is a single frequency CW laser rather than a linear FMCW laser at a very slow sweep rate of 92 GHz/s in [6]. Second, the signal light to be tested here is the linear FMCW light at sweep rate of 5000 THz/s that can be generated by our system, as shown in Figure 1, and the parameter to be tested here is the instantaneous frequency of the fast sweep light, rather than a linewidth of a typical CW laser diode in Ref. [6]. Third, there is no electronic low-pass filter in front of the OSC in our system (see Figure 4), while in COSA there is one. It is necessary to have an electronic low-pass filter between the balanced photodetectors and the OSC to improve the frequency resolution of COSA. The smaller the bandwidth of the filter, the higher the resolution is. However, a very narrow bandwidth low-pass filter cannot work well here in our detection system. The reason that it would not work well is because our sweep rate of 5000 THz/s is too fast compared to 92 GHz/s in the original COSA [6]. Therefore, we need to perform numerical filtering to retrieve the interference pattern in a small band centered at the optical carrier frequency $\nu_L$ of our local CW laser.

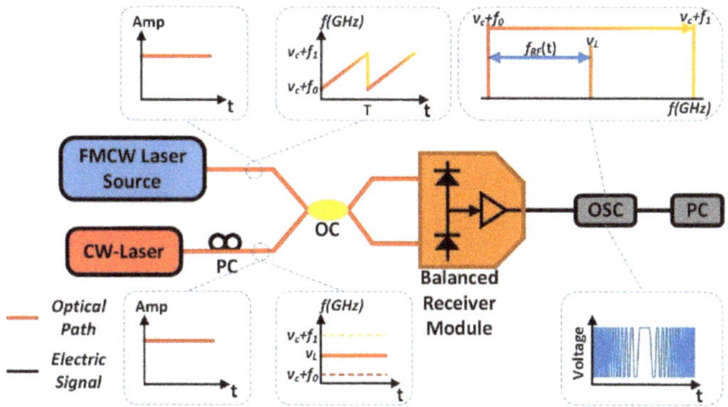


**Figure 4.** The measurement system of the under-test FMCW light wave.

The simulation results are shown in Figure 5. The interference pattern which is generated when the under-test FMCW light is swept over a non-filtered frequency band centered at the local CW laser in the frequency domain is recorded in real-time by an OSC and is presented in Figure 5a. Because both the FMCW light and the local single frequency laser are continuous waves with constant amplitude in an ideal simulation model, the interference signal should have a constant amplitude if there is no electronic low-pass filter in front of the OSC. However, because the FMCW light and the local CW laser are independent of each other in the time domain, the central part of the interference signal could present different shapes due to the various phase differences when two lights mix in the OC shown in Figure 4. Therefore, the data in the central part of the interference pattern is not reliable for retrieving the point in time when the FMCW sweeps across the point at the frequency of the local CW laser. Only the bilateral parts of the interference pattern which are formed when the FMCW is approaching to and leaving the local CW laser have consistency and are reliable. By performing a narrow-band numerical filtering operation, a profile of interference fringes can be clearly seen and are shown in yellow color in Figure 5b. The peak of the profile should be the point in time when the FMCW sweeps across the point at the frequency of the local CW laser. Thereby, it is easy to obtain the point in time precisely.

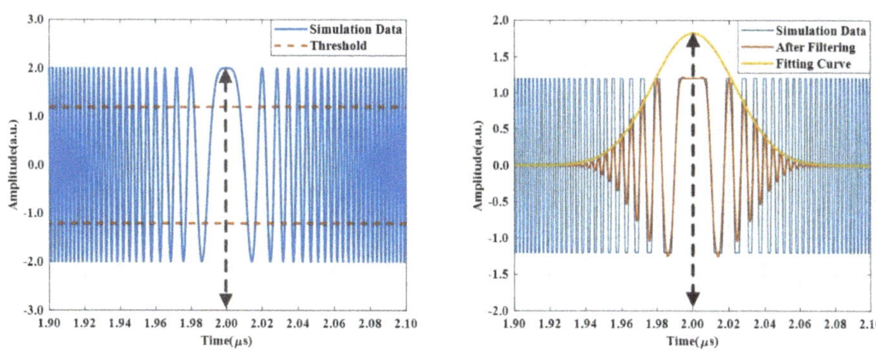


(a) (b)

**Figure 5.** Simulation of the measurement method for determining the moment of time when the under-test FMCW light crosses the local CW laser in frequency domain. (**a**) The heterodyne signal collected by the OSC, (**b**) performing a numerical filtering and chopping off the top and bottom parts in the amplitude to eliminate ripple effects in amplitude.

In reality, the interference signal which is shown in Figure 5a may be deformed in amplitude due to imperfect saturation effects of the balanced photodetectors. We can show that the imperfect effect does not interfere in recovering the profile of the interference pattern around the local CW laser. We deliberately performed an operation of chopping the top 'head' and bottom 'foot' parts of the interference signal along the dash lines shown in Figure 5a before the numerical filtering is taken. It can be seen in Figure 5b that the process of chopping off the 'head' and 'foot' just deforms parts of interference fringes. It does not change the shape and position of the interference profile in the time domain. In addition, this technique of chopping off the 'head' and 'foot' can increase the accuracy for determining the moment of time when the under-test FMCW sweep crosses the local CW laser in the frequency domain because the chopped interference signal becomes perfectly flat in amplitude so the symmetry of the profile can be improved. Furthermore, removing the 'head' and 'foot' parts can effectively reinforce the signals on left side and right side of the interference fringes which are shown in the orange color line in Figure 5b.

## 5. Experiments for Measuring the Instantaneous Frequency of a FMCW Light Source and Results Analysis

The experimental setup is shown in Figure 4. The wavelength $\lambda_L$ (or frequency $\nu_L$) of the local CW laser (TECLD) is tuned carefully to be $\lambda_L$ = 1546.5033 nm to make sure that the local frequency $\nu_L$ is in the middle of the sweep excursion of the under-test FMCW light which has been generated in the system shown in Figure 1. The devices and equipment are the same as what have been described in previous sections. The heterodyne signal in 5 successive periods is recorded by the same real-time OSC for which the sampling rate is 80 GSa/s and is shown in Figure 6a. It can be seen that the heterodyne signals in amplitudes are not constant. However, the insets of the heterodyne signals are flatter. It can be seen in the two insets of Figure 6a that the central parts of heterodyne signals have different shapes. This is because there is no fixed phase relationship between the FMCW light and the local CW Laser (TECLD). Finally, it also can be seen clearly in the two insets that the saturation effects of the balanced photodetectors cut off the top and bottom parts of the heterodyne signals. We have discussed in previous sections that the saturation effect does not bring negative effects for precisely determining the point in time at the peak of an interference fringe profile. On the contrary, we found that chopping off the top and bottom parts could remove the 'ripples' in amplitude of the heterodyne signal making the profile fit better with the filtered interference fringes, and it does not shift the position in time of the profile.

Figure 6b shows the time-frequency curve based on the data in Figure 6a by using the algorithm of STFT. The turning point can be clearly seen in the middle of each period of 4 μs. Actually, the time of the turning point is the time coordinate we want to measure, however the accuracy for determining the turning point in time is not satisfied because the uncertainty of the STFT is about 50 MHz in Figure 6b. We have explained this issue in the previous sections for the traditional STFT method. We will show that our new method can reduce the uncertainty to less than 100 kHz in following paragraph.

We now show how we precisely determine the point in time when the FMCW sweeps across the local CW laser. The five-period heterodyne signal formed by the FMCW light and the local CW laser is collected by the real-time OSC and is shown as the blue curve in Figure 7a. We perform a numerical filtering to the data of the blue curve. The numerical filter is a finite impulse response (FIR) low-pass filter with a bandwidth of 600 kHz Kaiser type window with a stopband attenuation of 60 dB and transition zone steepness factor of 0.9999. The filtered curves are shown in orange color in Figure 7a. The curves in the first period and the second period in Figure 7a are zoomed-in and shown in Figure 7b,c, respectively. The central parts of the two interference fringes are different due to the reasons we have explained in previous section, however the profiles look the same as we have used the technique of chopping off the head and foot before performing numerical filtering. Now, it is straightforward to get five central points in time of these five profiles which are listed in Table 1.

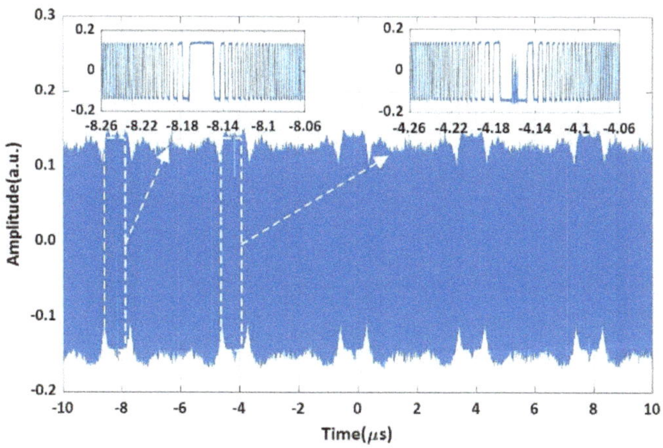


(a)

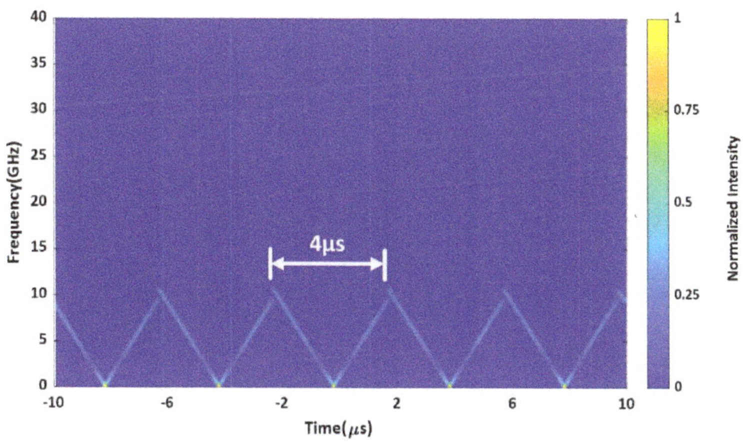


(b)

**Figure 6.** The experimental results for measuring the instantaneous frequency of a FMCW light source. (**a**) The heterodyne signal of the under-test FMCW light and a local CW laser. (**b**) The time-frequency curve of the heterodyne signal in (**a**) by the STFT algorithm.

**Table 1.** Central Points in Time of 5 Interference Profiles in Series.

| Number of Profile | Center Point in Time of Profile | Interval to the Center of 1st Profile | Equivalent Frequency Error |
|---|---|---|---|
| 1st | −8.16186146 μs | 0 μs | 0 kHz |
| 2nd | −4.16184358 μs | 4.00001788(0.00001788) μs | 89.4 kHz |
| 3rd | −0.16184241 μs | 8.00001905(0.00001905) μs | 95.25 kHz |
| 4th | 3.83792087 μs | 11.99978233(−0.00021767) μs | −1.08835 MHz |
| 5th | 7.83911479 μs | 16.00097625(0.00097625) μs | 4.88125 MHz |

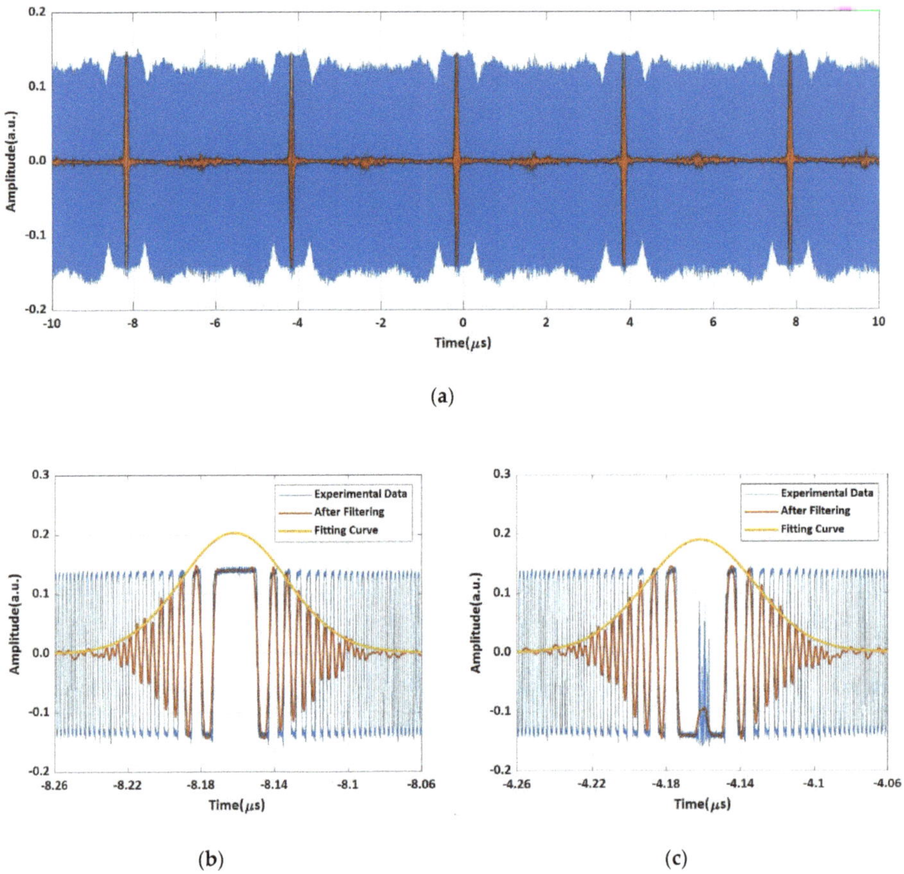


**Figure 7.** The multi-period heterodyne signal of the FMCW light and the local CW laser, (**a**) collected by the real-time OSA, (**b**) the 1st period, (**c**) the 2nd period.

The intervals of the 2nd profile, 3rd profile, 4th profile, and 5th profile to the 1st profile are listed in the third column in Table 1. They should be 0μs, 4.00000000 μs, 8.00000000 μs, 12.00000000 μs, and 16.00000000 μs if the AWG could be trusted to output signals exactly. Now the errors in time domain are obtained by subtracting these measured intervals with their ideal intervals, respectively. Then the equivalent instantaneous frequency errors (shown in the last column in Table 1) are obtained by multiplying the errors in time with the sweep rate of 5000 THz/s, respectively. It can be seen that the errors are increased by one order of magnitude from less than 100 kHz to around 1 MHz starting from the 4th profile. The reason could be the frequency drift of two lasers in our current system after 8 μs and the further drift after 12 μs. One is the local CW laser in the detection system shown in Figure 4, and the other is the seed DFB laser in the generation system of the linear FMCW light (see Figure 1). In our systems, we have not introduced optical frequency feedback servo systems to lock the local CW laser or the seed DFB laser to ultra-stable frequency reference sources so far [28–30]. If we had frequency feedback servo systems we believe that the uncertainty of this method for measuring the instantaneous frequency by one-to-one mapping time and frequency would be less than 100 kHz for which the accuracy can be increased by two orders of magnitude compared to 50 MHz obtained by other traditional methods.

## 6. Conclusions

In this paper, a new scheme is proposed for measuring the instantaneous frequency of a FMCW laser against a stable CW frequency comb over the full frequency sweep excursion of the FMCW laser. It is different from traditional methods based on femtosecond mode-locked pulsed frequency combs which measure the phase increment or phase decrement versus time in a short detection time window. However, the shorter the window is, the larger is the uncertainty of the frequency measurement. In our scheme, the issue of detection time window is avoided because we do not need to retrieve the phase function with time of the under-test FMCW laser, we just need to build a one-to-one mapping between the frequency of the each CW comb tooth and the point of time that the FMCW laser sweeps each comb tooth. Based on the one-to-one mapping, the time-frequency curve of the FMCW laser can be determined so that the instantaneous frequency of the FMCW laser over the full sweep excursion can be obtained precisely. We demonstrate how to determine the point in time that a FMCW laser sweeps one of the CW comb teeth by both simulation and experiments. It is shown by experiments that the uncertainty of measuring the instantaneous frequency of the FMCW at an ultrafast sweep rate of 5000 THz/s can reach less than 100 kHz, which is at least two orders of magnitude smaller in frequency compared to traditional methods. The method we developed in this paper has the potential to allow a precise characterization of an ultrafast sweep FMCW laser by combining a CW frequency comb with a large frequency spacing, such as 1 GHz, which is 10 times larger than traditional mode-locked pulsed frequency comb. So, the sweep rate of a FMCW laser could be potentially measured up to 10 times higher than 1500 THz/s while the uncertainty could be kept as low as 100 kHz, because measuring window issue in a traditional system can be avoided in our system in which instantaneous frequency is measured by a one-to-one mapping between the point in time of a FMCW laser and the known frequency of each CW comb tooth.

**Author Contributions:** T.Y. led the research group and wrote the manuscript. J.Y. completed the simulations, experiments, and data analysis in addition to drawing all figures and preparing a table. Z.W. managed the devices and advanced equipment. D.J. revised the manuscript and C.G. assisted in the group lab. All authors have read and agreed to the published version of the manuscript.

**Funding:** This research was funded by the Tianjin Science and Technology Commission (TSTC), grant number 19JCZDJC32500 and 18JCYBJC16800, and by the National Science Foundation of China, grant number 61975145.

**Acknowledgments:** We thank the Keysight Inc. for the use of their equipment.

**Conflicts of Interest:** The authors declare no conflict of interest.

## References

1. Yu, W.; Pfeiffer, P.; Morsali, A.; Yang, J.; Fontaine, J. Comb-calibrated frequency sweeping interferometry for absolute distance and vibration measurement. *Opt. Lett.* **2019**, *44*, 5069–5072. [CrossRef] [PubMed]
2. Baumann, E.; Giorgetta, F.R.; Coddington, I.; Sinclair, L.C.; Knabe, K.; Swann, W.C.; Newbury, N.R. Comb-calibrated frequency-modulated continuous-wave ladar for absolute distance measurements. *Opt. Lett.* **2013**, *38*, 2026–2028. [CrossRef] [PubMed]
3. Hariyama, T.; Sandborn, P.A.M.; Watanabe, M.; Wu, M.C. High-accuracy range-sensing system based on FMCW using low-cost VCSEL. *Opt. Express* **2018**, *26*, 9285–9297. [CrossRef]
4. DiLazaro, T.; Nehmetallah, G. Large-volume, low-cost, high-precision FMCW tomography using stitched DFBs. *Opt. Express* **2018**, *26*, 2891–2904. [CrossRef]
5. Zhang, X.; Pouls, J.; Wu, M.C. Laser frequency sweep linearization by iterative learning pre-distortion for FMCW LiDAR. *Opt. Express* **2019**, *27*, 9965–9974. [CrossRef] [PubMed]
6. Baney, D.M.; Szafraniec, B.; Motamedi, A. Coherent optical spectrum analyzer. *IEEE Photonic Technol. Lett.* **2002**, *14*, 355–357. [CrossRef]
7. Szafraniec, B.; Lee, A.; Law, J.Y.; McAlexander, W.; Pering, R.D.; Tan, T.S.; Baney, D.M. Swept coherent optical spectrum analysis. *IEEE Trans. Instrum. Meas.* **2004**, *53*, 203–215. [CrossRef]
8. Szafraniec, B.; Law, J.Y.; Baney, D.M. Frequency resolution and amplitude accuracy of the coherent optical spectrum analyzer with a swept local oscillator. *Opt. Lett.* **2002**, *27*, 1896–1898. [CrossRef] [PubMed]

9. Butler, T.; Slepneva, S.; O'Shaughnessy, B.; Kelleher, B.; Goulding, D.; Hegarty, S.P.; Lyu, H.C.; Karnowski, K.; Wojtkowski, M.; Huyet, G. Single shot, time-resolved measurement of the coherence properties of OCT swept source lasers. *Opt. Lett.* **2015**, *40*, 2277–2280. [CrossRef] [PubMed]
10. Wieser, W.; Biedermann, B.R.; Klein, T.; Eigenwillig, C.M.; Huber, R. Multi-Megahertz OCT: High quality 3D imaging at 20 million A-scans and 4.5 GVoxels per second. *Opt. Express* **2010**, *18*, 14685–14704. [CrossRef]
11. Kolb, J.P.; Pfeiffer, T.; Eibl, M.; Hakert, H.; Huber, R. High-resolution retinal swept source optical coherence tomography with an ultra-wideband Fourier-domain mode-locked laser at MHz A-scan rates. *Biomed. Opt. Express* **2018**, *9*, 120–130. [CrossRef] [PubMed]
12. Del'Haye, P.; Arcizet, O.; Gorodetsky, M.L.; Holzwarth, R.; Kippenberg, T.J. Frequency comb assisted diode laser spectroscopy for measurement of microcavity dispersion. *Nat. Photonics* **2009**, *3*, 529–533. [CrossRef]
13. Liu, T.-A.; Newbury, N.R.; Coddington, I. Sub-micron absolute distance measurements in sub-millisecond times with dual free-running femtosecond Er fiber-lasers. *Opt. Express* **2011**, *19*, 18501–18509. [CrossRef] [PubMed]
14. Behroozpour, B.; Sandborn, P.A.M.; Quack, N.; Seok, T.J.; Matsui, Y.; Wu, M.C.; Boser, B.E. 11.8 Chip-scale electro-optical 3D FMCW lidar with 8μm ranging precision. In Proceedings of the 2016 IEEE International Solid-State Circuits Conference (ISSCC), San Francisco, CA, USA, 31 January–4 February 2016; pp. 214–216.
15. Lyu, Y.; Yang, T.; Lu, Z.; Guo, C.; Ge, C.; Wang, Z.; Jia, D.; Yin, H. External Modulation Method for Generating Accurate Linear Optical FMCW. *IEEE Photonic. Technol. Lett.* **2017**, *29*, 1560–1563. [CrossRef]
16. Satyan, N.; Vasilyev, A.; Rakuljic, G.; Leyva, V.; Yariv, A. Precise control of broadband frequency chirps using optoelectronic feedback. *Opt. Express* **2009**, *17*, 15991–15999. [CrossRef] [PubMed]
17. Lu, Z.; Yang, T.; Li, Z.; Guo, C.; Wang, Z.; Jia, D.; Ge, C. Broadband linearly chirped light source with narrow linewidth based on external modulation. *Opt. Lett.* **2018**, *43*, 4144–4147. [CrossRef]
18. Zhang, F.; Guo, Q.; Pan, S. Photonics-based real-time ultra-high-range-resolution radar with broadband signal generation and processing. *Sci. Rep.* **2017**, *7*, 13848. [CrossRef]
19. Barber, Z.W.; Giorgetta, F.R.; Roos, P.A.; Coddington, I.; Dahl, J.R.; Reibel, R.R.; Greenfield, N.; Newbury, N.R. Characterization of an actively linearized ultrabroadband chirped laser with a fiber-laser optical frequency comb. *Opt. Lett.* **2011**, *36*, 1152–1154. [CrossRef]
20. Giorgetta, F.R.; Coddington, I.; Baumann, E.; Swann, W.C.; Newbury, N.R. Fast high-resolution spectroscopy of dynamic continuous-wave laser sources. *Nat. Photonics* **2010**, *4*, 853–857. [CrossRef]
21. Coddington, I.; Giorgetta, F.R.; Baumann, E.; Swann, W.C.; Newbury, N.R. Characterizing fast arbitrary cw waveforms with 1500 THz/s instantaneous chirps. *IEEE J. Sel. Top. Quantum Electron.* **2012**, *18*, 228–238. [CrossRef]
22. Fortier, T.; Baumann, E. 20 years of developments in optical frequency comb technology and applications. *Commun. Phys.* **2019**, *2*, 1–16. [CrossRef]
23. Coddington, I.; Swann, W.C.; Newbury, N.R. Coherent linear optical sampling at 15 bits of resolution. *Opt. Lett.* **2009**, *34*, 2153–2155. [CrossRef] [PubMed]
24. Duran, V.; Djevarhidjian, L.; de Chatellus, H.G. Bidirectional frequency-shifting loop for dual-comb spectroscopy. *Opt. Lett.* **2019**, *44*, 3789–3792. [CrossRef]
25. Li, R.; Chen, H.; Yu, Y.; Chen, M.; Yang, S.; Xie, S. Multiple-frequency measurement based on serial photonic channelization using optical wavelength scanning. *Opt. Lett.* **2013**, *38*, 4781–4784. [CrossRef] [PubMed]
26. Wang, Z.Y.; Liu, S.Y.; Luo, S.; Yuan, Q.; Ma, R.; Ge, C.F.; Yang, T.X. Broadband lightwave synthesized frequency sweeper using self-induced auto-tracking filter. *Opt. Express* **2015**, *23*, 22134–22140. [CrossRef] [PubMed]
27. Wang, Z.; Qin, X.; Gao, C.; Wang, T.; Jia, D.; Sang, M.; Yang, T. Precise simultaneous multiwavelength tuning by electrical RF signals. *IEEE Photonic Technol. Lett.* **2013**, *25*, 914–916. [CrossRef]
28. Black, E.D. An introduction to Pound–Drever–Hall laser frequency stabilization. *Am. J. Phys.* **2001**, *69*, 79–87. [CrossRef]
29. Drever, R.W.P.; Hall, J.L.; Kowalski, F.V.; Hough, J.; Ford, G.M.; Munley, A.J.; Ward, H. Laser phase and frequency stabilization using an optical resonator. *Appl. Phys. B* **1983**, *31*, 97–105. [CrossRef]
30. Ming, M.; Luo, Y.; Liang, Y.-R.; Zhang, J.-Y.; Duan, H.-Z.; Yan, H.; Jiang, Y.-Z.; Lu, L.-F.; Xiao, Q.; Zhou, Z.; et al. Ultraprecision intersatellite laser interferometry. *Int. J. Extrem. Manuf.* **2020**, *2*, 022003. [CrossRef]

sensors

MDPI

Article

# Bipolar Optical Code Division Multiple Access Techniques Using a Dual Electro-Optical Modulator Implemented in Free-Space Optics Communications

**Shin-Pin Tseng [1], Eddy Wijanto [2], Po-Han Lai [2] and Hsu-Chih Cheng [2,*]**

[1] Department of Electronic Engineering, National United University, Miaoli 36003, Taiwan; sptseng@nuu.edu.tw
[2] Department of Electro-Optical Engineering, National Formosa University, Yunlin 632, Taiwan; d0777106@gm.nfu.edu.tw (E.W.); carl60296@gmail.com (P.-H.L.)
* Correspondence: chenghc@nfu.edu.tw

Received: 2 May 2020; Accepted: 22 June 2020; Published: 24 June 2020

**Abstract:** This study developed a bipolar optical code division multiple access (Bi-OCDMA) technique based on spectral amplitude coding for the formation and transmission of optical-polarized and coded signals over wireless optical channels. Compared with conventional Bi-OCDMA schemes, the proposed free-space optics communication system that uses a dual electro-optical modulator design improves the transmission rate. In theory, multiple access interference can be removed by using correlation subtraction schemes. The experiment results revealed that the proposed system can be employed to accurately extract codewords from an *M*-sequence and subsequently reconstruct the desired original data. Moreover, the proposed architecture can be implemented easily in simple and cost-effective designs and may be beneficial for broadening the use of Bi-OCDMA schemes in wireless optical communications.

**Keywords:** bipolar; optical code division multiple access; electro-optical modulator; free-space optics communication

## 1. Introduction

Future fifth-generation (5G) networks require high bandwidth, low latency, accurate synchronization, and high reliability because they use key 5G technologies—namely enhanced mobile broadband that provides a peak data rate of $\geq$10 Gbps, massive machine-type communications (mMTCs) that transmit data among Internet of things (IoT) devices, and ultrareliable low-latency communications—with reliability and latency in the millisecond range. These requirements present numerous challenges in communication systems. Optical communication techniques are also promising candidates for overcoming such challenges because they can provide high bandwidth and a small latency. Optical communications have a small time delay because light provides high-speed transmission that improves the propagation delay and optical fibers have low attenuation that reduces the need for repeating and processing transmission signals. Furthermore, all current IoT applications, including e-health, telemedicine, surveillance systems, autonomous vehicles, and virtual reality platforms, require high bandwidth; therefore, wireless optical communications (WOC) have received considerable research attention [1–4]. Compared with conventional wireless communication techniques, WOC schemes can substantially resist electromagnetic wave interference (EMI).

A fundamental part of a WOC system is multiplexing techniques, which entail multiple users transmitting data by using a single link. In optical communication environments, the most widely used multiplexing technique is wavelength division multiplexing (WDM). WDM is advantageous for its configuration simplicity; however, it has disadvantageous spectral efficiency. In 2019, Ahmed et al. used WDM in free-space optical (FSO) communications to improve the performance of a system with a frequency range in the visible light spectrum [1].

Time division multiplexing (TDM) is another multiplexing scheme that allows users to simultaneously access the same channel by assigning time slots to all users. Although TDM schemes have sufficient spectral efficiency, they are subject to nonlinear fiber effects. In 2014, Mahloo et al. proposed a hybrid WDM–TDM approach for passive optical networks to increase the number of users in FSO communication systems while maintaining bandwidth [2]. Hybrid WDM–TDM combines the advantages of WDM and TDM to increase the number of users and achieve long-range communication. In addition, space division multiplexing (SDM) involves the application of beam separation to simultaneously deliver different data streams. In 2019, Rommel et al. proposed SDM with multicore fibers, and they used optical beamforming to access high-capacity millimeter-wave radios [3]. In 2019, Li et al. designed a novel quadrant detector to improve WOC transmission [4].

Some studies have reported a novel multiplex technique, namely optical code division multiple access (OCDMA) [5–21]. OCDMA employs CDMA techniques in optical fiber communication environments. This multiplexing scheme uses an optical coding technique in which a channel assigns each user a unique codeword to prevent mutual interference in the same channel. This technique allows the simultaneous transmission of unsynchronized data from multiple users of the same channel and bandwidth [5,6]. Therefore, OCDMA has favorable antijamming properties and moderate security with high-capacity processing. Among OCDMA schemes, spectral amplitude coding (SAC) is the most effective for alleviating multiple access interference (MAI).

On the basis of optical signal demodulation, OCDMA techniques can be divided into two categories. First, an incoherent OCDMA system uses optical field intensity to encode optical signals. These systems mainly use unipolar encoding (0, 1), which has a simple system structure and cost-effective design. However, the number of codewords that can be obtained through unipolar encoding is considerably smaller than that obtained through bipolar coding. To increase the number of simultaneous users, the code length must be increased, but this increases the system cost. Second, coherent OCDMA systems use the spectral phase of light to encode signals and a matching filter to control the optical phase [7]. These systems use bipolar encoding $(-1, 1)$. Because bipolar codes have pseudo-orthogonality, the value of a cross-correlation function between any two codewords can be approximated to 0, which results in low MAI and considerably enhances system performance. In addition, Zefreh et al. introduced a power-cubic nonlinear preprocessor for improving the coherent SAC OCDMA system performance; through numerical calculations, they demonstrated that MAI is the dominant noise in high-power scenarios [8].

Furthermore, unipolar and bipolar OCDMA techniques increase the security of communication networks [9,10].

In 2006, Chang et al. developed a spectral polarization coding approach for implementing complementary bipolar optical correlation in an incoherent bipolar OCDMA (Bi-OCDMA) network [11]. Each decoder employed several fiber Bragg gratings (FBGs) and polarization beam splitters to construct differential photodetectors. The spectral amplitude was incorporated into polarization coding as a specific address code. Their complementary bipolar spectral polarization coding scheme used Hadamard codes as optical codewords for each user, and the coded optical signals were then assigned to either a vertical or horizontal polarization state for polarization coding. Although MAI could be eliminated through correlation subtraction for differential photodetectors, their system had high complexity.

In 2007, Zeng et al. implemented a unipolar-encoding/bipolar-decoding OCDMA scheme that used an electro-optic phase modulator and two FBG arrays in system design [12]. On the transmitter side, a data sequence was used to modulate the phase of the optical carriers, and an FBG encoder array was used for wavelength mapping to an optical phase sequence. On the receiver side, an FBG decoder array was employed as a frequency discriminator to convert phase-modulated optical signals into intensity-modulated signals for optical decoding. However, the decoders used a series of FBGs, which further limited the rate of signal transmission.

In 2018, Patel et al. developed a double-weight code pattern for bipolar codes by using a reconfigurable encoder design [13]. The design increased security against eavesdroppers at the transmitting end. To reconstruct the information of a desired user, the receiver employed complementary subtraction and single photodiode detection. However, the pattern of the code used was straightforward.

In 2019, Filho et al. compared the encoding and decoding of both bipolar and unipolar sequences by using a superstructure FBG (SFBG) [14]. They evaluated SFBG performance in autocorrelation and cross-correlation. This enabled the measurement of unipolar and bipolar coding quality and the effect of multiple users on a network.

In 2020, Ghoumid et al. developed a Bi-CDMA system with a phase shift by applying an E-beam technique to a $H_xLi_{1-x}NbO_3$ transmission channel by using several cascaded Bragg filters and Hadamard codes to conduct several experiments [15]. Because phase shifts must be identified on a coder spectral response, their entire structure is relatively complex. To improve system performance, some researchers have proposed an intelligently structured receiver to suppress noise effects and a semiconductor optical amplifier (SOA) to mitigate temperature variation effects on links [16,17]. In 2018, Yen et al. presented Walsh–Hadamard-code-based OCDMA techniques with moderate security for applications in WOC environments [18].

We developed a simple Bi-OCDMA FSO system by using a family of $M$-sequences and dual electro-optical modulator (EOM) schemes. In this study, the corresponding system was simplified and had the advantages of small size, cost effectiveness, and moderate security. Furthermore, compared with conventional SAC OCDMA schemes, the proposed Bi-OCDMA techniques retained the benefits of the same SAC codec design, MAI alleviation, and complementary keying to enhance overall transmission performance. However, there is a restriction in keeping the properties of two EOMs as similar as possible. Subsequently, we conducted an experiment to test the proposed scheme. The experimental results revealed that the transmission rates for each user can be improved.

The rest of this paper is organized as follows: Section 2 describes the proposed FSO communication system that uses Bi-OCDMA schemes and includes explanation of the coding theory, corresponding system design, and operation. Section 3 details the experimental setup and the results of Bi-OCDMA encoding and decoding. Section 4 describes the FSO system and provides conclusions.

## 2. Development of the Proposed FSO Bi-OCDMA System

$M$-sequences have been used to develop SAC and all-fiber loop vibration sensor systems [19,20]. A family of $M$-sequences that forms all sequences of the same length is used in Bi-OCDMA schemes. Let $\boldsymbol{X}_1$ be a codeword from $M$-sequences as follows:

$$\boldsymbol{X}_1 = [x_1(1),\, x_1(2), \ldots,\, x_1(N)] \tag{1}$$

where $x_k(i)$ is the $i$th element of the $k$th codeword of the $M$-sequence and $N$ is the code length of the $M$-sequence. Subsequently, the cyclic property of $M$-sequences is used to easily generate codewords of the same length $N$ through an operation of $\boldsymbol{X}_{(k+1)} = T^k\boldsymbol{X}_1$, where $k$ is the number of cyclic shifts to the right side. Consider the $M$-sequence for which $N = 3$. The codeword then assigned to each user can be explained as follows:

(1) Codeword assigned to the first user: $\boldsymbol{X}_1 = [101]$;
(2) Codeword assigned to the second user: $\boldsymbol{X}_2 = T^1\boldsymbol{X}_1 = [110]$;
(3) Codeword assigned to the third user: $\boldsymbol{X}_3 = T^2\boldsymbol{X}_1 = [011]$.

With polarization coding and modulation techniques, Bi-OCDMA schemes using $M$-sequence codes can be implemented as follows: The optical signal corresponding to the assigned codeword with a vertical polarization state is transmitted when the data bit of the $k$th user is 1, and that corresponding to the assigned codeword with a horizontal polarization state is transmitted when the data bit of the $k$th user is 0. Table 1 presents $M$-sequences for which $N = 3$ with a bipolar scheme. The subscripts V and H represent optical signals with vertical and horizontal polarization states, respectively. To employ $M$-sequences in the proposed Bi-OCDMA schemes, the results of a correlation with length $N$ could be obtained as follows:

$$R_{XX}^{(nm)}(j,k) = \sum_{i=1}^{N} x_j^{(n)}(i)x_k^{(m)}(i) = \begin{cases} (N+1)/2 \quad , & \text{for } j = k \text{ and } n = m \\ (N+1)/4 \quad , & \text{for } j \neq k \text{ and } n = m \\ 0 \quad , & \text{otherwise} \end{cases} \tag{2}$$

and

$$R_{XX}^{(nm)}(j,\bar{k}) = \sum_{i=1}^{N} x_j^{(n)}(i)\left[1 - x_k^{(m)}(i)\right] = \begin{cases} (N+1)/4 \quad , & \text{for } j \neq k \text{ and } n = m \\ 0 \quad , & \text{otherwise} \end{cases} \tag{3}$$

where $n$ and $m$ represent the optical codewords with individual horizontal and vertical polarization states, respectively. Theoretically, similar to SAC techniques, the following equations can be employed to prevent the influence of MAI.

$$\begin{aligned} &\left[R_{XX}^{(nV)}(j,k) + R_{XX}^{(nH)}(j,\bar{k})\right] - \left[R_{XX}^{(nV)}(j,\bar{k}) + R_{XX}^{(nH)}(j,k)\right] \\ &= \begin{cases} (N+1)/2 \quad , & \text{for } j = k \text{ and } n = \mathrm{V} \\ -(N+1)/2 \quad , & \text{for } j = k \text{ and } n = \mathrm{H} \\ 0 \quad , & \text{otherwise} \end{cases} \end{aligned} \tag{4}$$

**Table 1.** $M$-sequence for which $N = 3$ with a bipolar scheme.

| | Codeword X | Data Bit (D) | Transmitting Optical Signal $\lambda_1$ $\lambda_2$ $\lambda_3$ |
|---|---|---|---|
| User 1 | 1 0 1 | 0<br>1 | $[1\ 0\ 1]_H$<br>$[1\ 0\ 1]_V$ |
| User 2 | 1 1 0 | 0<br>1 | $[1\ 1\ 0]_H$<br>$[1\ 1\ 0]_V$ |
| User 3 | 0 1 1 | 0<br>1 | $[0\ 1\ 1]_H$<br>$[0\ 1\ 1]_V$ |

On the basis of these deductions, the corresponding FSO system using Bi-OCDMA schemes can be developed.

Figure 1 illustrates the design of the proposed FBG encoder with the $M$-sequence for which $N = 3$. The proposed encoder comprises a superluminescent diode (SLD) light source, an optical circulator, a series of FBGs, a 1 × 2 optical splitter, a pattern generator, two polarizers (0° and 90°), a beam splitter (BS), and two EOMs for bipolar coding. First, an SLD output is inserted into Port 1 of the optical circulator and then used as an input for the series of FBGs through Port 2 of the optical circulator. Subsequently, the series of FBGs is employed to reflect specific wavelengths according to the codeword assigned using the $M$-sequence. For example, when $N = 3$, the FBG resonance wavelengths are $\lambda_1$ and $\lambda_3$, corresponding to codeword $X_1 = [1\ 0\ 1]$ for the first user.

Subsequently, the reflected optical signals are entered as an input into Port 2 of the optical circulator, which then provides output from Port 3. Subsequently, these optical signals are coupled into the 1 × 2 optical splitter through an erbium-doped fiber amplifier (EDFA) to compensate for the attenuation of devices in the encoder. The amplified signals are distributed in a parallel manner in the two EOMs for dual EOM modulation. The output signals of the two EOMs are then determined according to the normal (D) and complementary ($\bar{D}$) outputs of the pattern generator, where D is the data bit of the user, represented as "0" or "1". No optical signal appears at the output port of EOM 2 for user data bits of "1". By contrast, no output signal appears at the output port of EOM 1 for user data bits of "0". The outputs of EOMs 1 and 2 are entered as inputs into the vertical and horizontal polarizers, respectively, for polarization coding and are then combined through the BS. For example, if the data bit of user 1 is "1," the BS output corresponds to $[1\ 0\ 1]_V$ and $[0\ 0\ 0]_H$; however, if the data bit of user 1 is "0," the BS output corresponds to $[0\ 0\ 0]_V$ and $[1\ 0\ 1]_H$. Finally, the output of each encoder is coupled into a fiber collimator and transmitted via a wireless optical channel.

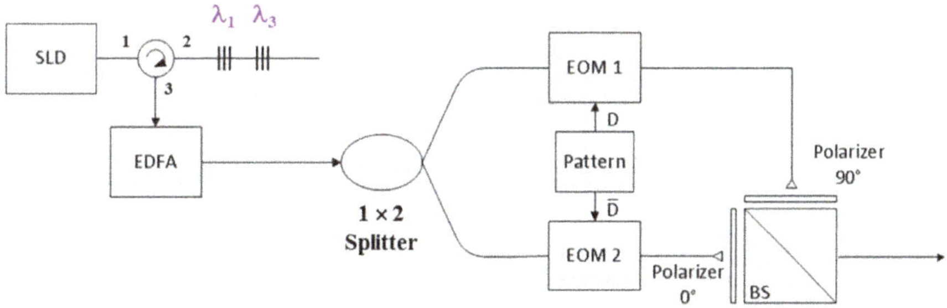


**Figure 1.** Proposed bipolar optical code division multiple access encoder.

During the receiving process, the wireless optical signal is received through a fiber collimator and then distributed to the input port of each decoder. Figure 2 illustrates the structure of the proposed FBG-based Bi-OCDMA decoder, which contains a polarization beam splitter (PBS), two optical circulators, two series of FBGs, two 2 × 1 optical couplers, an attenuator, and a balanced photodetector (BPD) to subtract upper and lower signals and mitigate potential MAI.

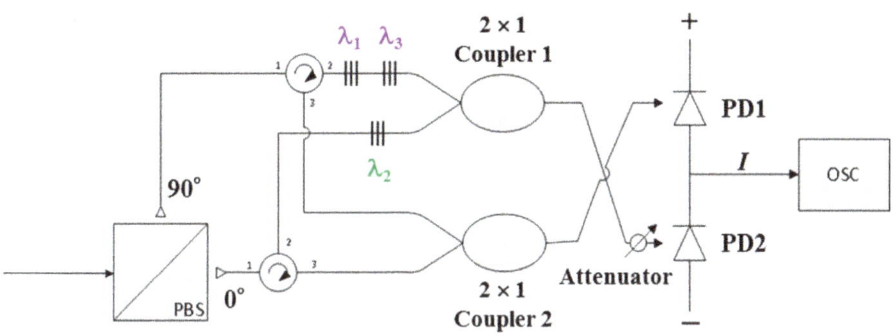


**Figure 2.** Structure of the proposed bipolar optical code division multiple access decoder.

First, the optical signals received from the collimator output are depolarized through the PBS and then used as input for the first ports of two circulators. The received optical signals with vertical and horizontal polarization components appear in the upper arm and lower branch paths, respectively. The two ports of the two optical circulators are connected to the two series of FBGs, which correspond to the normal ($X$) and complementary ($\overline{X}$) codewords. For example, when $N = 3$, the series of upper FBGs reflects the central wavelengths of $\lambda_1$ and $\lambda_3$ that correspond to the normal codeword $X_1 = [1\ 0\ 1]$ in FBG Decoder 1. Similarly, the second series of FBGs reflects the central wavelength of $\lambda_2$, which corresponds to the complementary codeword $\overline{X}_1 = [0\ 1\ 0]$ in FBG Decoder 1. The output signals of the two series of FBGs are coupled into the upper 2 × 1 optical coupler. The lower optical coupler is used to collect optical signals from Port 3 of each of two optical circulators. The output signal received from the upper coupler passes through the attenuator and arrives at the second input port of the BPD. The purpose of this process is to alleviate the influence of unwanted spectral outputs caused by imperfect reflections in the FBG decoder. The output signal of the lower coupler is used as an input to the first input port of the BPD. On the basis of Equations (2) and (3), we developed two models for the outputs of the upper and lower couplers of FBG Decoder 1, namely $F_{11}$ and $F_{21}$, when the codeword of $X_j$ is received, expressed as

$$F_{11} = \left[R_{XX}^{(nH)}(j,1) + R_{XX}^{(nV)}\left(j,\bar{1}\right)\right] = \begin{cases} 0, & \text{for } j = 1 \text{ and } n = \text{V} \\ 2, & \text{for } j = 1 \text{ and } n = \text{H} \\ 1, & \text{for } j = 2,3 \end{cases} \tag{5}$$

and

$$F_{21} = \left[R_{XX}^{(nV)}(j,1) + R_{XX}^{(nH)}\left(j,\bar{1}\right)\right] = \begin{cases} 2, & \text{for } j = 1 \text{ and } n = \text{V} \\ 0, & \text{for } j = 1 \text{ and } n = \text{H} \\ 1, & \text{for } j = 2,3 \end{cases} \tag{6}$$

where $R_{XX}^{(nV)}\left(j,\bar{1}\right)$ and $R_{XX}^{(nH)}(j,1)$ are the optical signals at the upper and lower input ports of the upper coupler, respectively. The expressions $R_{XX}^{(nV)}(j,1)$ and $R_{XX}^{(nH)}\left(j,\bar{1}\right)$ are the optical signals at the upper and lower input ports of the lower coupler, respectively. These optical signals arrive at the input ports of the BPD for correlation subtraction and MAI removal according to the results of Equation (4) in Section 2 and are then converted into electrical signals. Therefore, the output of the BPD can be converted into model ($F$), as follows, when the codeword of $X_j$ is received:

$$F = F_{21} - F_{11} = \begin{cases} 2, \text{ for } j = 1 \text{ and } n = \text{V} \\ -2, \text{ for } j = 1 \text{ and } n = \text{H} \\ 0, \text{ otherwise} \end{cases} \tag{7}$$

Finally, a decision current ($I$) is used to determine the data bit of the desired user from the wireless optical channel.

## 3. Experimental Setup and Results

On the basis of the structure illustrated in Figure 3 with $N$ = 3, the feasibility of the proposed FSO communication system was verified through several experiments by using a model with the following specifications: (1) for the light source, the NXTAR SLD-2000 was adopted. (2) Couplers (Fiber Optic Communications, Inc., Hsinchu, Taiwan) were used as 1 × 2 splitters and 2 × 1 couplers. (3) A Pirelli 10-Gbps integrated optic intensity modulator—which used two EOMs—was used to modulate the signal of the pattern output. (4) An Agilent 81130A pulse pattern generator was used to generate desired patterns for transmission. (5) A left-handed plastic circular polarizer (CP42HE) 12.5 mm in diameter—which used two polarizers (0° and 90°)—was used to assign the light signal to the specific polarization state. (6) A 25-mm nonpolarizing cube BS with a wavelength range of 1100–1620 nm was used to combine optical signals from different paths. (7) A 5-mm VIS polarizing cube BS was used to split two polarization states (0° and 90°) from the input optical signal. (8) A single-mode circulator (1550 nm and 500 mW; FCIR-1550-3-3-A-0-1-2-1-2) was used as the optical circulator. (9) The attenuator range was adjusted from 0 to −30 dB to alleviate the influences of noise and the lower branch signal in the front of PD2. (10) The BPD (Model-1817, New Focus Inc., USA) was used for optical signal subtraction and the conversion of the results into electrical signals. (11) A Tektronix oscilloscope (OSC, model TDS2102B) was used to monitor BPD outputs. (12) A limited power supply of 15 V was provided to the BPD. (13) An Anritsu MS9710C optical spectrum analyzer was employed to assess the accuracy of the spectral output acquired from a codec.

Assume that the first user is the desired user. First, the FBG resonance wavelengths used for the FBG codec are 1543, 1546, and 1549 nm for $\lambda_1$, $\lambda_2$, and $\lambda_3$, respectively. Therefore, user 1 is assigned the codeword $X_1$ = [$\lambda_1$ 0 $\lambda_3$].

Figure 4 presents the measured reflected spectra ($\lambda_1$, $\lambda_3$) for user 1 with corresponding central wavelengths of 1543 and 1549 nm and light intensities of −22.96 and −23.02 dBm, which appeared at the circulator port in Encoder 1. After the reflected spectra passed through the EDFA and 1 × 2 splitter, they were modulated by EOMs 1 and 2 according to the normal (D) and complementary ($\overline{D}$) outputs of the pattern generator, and the spectra were then entered as inputs that were parallel to the input ports of the two polarizers (90° and 0°) to determine suitable polarization states.

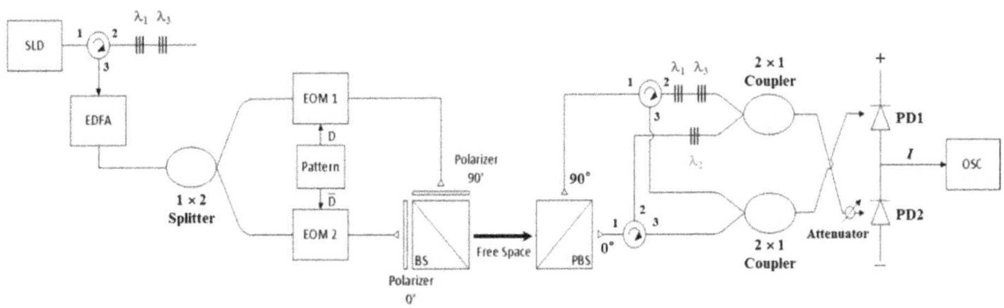


**Figure 3.** Proposed free-space optical communication system using bipolar optical code division multiple access and dual electro-optical modulator schemes.

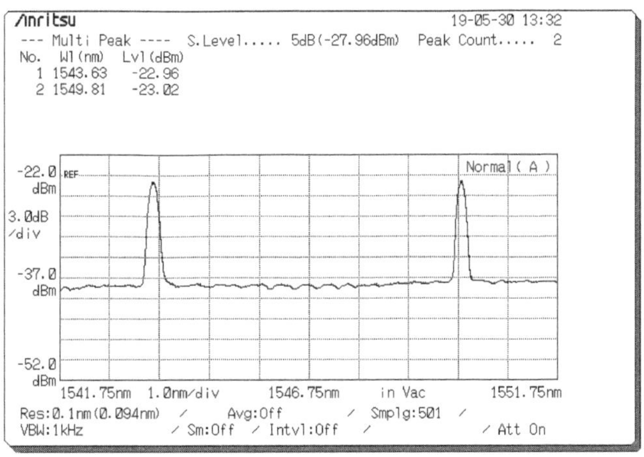


**Figure 4.** Reflected spectra ($\lambda_1$, $\lambda_3$) used for user 1.

Figure 5 presents the spectra obtained at the output ports of the two EOMs operating with different data bits. Figure 5A presents the output spectrum obtained at the output port of EOM 1 when the data bit (D) of user 1 is "1". The central wavelengths of $\lambda_1$ and $\lambda_3$ are 1543 and 1549 nm, respectively, and the corresponding light intensities are −8.82 and −8.77 dBm, respectively. Figure 5B presents the output spectrum obtained at the output port of EOM 2 when the data bit (D) of user 1 is "0". The central wavelengths of $\lambda_1$ and $\lambda_3$ are 1543 and 1549 nm, respectively, and the corresponding light intensities are −10.39 and −10.55 dBm, respectively. Figure 6 indicates that a signal frequency of 500 Hz was acquired from pattern generation and entered as an input into EOMs 1 and 2 of Encoder 1, where the input signals of the two EOMs complemented each other. The output signals of the EOM 1 and EOM 2 were assigned polarization states of 0° and 90°, respectively, and they were then combined into a free-space channel passing through the BS and collimator.

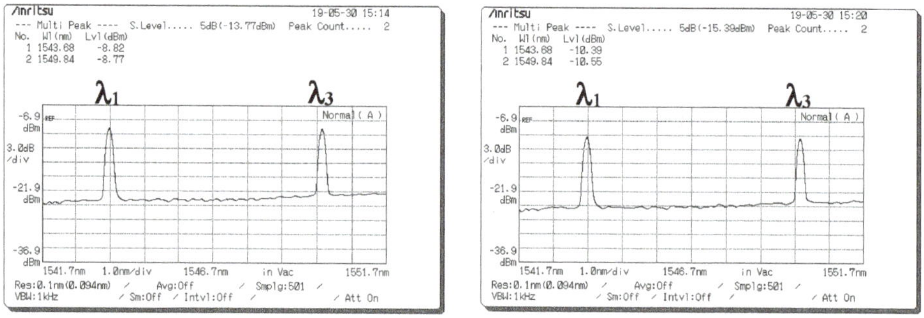


(A) (B)

**Figure 5.** Output spectrum acquired at the output ports of the two electro-optical modulators (EOMs) for different data bits. (**A**) Optical spectra of EOM 1 for a data bit of "1". (**B**) Optical spectra of EOM 2 for a data bit of "0".

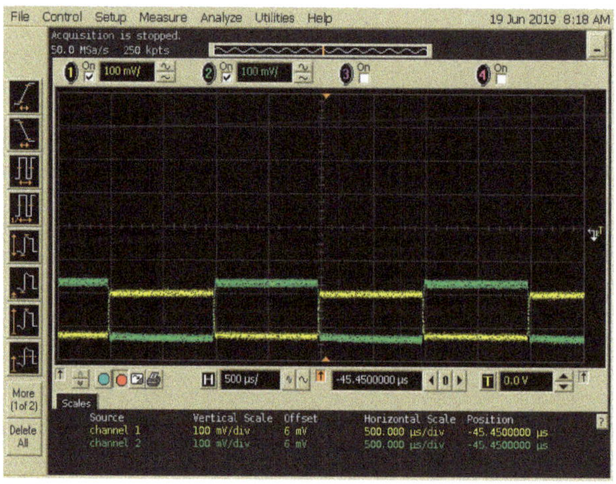


**Figure 6.** Input signals of the two electro-optical modulators (EOMs) when the signals of Channels 1 and 2 were entered as inputs into EOM 1 and EOM 2, respectively.

Figure 7 presents the depolarized spectra obtained at the output ports of the PBS in Decoder 1 when the encoded signal with different data bits was received from Encoder 1. Figure 7A,B presents the depolarized spectra acquired at the output ports of the PBS in Decoder 1 when a data bit (D) of "1" is transmitted. In the depolarized spectra ($\lambda_1$, $\lambda_3$) with the 90°polarization state at the first output port of the PBS, the corresponding central wavelengths were 1543 and 1549 nm, and the light intensities were −25.31 and −24.68 dBm, respectively (Figure 7A). Figure 7B indicates that no signal appeared at the second output port of the PBS when the data bit (D) of user 1 was "1". Figure 7C,D presents the depolarized spectra acquired at the output ports of the PBS in Decoder 1 when a data bit (**D**) of "0" is transmitted. Figure 7C indicates that no signal appeared at the first output port of the PBS when the data bit (D) of user 1 was "0". In the depolarized spectra ($\lambda_1$, $\lambda_3$) with the 0° polarization state at the second output port of the PBS, the corresponding central wavelengths were 1543 and 1549 nm, and the light intensities were −24.06 and −23.9 dBm, respectively (Figure 7D). Therefore, in the experiment, the degree of light intensity demonstrated an approximate 13.35–16.5-dB loss from the EOM output in Encoder 1 to the PBS output in Decoder 1 through the wireless optical channel.

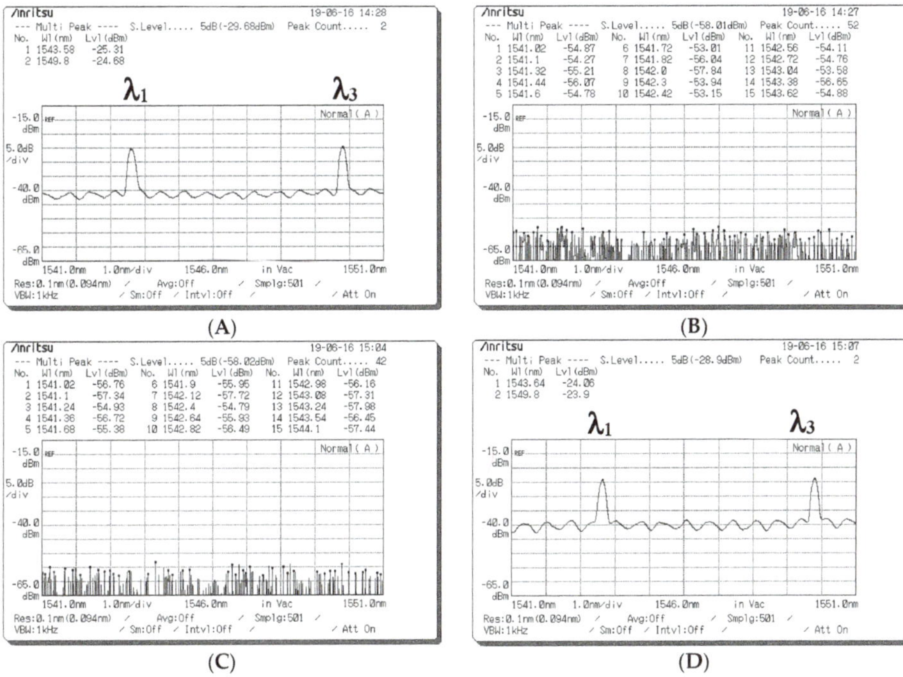


**Figure 7.** Optical spectrum acquired at the output ports of the polarization beam splitter (PBS) in Decoder 1 when different data bits (D) are sent from Encoder 1. Depolarized spectra acquired at (**A**) Output Port 1 and (**B**) Output Port 2 of the PBS when a data bit of "1" is sent. Depolarized spectra acquired at (**C**) Output Port 1 and (**D**) Output Port 2 of the PBS when a data bit of "0" is sent.

Figure 8 presents the measured spectra before the optical signals entered the upper and lower couplers when a data bit (D) of "1" is sent from Encoder 1. Figure 8A presents the spectra measured using the upper optical circulator and FBG Decoder 1 before input was entered in Port 1 of the upper coupler when the a data bit (D) of user 1 was "1". The corresponding central wavelengths were 1543 and 1549 nm, and the light intensities were −34.29 and −35.64 dBm, respectively. Figure 8B indicates that no signal appeared at Port 2 of the upper coupler for user 1's data bit (D) of "1" when the spectra passed through the horizontal (0° component) path; the corresponding central wavelengths were 1543, 1546, and 1549 nm, and all light intensities were less than −53 dBm. Figure 8C presents the decoded spectra that passed through the upper optical circulator and FBG Decoder 1 before it was entered as an input into Port 1 of the lower coupler when the data bit (D) of user 1 was "1". The corresponding central wavelengths were 1543 and 1549 nm, and the light intensities were −28.24 and −27.57 dBm, respectively. Figure 8D indicates that no signal appeared at Port 2 of the lower coupler for user 1's data bit (D) of "1" when the spectra passed through the horizontal (0° component) path; the corresponding central wavelengths were 1543, 1546, and 1549 nm, and all light intensities were less than −53 dBm. Some depolarized spectral outputs were generated because of the imperfect reflection obtained from FBG Decoder 1 (Figure 8A).

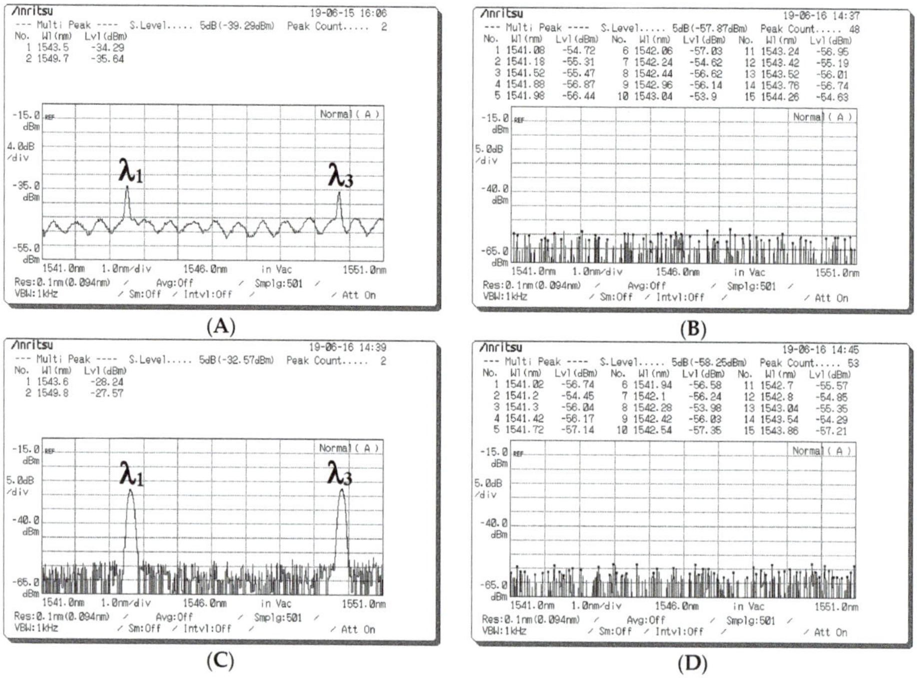


**Figure 8.** Optical spectrum before the optical signals entered upper and lower couplers when a data bit (D) of "1" was sent from Encoder 1. (**A**) Optical spectra obtained at Port 1 of the upper coupler when the optical signals passed through fiber Bragg grating (FBG) Decoder 1. (**B**) Optical spectra obtained at Port 2 of the upper coupler when the optical signals passed through the 0° component path. (**C**) Optical spectra obtained at Port 1 of the lower coupler when the optical signals passed through FBG Decoder 1. (**D**) Optical spectra obtained at Port 2 of the lower coupler when the optical signals passed through the 0° component path.

Figure 9 presents the spectra measured before the optical signals entered the upper and lower couplers when a data bit (D) of "0" was sent from Encoder 1. Figure 9A indicates that no signal appeared at Port 1 of the upper coupler for user 1's data bit (D) of "0" when the spectra passed through the vertical (90° component) path; the corresponding central wavelengths were 1543, 1546, and 1549 nm, and all light intensities were less than −53 dBm. Figure 9B presents the spectra decoded using the lower optical circulator and FBG Complement Decoder 1 before they were input into Port 2 of the upper coupler for user 1's data bit (D) of "0". The corresponding central wavelengths were 1543 and 1549 nm, and the light intensities were −28.97 and −28.56 dBm, respectively. Figure 9C indicates that no signal appeared at Port 1 of the lower coupler for user 1's data bit (D) of "0" when the spectra passed through the vertical (90° component) path; the corresponding central wavelengths were 1543, 1546, and 1549 nm, and all light intensities were less than −53 dBm. Figure 9D presents the optical spectra obtained through the lower optical circulator and FBG Complement Decoder 1 before they were input into Port 2 of the lower coupler for user 1's data bit (D) of "0". The corresponding central wavelengths were 1543, 1546, and 1549 nm, and all light intensities were less than −47 dBm. Some unwanted spectral outputs were produced because of the imperfect upper coupler connection and lower circulator (Figure 9D). However, the decoded output was unaffected by leakage intensities (Figure 9D).

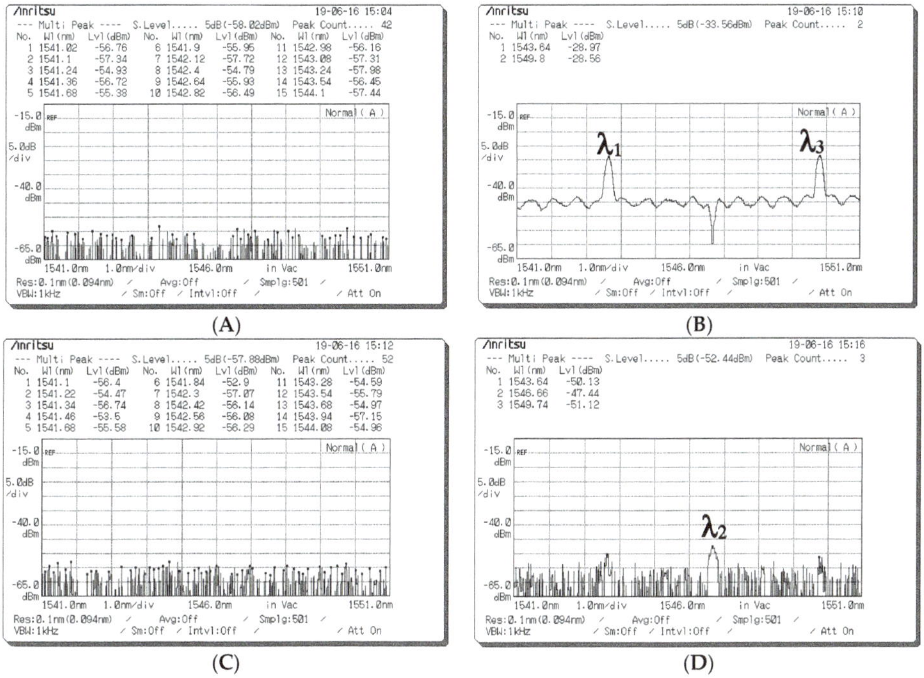


**Figure 9.** Output spectrum before optical signals entered the upper and lower couplers when a data bit (D) of "0" was sent from Encoder 1. (**A**) Optical spectra acquired at Port 1 of the upper coupler when the signal passed through the 90° component path. (**B**) Optical spectra acquired at Port 2 of the upper coupler when the signal passed through fiber Bragg grating (FBG) Complement Decoder 1. (**C**) Optical spectra acquired at Port 1 of the lower coupler when the signal passed through the 90° component path. (**D**) Optical spectra acquired at Port 2 of the lower coupler when the signal passed through FBG Complement Decoder 1.

Figure 10 presents the output spectrum obtained at the output ports of the upper and lower couplers in Decoder 1 when the different data bits (D) of user 1 were sent. Figure 10A,B presents the optical spectra appearing at the output ports of the upper and lower couplers in Decoder 1, respectively, and Figure 10C,D presents those appearing for data bits (D) of "1" and "0" entered as inputs into Encoder 1. Subsequently, the BPD converted the decoded spectra corresponding to its input ports into electrical signals.

Figure 11 presents the decoding results of changing frequencies when data were transmitted from Encoder 1. A digital OSC was used to access the transmitted signal from Encoder 1. In Figure 11A–D, input frequencies of 0.5, 50, 5000, and 10,000 kHz were used as inputs for FBG Encoder 1. Compared with previous systems [21], the novel FSO communication system with the proposed Bi-OCDMA scheme was implemented successfully and further enhanced the overall transmission rate.

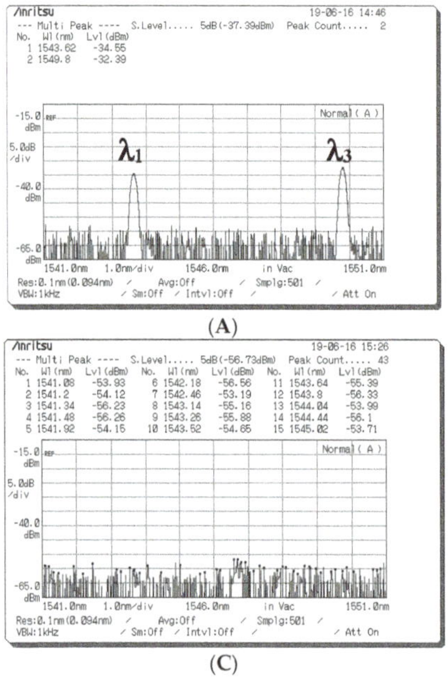


**Figure 10.** Decoded spectra appearing at the output ports of the upper and lower couplers when different data bits (D) of user 1 were sent. Optical spectra obtained at the output port of the (**A**) upper coupler and (**B**) lower coupler for a data bit (D) of "1" for user 1. Optical spectra obtained at the output port of the (**C**) upper coupler and (**D**) lower coupler for a data bit (D) of "0" for user 1.

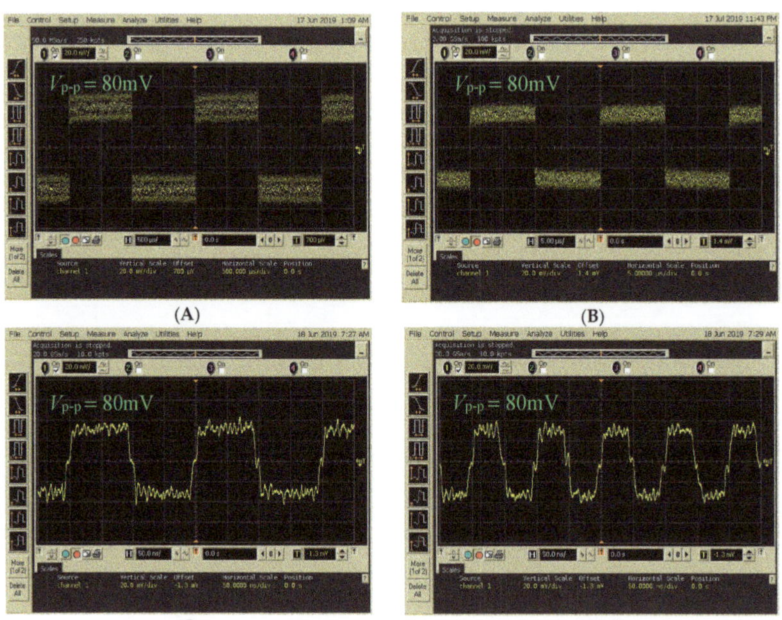


**Figure 11.** Decoding results for Decoder 1 acquired by the balanced photodetector at signal frequencies of (**A**) 500 Hz, (**B**) 50 kHz, (**C**) 5 MHz, and (**D**) 10 MHz input to Encoder 1.

## 4. Conclusions

In this study, the use of Bi-OCDMA with a dual EOM scheme implemented in WOC environments was proposed and successfully demonstrated at normal atmospheric temperatures. FBGs were employed as primary devices for developing the codec. The measurement results of the signal transmission rate revealed that switching limitations of previous systems using an optical switch [21] can be improved through use of the proposed design with a dual EOM structure.

The proposed Bi-OCDMA method is based on original SAC OCDMA techniques, which theoretically alleviate the MAI effect and reduce crosstalk from other FBG encoders. When deployed, the proposed FSO system exhibited excellent properties in terms of its light weight, cost effectiveness, moderate security, and EMI resistance. These properties may further enhance the overall transmission rates of WOC applications in the near future.

Future work can apply the proposed Bi-OCDMA technique to multiuser and long-distance WOC scenarios that involve MAI mitigation and performance measurement by using parameters such as the bit error rate, Q-factor, and eye diagrams.

**Author Contributions:** Conceptualization, S.-P.T. and H.-C.C.; methodology, S.-P.T. and H.-C.C.; validation, S.-P.T. and H.-C.C.; resources, H.-C.C.; data curation, P.-H.L.; writing—original draft preparation, E.W.; writing—review and editing, S.-P.T. and H.-C.C. All authors have read and agreed to the published version of the manuscript.

**Funding:** This work was supported by the Ministry of Science and Technology, Taiwan, under Grant MOST 108-2221-E-150-041 and Grant MOST 108-2221-E-239-003.

**Conflicts of Interest:** The authors declare no conflict of interest.

## References

1. Ahmed, A.; Singh, A.; Kaur, S. Performance analysis of WDM-MIMO free space optical system under atmospheric turbulence. In Proceedings of the 6th International Conference on Signal Processing and Integrated Networks (SPIN), Noida, India, 7–8 March 2019; pp. 820–825.
2. Mahloo, M.; Chen, J.; Wosinska, L.; Dixit, A.; Lannoo, B.; Colle, D.; Machuca, C.M. Toward reliable hybrid WDM/TDM passive optical networks. *IEEE Commun. Mag.* **2014**, *52*, S14–S23. [CrossRef]
3. Rommel, S.; Perez-Galacho, D.; Fabrega, J.M.; Muñoz, R.; Sales, S.; Monroy, I.T. High-capacity 5G fronthaul networks based on optical space division multiplexing. *IEEE Trans. Broadcast.* **2019**, *65*, 434–443. [CrossRef]
4. Li, Q.; Xu, S.; Yu, J.; Yan, L.; Huang, Y. An improved method for the position detection of a quadrant detector for free space optical communication. *Sensors* **2019**, *19*, 175. [CrossRef] [PubMed]
5. Cheng, H.C.; Wu, C.H.; Yang, C.C.; Chang, Y.T. Wavelength division multiplexing/spectral amplitude coding applications in fiber vibration sensor systems. *IEEE Sens. J.* **2011**, *11*, 2518–2526. [CrossRef]
6. Hunter, D.B.; Minasian, R.A. Programmable high-speed optical code recognition using fiber Bragg grating arrays. *Electron. Lett.* **1999**, *35*, 412–414. [CrossRef]
7. Salehi, J.A.; Weiner, A.M.; Heritage, J.P. Coherent ultrashort light pulse code-division multiple access communication systems. *IEEE J. Light. Technol.* **1990**, *8*, 478–491. [CrossRef]
8. Zefreh, M.R.; Salehi, J.A. Theoretical studies of ultrashort light pulse spectrally-phase-encoded OCDMA system using power-cubic optical nonlinear preprocessor. *IEEE J. Light. Technol.* **2015**, *33*, 5062–5072. [CrossRef]
9. Sikder, S.; Sarkar, M.; Ghosh, S. Optical network security using unipolar Walsh code. In AIP Conference Proceedings, Secunderabad, India, 22–23 December 2017; pp. 1–4.
10. Farghal, A.E.A.; Shalaby, H.M.H. Reducing inter-core crosstalk impact via code-interleaving and bipolar 2-PPM for core-multiplexed SAC OCDMA PON. *J. Opt. Commun. Netw.* **2018**, *10*, 35–45. [CrossRef]
11. Chang, Y.T.; Huang, J.F. Complementary bipolar spectral polarization coding over fiber-grating-based differential photo detectors. *Opt. Eng.* **2006**, *45*, 045004. [CrossRef]
12. Zeng, F.; Wang, Q.; Yao, J. Sequence-inversion-keyed optical CDMA coding/decoding scheme using an electrooptic phase modulator and fiber Bragg grating arrays. *IEEE J. Sel. Top. Quant.* **2007**, *13*, 1508–1515. [CrossRef]

13. Patel, S.P.; Gupta, S. Novel bipolar reconfigurable code for OCDMA network. *Int. J. Adv. Eng. Res. Dev.* **2018**, *5*, 493–500.
14. Filho, A.; Sales, J.C.; Sousa, J.R.; Freire, M.M.; Alves, G.I.; Sombra, S. Sensors of encoding and decoding based on superstructure fiber Bragg gratings modulated in amplitude and phase for applications in systems OCDMA-63 chip. *Fiber Integr. Opt.* **2019**, *38*, 349–361. [CrossRef]
15. Ghoumid, K.; Ghadban, A.; Boukricha, S.; Ar-reyouchi, E.M.; Yahiaoui, R.; Mekaoui, S.; Raschetti, M.; Lepers, C. Spectral coded phase bipolar OCDMA technological implementation thanks to low index modulation filters. *Telecommun. Syst.* **2020**, *73*, 433–441. [CrossRef]
16. Ahmed, M.S.; Glesk, I. Application of semiconductor optical amplifier (SOA) in managing chirp of optical code division multiple access (OCDMA) code carriers in temperature affected fibre link. *App. Sci.* **2018**, *8*, 715. [CrossRef]
17. Chen, K.S.; Chen, Y.C.; Liao, L.G. Advancing high-speed transmissions over OCDMA networks by employing an intelligently structured receiver for noise mitigation. *Appl. Sci.* **2018**, *8*, 2408. [CrossRef]
18. Yen, C.T.; Huang, J.F.; Zhang, W.Z. Hiding stealth optical CDMA signals in public BPSK channels for optical wireless communication. *Appl. Sci.* **2018**, *8*, 1731. [CrossRef]
19. Tseng, S.P.; Wu, J. Extended M-sequence codes for SAC FO-CDMA PONs applications. *Electron. Lett.* **2008**, *44*, 488–490. [CrossRef]
20. Tseng, S.P.; Yen, C.T.; Su, R.S.; Cheng, H.C. Employing optical code division multiple access technology in the all fiber loop vibration sensor system. *Opt. Fiber Technol.* **2013**, *19*, 627–637. [CrossRef]
21. Cheng, H.C.; Wijanto, E.; Lien, T.C.; Lai, P.H.; Tseng, S.P. Multiple access techniques for bipolar optical code division in wireless optical communications. *IEEE Access* **2020**, *8*, 83511–83523. [CrossRef]

*sensors*

MDPI

*Letter*

# Compressed Nonlinear Equalizers for 112-Gbps Optical Interconnects: Efficiency and Stability

**Wenjia Zhang *, Ling Ge, Yanci Zhang, Chenyu Liang and Zuyuan He**

State Key Laboratory of Advanced Optical Communication Systems and Networks, Shanghai Jiao Tong University, Shanghai 200240, China; mrgreen@sjtu.edu.cn (L.G.); zhangyanci@sjtu.edu.cn (Y.Z.); lcy1993@sjtu.edu.cn (C.L.); zuyuanhe@sjtu.edu.cn (Z.H.)

* Correspondence: wenjia.zhang@sjtu.edu.cn; Tel.: +86-21-34204316

Received: 12 May 2020; Accepted: 17 August 2020; Published: 19 August 2020

**Abstract:** Low-complexity nonlinear equalization is critical for reliable high-speed short-reach optical interconnects. In this paper, we compare the complexity, efficiency and stability performance of pruned Volterra series-based equalization (VE) and neural network-based equalization (NNE) for 112 Gbps vertical cavity surface emitting laser (VCSEL) enabled optical interconnects. The design space of nonlinear equalizers and their pruning algorithms are carefully investigated to reveal fundamental reasons of powerful nonlinear compensation capability and restriction factors of efficiency and stability. The experimental results show that NNE has more than one order of magnitude bit error rate (BER) advantage over VE at the same computation complexity and pruned NNE has around 50% lower computation complexity compared to VE at the same BER level. Moreover, VE shows serious performance instability due to its intricate structure when communication channel conditions become tough. Moreover, pruned VE presents more consistent equalization performance within varying bias values than NNE.

**Keywords:** VCSEL; neural network-based equalization; Volterra series-based equalization

---

## 1. Introduction

Recent decades have witnessed the explosion of data traffic especially in regional spaces like data centers and supercomputers. The hardware infrastructure for supporting such massive connectivity has been turning to optical components and fibers even over a distance less than 100 m. The data rate of optical interconnects have exceeded 100 Gbps per lambda in commercialized products and will soon upgrade to 200 Gbps per lambda for 800 Gbps or 1 Tbps optical module [1]. Vertical cavity surface emitting laser (VCSEL)-based optical interconnect is a typical and competitive candidate because of interesting features of low cost and power consumption [2]. To meet capacity, cost and power consumption requirements, intensity modulation and direct detection (IM-DD) with more advanced modulations like four- or eight-level pulse amplitude modulation (PAM4/8) [3], carrier-less amplitude phase modulation (CAP) [4] and discrete multi-tone modulation (DMT) [5] has been proposed for 100/200 Gbps per lambda optical interconnects, in which PAM4 has become a widely accepted modulation format for 400G products. It is of necessity rather than an option to employ advanced modulations for improving spectral efficiency due to severe bandwidth limitation and noise accumulation of devices and channels. However, advanced modulations, though mitigating the system requirement on frequency response of critical devices, will also bring serious modulation non-linearity [6,7], reduced signal–noise ratio (SNR), and level-dependent noise accumulation [8]. In particular, VCSEL-based interconnect solution confronts large challenges to realize reliable 100/200 Gbps per lambda communications with advanced modulations mostly because of bandwidth limitation [9] and complex impairment combination of relative intensity noise (RIN) [10], mode partition noise (MPN) [11] and fiber mode dispersion. The complicated signal distortions in time, amplitude and frequency domain need advanced nonlinear equalization for compensation and adaption [3,6,8,11–14].

Advanced digital signal processing (DSP) technologies-based nonlinear equalization, such as Volterra series-based equalization (VE) [3,15–18] and neural network-based equalization (NNE) [12,19–22], have been proposed to realize powerful nonlinear compensation capability in the VCSEL enabled optical interconnect. However, high computation complexity of these nonlinear equalization hinders their practical implementation in the cost and power-sensitive module. Fortunately, it has been proved that there are various design spaces for nonlinear equalization to reduce computation complexity without sacrificing communication quality [3,12]. Therefore, only by marrying strong algorithms with limited computing and storage resources of underlying hardware, nonlinear equalization can find their values in the next wave of hardware evolution for a high-end optical module. There are various low-complexity designs of VE and NNE in the literature. In [23], coefficients in Volterra series that are not on the diagonal have been intentionally abandoned to construct a memory polynomial Volterra equalizer. However, this approach is not able to adapt to various transmission conditions since that simplification process runs before training. Moreover, in [12,17,24], the authors employ an approach of weight-pruning and retraining to reach a sparse but optimized equalizer structure for VE or NNE. Additionally, more aggressive methods [3,16,18] by introducing $\ell_0$ and $\ell_1$-regularization penalty term into original cost function for pruning Volterra nonlinear equalizer are proposed to force large amount of coefficients to zeros, demonstrating a over 93% complexity reduction in the 112 Gbps VCSEL-based optical interconnects. However, equalization architectures for Volterra series and neural network are fundamentally different in the nonlinear construction. In addition, there is tremendous difference in the growth pattern of computation complexity with construction parameters. So, pruning strategies should be carefully designed for two equalization architectures for exploiting best potentials. Questions are raised on which nonlinear equalization prevails over others of realizing lowest complexity equalization while keeping the reliable transmission performance [25]. More importantly, it is still unclear the window of key parameter variation for a working channel that equalization performance gain can keep consistent.

In this paper, we compare the efficiency and stability of a pruned three-layer neural network-based equalizer and a pruned three-order Volterra series-based equalizer for VCSEL enabled 112 Gbps optical interconnects. A threshold-based pruning and retraining algorithm, which is proposed in our previous publications [3,12], is used to compare the pruning efficiency of nonlinear equalizers. In this comparison study, we extend the conference paper [25] by elaborating design space of nonlinear equalizers, including engineering parameters on memory length, neural network layer and activation function, in order to reveal fundamental reasons of powerful nonlinear compensation capability and restriction factors of efficiency and stability. In addition, signaling performance under different bias voltages for a VCSEL is evaluated by using pruned VE and pruned NNE, respectively.

## 2. Principle of Pruned Nonlinear Equalizers

### 2.1. Volterra Series and Neural Network-Based Equalizer

The mathematical expression of $P$-order VE with memory length of $M_r$ can be expressed as Equation (1),

$$y(k) = W_{dc} + \sum_{r=1}^{P} \sum_{k_1=0}^{M_r-1} \cdots \sum_{k_r=k_{r-1}}^{M_r-1} W_r(k_1, k_2, ..., k_r) \cdot x(k-k_1) \cdots x(k-k_r). \tag{1}$$

where $x(k)$ is $k^{th}$ sampled data from received signals and $y(k)$ is output data through equalization. $W_r$ is the $r^{th}$-order Volterra kernel. $W_{dc}$ is responsible for $DC$ component, which is not included in the final model of an AC-coupled circuit. This model presented in Equation (1) can approximate any nonlinear system in theory. However, it is physically impossible in many cases due to high computation complexity. As low-order part of Volterra series model indicates most features of nonlinear system, approximation approaches are often used to truncate theoretically infinitely long expressions.

It can be seen from Equation (1), the number of coefficients, indicating computation burden of multiplication and addition, will grow very fast with increase of $P$ and memory length $M_r$. It has been known that multiplication contributes most of computation resources compared to addition and will be main reason of complexity effect of pruning algorithms [18]. Therefore, computation complexity for a $P$-order VE with memory length of $M_r$ can be defined as the number of multiplication operations [18] and expressed as Equation (2),

$$Complexity = M_1 + M_2(M_2+1) + M_3(M_3+1)(M_3+2)/2 + \cdots \tag{2}$$

According to [3,26], a three-order structure is sufficient for short-reach optical interconnects. By expanding Equation (1), a three-order VE, named by $VE(M_1, M_2, M_3)$ where $M_1, M_2, M_3$ are memory lengths of first, second and third order of VE, is described by

$$\begin{aligned} y(k) = & \sum_{k_1=0}^{M_1-1} w_1(k_1)x(k-k_1) \\ & + \sum_{k_1=0}^{M_2-1}\sum_{k_2=k_1}^{M_2-1} w_2(k_1,k_2)x(k-k_1)x(k-k_2) \\ & + \sum_{k_1=0}^{M_3-1}\sum_{k_2=k_1}^{M_3-1}\sum_{k_3=k_2}^{M_3-1} w_3(k_1,k_2,k_3)x(k-k_1)x(k-k_2)x(k-k_3). \end{aligned} \tag{3}$$

A basic $VE(2,2,2)$ structure with 9 coefficients is shown in Figure 1a. In this work, a three-order VE ($P = 3$) is applied for further experimental investigation.

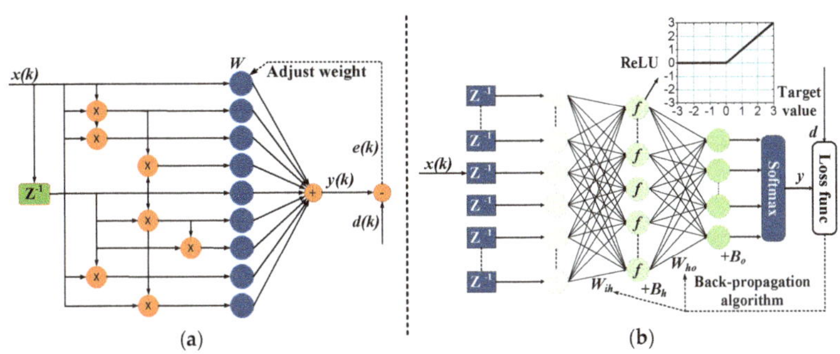


**Figure 1.** The structure of (**a**) three-order $VE(2,2,2)$ and (**b**) three-layer $NNE$.

Artificial neural networks have been widely deployed in various areas such as image classification and natural language processing, showing much better performance than traditional algorithms. Recently, it has been intensively researched in the field of optical communication, in which nonlinear equalization is one of hottest topic [13]. To equalize PAM-N communication signals, the received analog signals are taken to a neural network with N possible outputs, representing N levels. The signal stream is first delayed through a delay array, and then input into an input layer with designed neuron number. The numbers of input neuron and layer number are also of great significance to balance the equalization performance and computation complexity. According to universal approximation theorem, a three-layer neural network will provide strong nonlinear equalization capability for the communication systems.

The mathematical formula of NNE process can be expressed as Equation (4),

$$y = argmax[softmax(f(x^T(k) \times W_{ih} + B_h) \times W_{ho} + B_o)] \tag{4}$$

where $x(k)$ is a sampled signal sequence, $W_{ih}$ and $W_{ho}$ are weight matrices of input layer to hidden layer and hidden layer to output layer respectively. $B_h$ and $B_o$ are bias vectors of hidden layer and output layer. $f$ means activation function of hidden layer. Rectified linear unit (ReLU), tangent hyperbolic (Tanh) and sigmoid functions are considered in this work. Function $softmax(\cdot)$ is used to convert the results of output layer to probability distribution for each class. Finally, $argmax[\cdot]$ is decision function, which returns an index of the maximum value of output probability vector. $y$ is final result indicating a specific symbol of information. Loss functions of ANN, including mean square error (MSE) loss function, logarithmic loss function and cross-entropy loss function, are important for training model and final performance. The MSE loss function usually shows slow convergence and is seldom used in classification task. In this work, back propagation (BP) algorithm combined with cross-entropy loss function are used to train NNE to achieve better classification performance.

The structure of the 3-layer NNE is shown in Figure 1b, where $NNE(N_1, N_2, N_3)$ represents a 3-layer NNE with $N_1, N_2, N_3$ as neuron number of input, hidden and output layer. Similar to VE, computation complexity for NNE is defined as number of multiplication operation as calculated in Equation (5),

$$Complexity = N_1 \times N_2 + N_2 \times N_3 + N_3 \times N_4 + \cdots \tag{5}$$

### 2.2. Pruning Algorithm

It has been proved that most of multiplication operations in the nonlinear equalizers are redundant [23] as either weight value is small or connections in network level contribute very little to final results. Therefore, there is possibility of best pruning algorithm that can realize a lightest equalizer without sacrificing transmission performance. The choice of strategy to prune the complex networks as shown in Figure 1 become very important to reduce computation complexity for each architecture while maintaining equalization performance. We proposed a threshold-based pruning and retaining approach for equalizers, where coefficients after initial training are intentionally discarded based on a threshold and damages caused by pruning process are recovered as much as possible through retraining [3,12]. The pruning process can be easily expressed as Equation (6), where $S(\cdot)$ represents weight setting and $T$ is threshold. With weight value setting to zero, connections relating to this value is cut in these complex networks. Due to different features of two architectures, there also needs special design considerations for each equalizer. For VE, second- and third-order coefficients are pruned because we find linear terms only occupy a small amount of computation complexity while contribute most on the performance gain due to bandwidth limitation of the experimental system. For NNE, all weights both from input layer to hidden layer and hidden layer to output layer are run for pruning based on the threshold. An iterative pruning process is developed to achieve a more efficient network structure.

$$S(W) = 0, \quad when \quad S(\cdot) < T. \tag{6}$$

## 3. Experiment and Results

### 3.1. Experiment Setup

Figure 2 shows experimental setup for 112 Gbps optical interconnects. In the transmitter, 112 Gbps pseudo-random binary sequence (PRBS)-11 PAM-4 signals are generated by an arbitrary waveform generator (AWG), which is Keysight M8195A with 64-GSa/s sampling rate and 25-GHz 3dB bandwidth. Please noted that the waveform generator is bandwidth limited for generating 112 Gbps PAM4 signals, but this inherent bandwidth limitation can be mitigated by pre-distortion and equalization. PRBS-11 is used due to memory limitation of our AWG. High-speed data-streaming combined with DC bias are applied on an 850 nm multimode VCSEL bare die through a high-speed probe. Light-current-voltage

(LIV) curve and 3-dB bandwidth of VCSEL can be found in the inset of the Figure 2. Lights from VCSEL are coupled into a 100 m OM3 multimode fiber through a three-dimension coupling station. Through a variable optical attenuator (VOA), lights are detected by a photodetector (PD, New Focus 1484-A-50) integrated with a transimpedance amplifier (TIA). At the receiver, detected signals are sampled and captured by a real-time digital storage oscilloscope (DSO, Keysight DSOZ592A) with 160-GSa/s sampling rate and 59-GHz analog bandwidth for offline digital signal processing. The captured data are re-sampled to one sample per symbol, followed by nonlinear equalization and bit error rate (BER) calculation.

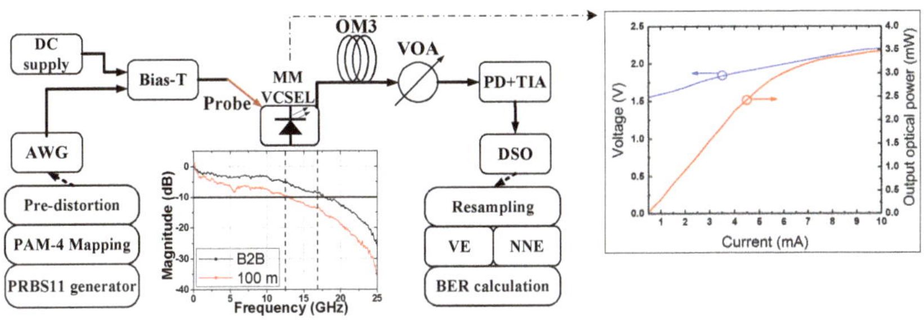


**Figure 2.** Experimental setup.

To avoid the risk of performance over-estimation from nonlinear equalization, we randomly disrupt the signals to create a sequence without periodicity before the received signals are fed into equalizer and conducted mini-batch training [12]. After such process, a disrupted sequence, with an ultra-long unrepeated pattern, is used to train equalizer, which enables equalizer to learn system characteristics instead of generation rules of PRBS.

### 3.2. *Efficiency Comparison*

First, memory length should be carefully tuned for VE and NNE to better understand the efficiency of pruning algorithms. If these parameters are intentionally set with large numbers, complexity reduction ratio will be exaggerated through pruning due to expanded redundancy, which is not fair for comparison. Therefore, we have tuned memory length of first order in VE by setting high-order memory length to zeros. Figure 3a shows that BER keeps stable when memory length of this order is from 51 to 251 for 100 Gbps PAM4 transmission, which is similar for back-to-back (B2B) and 100 m transmission with various received optical powers. Therefore, 51 is set for the first-order memory length. For nonlinear terms selection, second- and third-order memory lengths are increased simultaneously while trying to make their coefficient number in a similar level in order to balance performance gain introduced by second- and third-order term. For NNE, we fix the number of hidden layer neurons as 51 and only change the number of input symbols from 9 to 51. Figure 3b shows BER is reduced with increasing input symbol. The smooth BER change, without large-scale jump between any adjacent two points, also indicates that NNE in our experiment does not obtain over-estimation gain thanks to the use of randomization process, since that over-estimation problem for NNE will lead to stair-like curve of BER versus input symbol number [27]. Moreover, design space for NNE also includes activation function and layer number, which influence equalization performance. As shown in Figure 4, ReLU and Tanh activation function show better BER performance than Sigmoid. Tanh has similar BER with ReLU when optical power is smaller than 2-dBm while ReLU will achieve better performance when received optical power is more than 2-dBm. We can also learn from Figure 4 that BER performance can be slightly improved by extending the layer number of NNE from three to four with the same activation function of ReLU through comparing a three-layer $NNE(51, 51, 4)$ with a four-layer $NNE(51, 51, 51, 4)$. However, more BER performance improvement can be obtained by increasing input neuron number to 71 by comparing three-layer $NNE(51, 51, 4)$ with $NNE(71, 71, 4)$.

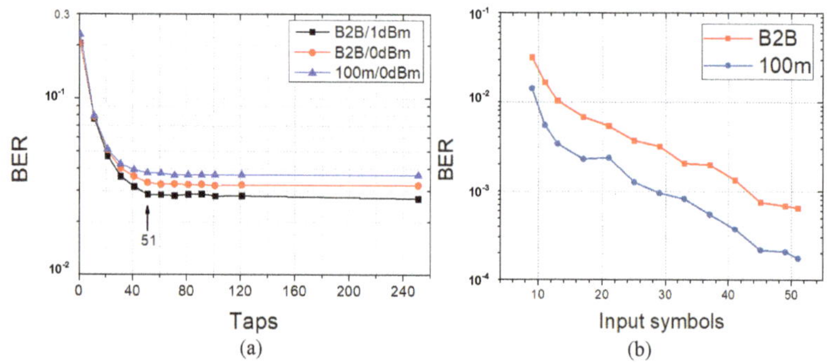


**Figure 3.** (**a**) BER versus memory length of first-order VE with received optical power of 1 dBm and 0 dBm (B2B) and 0 dBm (100 m); (**b**) BER versus input symbol number with 51 hidden neurons and three layers and received optical power of 1 dBm.

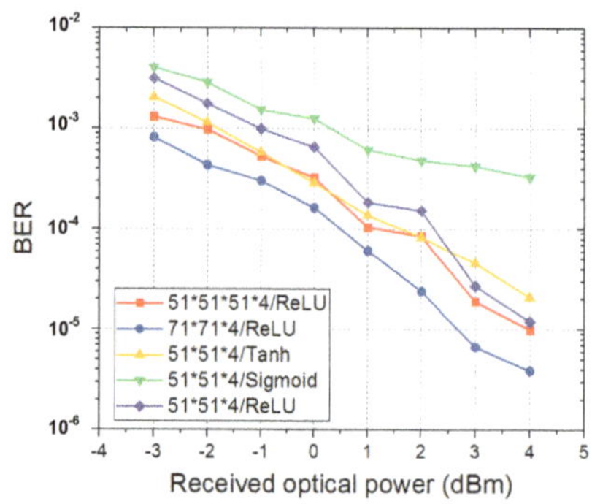


**Figure 4.** Impact of activation function and layer number of NNE on equalization performance.

Second, 100 Gbps PAM4 signaling at a typical 1-dBm received optical power is used for comparing efficiency of VE and NNE and their pruning version. Figure 5 shows the experimental results at B2B and 100 m MMF cases. We carefully set the initial equalizer configurations of $VE(51, 31, 15)$ and $NNE(51, 56, 4)$ with the initial complexity 3083 for VE and 3080 for NNE in order to balance the initial complexity of VE and NNE for fair comparison. As we can see from Figure 5, NNE presents more than one order of magnitude BER advantage over VE for both B2B and 100 m MMF cases at the same computation complexity. By using pruning algorithms, lower computation complexity could be achieved for both VE and NNE. At 7% HD FEC limit of 100 m MMF case, complexity values are 1144 and 603 for general VE and NNE and is reduced to 743 and 386 with pruning algorithms, in which pruned NNE shows 48% complexity performance improvement than pruned VE.

Third, in order to better understand behavior of VE and NNE, we increase the data rate to 112 Gbps and eliminate the pre-distortion process for pressuring more on the bandwidth limitation as indicated in the inset of Figure 1. Figure 6 illustrates experimental results with received optical power of 0-dBm. Noted that the computation complexity of equalizers without pruning process is changing with memory lengths of equalization architecture. From the Figure 6, VE shows serious numerical instability due to its intricate structure. With memory length increasing, BER performance powered by VE shows a slow reduce and even starts to increase until misconvergence with BER of 0.5. On the other hand, NNE is still able to achieve remarkable equalization performance for B2B and 100 m cases. Therefore, NNE has significantly more robust performance than VE in a harsh link situation. With pruning algorithms, NNE attains around 50% complexity reduction compared to conventional NNE at the FEC limit of BER with $3.8 \times 10^{-3}$. In addition, at the same complexity, equalizer without pruning algorithms shows worse BER performance due to reduced memory length. From Figure 6, we also learn that VE outperforms NNE when complexity is lower than 800 since that VE can always maintain the least linear equalizer. However, the performance of NNE, on the other hand, will decline rapidly after large-scale pruning, as NNE fails to have the property of separate order like VE.

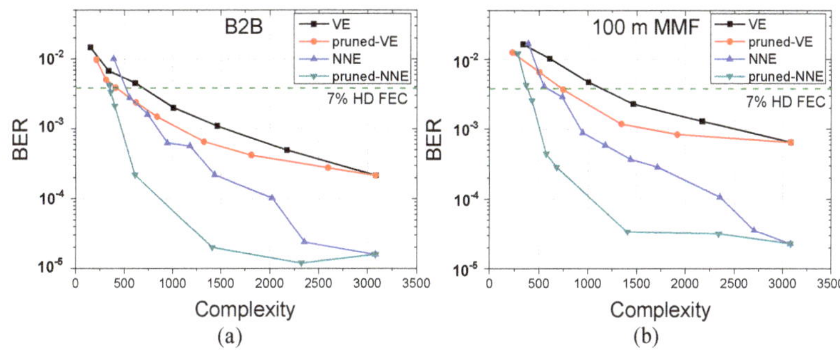


**Figure 5.** The efficiency comparison for (**a**) B2B case and (**b**) 100 m MMF case at 100 Gbps PAM-4 with pre-distortion.

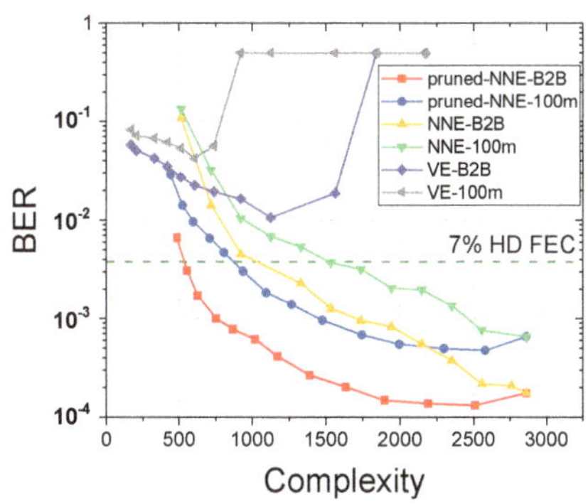


**Figure 6.** The efficiency comparison at 112 Gbps PAM-4 without pre-distortion.

### 3.3. Stability Comparison

Performance stability for equalization is of great significance in the real application where output optical power or 3-dB bandwidth of critical devices in the optical interconnect will change within a life-cycle because of component aging or ambient temperature fluctuation. It is likely that performance stability will become worse with pruning algorithms since that the redundancy of connections in the equalization will be useful in changing scenarios. To verify the stability of pruning algorithm with changing link conditions, we choose DC bias of VCSEL as a key parameter for further evaluation because DC bias variation of VCSEL will not only change the output power but also bandwidth of directly modulated laser. The transmission experiment is run with a 100 Gbps PAM4 pre-distorted link. We set $VE(51, 23, 11)$ with complexity of 1461 and $NNE(31, 41, 4)$ with complexity of 1435, respectively. BER performance under different bias is shown in Figure 7. The BER bathtub curve indicates that equalization performance variation, instead of bandwidth variation with bias current, is the main reason of BER deviation. First, BER is calculated as function of bias for NNE and VE with B2B and 100 m condition. Figure 7a shows stable BER level with bias from 4 mA to 10 mA. 6 mA, the optimal operating point of VCSEL, is chosen as a benchmark point. Second, pruning algorithms for VE and NNE are running and optimized at the bias of 6 mA. The same pruned equalizers are then applied to data received with a different bias current for VCSEL. As we can see from Figure 7b, the pruned VE optimized for 6 mA bias also shows excellent tolerance for bias from 7 mA to 9 mA. This means that sparse VE structure has consistent equalization performance within a bias range of 2.5 mA in this experiment. However, this is not case for NNE. Pruned NNE, optimized using data of 6 mA bias, seems less effective and more fluctuating for cases with a different bias. The experimental results in Figure 7b of fluctuation BER with pruned NNE shows that the iterative pruning and retraining approach has broken the equalization resilience due to reduced connections between layers.

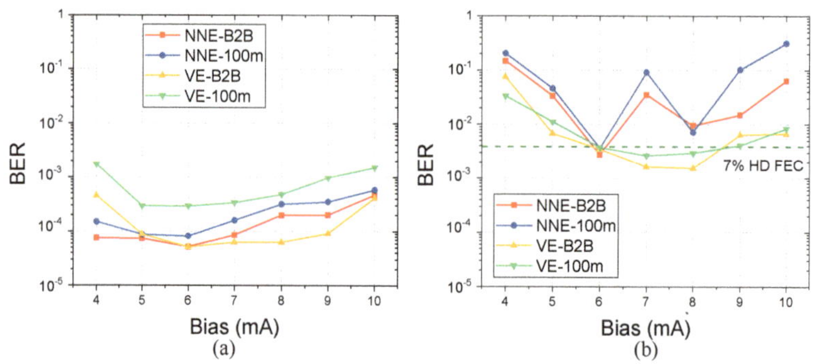


**Figure 7.** BER versus VCSEL DC bias for (**a**) VE and NNE (**b**) Pruned VE and NNE.

## 4. Conclusions

In this paper, we compare the efficiency and stability performance of pruned Volterra series-based equalization and neural network-based equalization for 112 Gbps VCSEL enabled optical interconnects. From the 112 Gbps PAM4 transmission experiment, we can conclude that NNE has more than one order of magnitude BER advantage over VE at the same computation complexity. By using pruning algorithms, NNE has around 50% lower computation complexity compared to VE at the same BER level. Moreover, when communication channel condition become tough, VE shows serious performance instability due to its intricate structure. Through performance stability evaluation by changing bias current of a VCSEL, pruned VE shows more consistent equalization performance than NNE.

**Author Contributions:** Conceptualization, W.Z. and Z.H.; methodology, L.G.; validation, Y.Z. and C.L.; writing–original draft preparation, W.Z.; writing–review and editing, Z.H. All authors have read and agreed to the published version of the manuscript.

**Funding:** This research was funded by National Key Research and Development Program of China (2019YFB1802903) and Open Projects Foundation of Yangtze Optical Fiber and Cable Joint Stock Limited Company (YOFC) (grant No.SKLD1903).

**Conflicts of Interest:** The authors declare no conflict of interest.

## Abbreviations

The following abbreviations are used in this manuscript:

| | |
|---|---|
| VE | Volterra series-based equalization |
| NNE | Neural network-based equalization |
| VCSEL | Vertical cavity surface emitting laser |
| BER | Bit error rate |
| IM-DD | Intensity modulation and direct detection |
| PAM | Pulse amplitude modulation |

| | |
|---|---|
| CAP | Carrier-less amplitude phase modulation |
| DMT | Discrete multi-tone modulation |
| SNR | Signal–noise ratio |
| RIN | Relative intensity noise |
| MPN | Mode partition noise |
| DSP | Digital signal processing |
| PRBS | Pseudo-random binary sequence |
| VOA | Variable optical attenuator |
| PD | Photodetector |
| TIA | Transimpedance amplifier |
| DSO | Digital storage oscilloscope |
| B2B | Back-to-back |
| Tanh | Tangent hyperbolic |
| ReLU | Rectified linear unit |

## References


1. Wu, X.; van den Borne, D.; Maki, J.J.; Alleston, S.; Di Mola, D. Interoperable Coherent Pluggables Beyond 400ZR. In Proceedings of the 2019 Asia Communications and Photonics Conference (ACP), Chengdu, China, 2–5 November 2019; pp. 1–3.
2. Zhang, W.; Liang, C.; Ge, L.; He, Z. Multimode optical interconnects based on VCSEL and MMF for more than 100-Gb/s/lane and 100 m transmission. In *Semiconductor Lasers and Applications VIII*; SPIE/COS Photonics Asia; SPIE: Paris, France, 2018; Volume 10812, p. 10.
3. Ge, L.; Zhang, W.; Liang, C.; He, Z. Threshold-Based Pruned Retraining Volterra Equalization for 100 Gbps/Lane and 100-m Optical Interconnects Based on VCSEL and MMF. *J. Lightwave Technol.* **2019**, *37*, 3222–3228. [CrossRef]
4. Dong, X.; Bamiedakis, N.; Cunningham, D.G.; Penty, R.V.; White, I.H. 112 Gb/s CAP-based Data Transmission over 100 m MMF Links using an Artificial Neural Network Equalizer. In Proceedings of the Conference on Lasers and Electro-Optics, San Jose, CA, USA, 5–10 May 2019; Optical Society of America: Rochester, NY, USA, 2019; p. SM1G.3. [CrossRef]
5. Diamantopoulos, N.P.; Yamazaki, H.; Yamaoka, S.; Nagatani, M.; Nishi, H.; Tanobe, H.; Nakao, R.; Fujii, T.; Takeda, K.; Kakitsuka, T.; et al. Net 321.24-Gb/s IMDD Transmission Based on a >100-GHz Bandwidth Directly-Modulated Laser. In Proceedings of the Optical Fiber Communication Conference Postdeadline Papers, San Diego, CA, USA, 8–12 March 2020; Optical Society of America: Rochester, NY, USA, 2020; p. Th4C.1. [CrossRef]
6. Zhang, K.; Zhuge, Q.; Xin, H.; Hu, W.; Plant, D.V. Performance comparison of DML, EML and MZM in dispersion-unmanaged short reach transmissions with digital signal processing. *Opt. Express* **2018**, *26*, 34288–34304. [CrossRef] [PubMed]
7. Sun, L.; Du, J.B.; Liu, J.C.; Chen, B.; Xu, K.; Liu, B.; Lu, C.; He, Z.Y. Intelligent 2-Dimensional Soft Decision Enabled by K-Means Clustering for VCSEL-Based 112-Gbps PAM-4 and PAM-8 Optical Interconnection. *J. Lightwave Technol.* **2019**, *37*, 6133–6146. [CrossRef]
8. You, Y.; Zhang, W.; Sun, L.; Du, J.; Liang, C.; Yang, F.; He, Z. Time skewing and amplitude nonlinearity mitigation by feedback equalization for 56 Gbps VCSEL-based PAM-4 links. *Opt. Commun.* **2018**, *410*, 909–915. [CrossRef]
9. Tatum, J.A.; Landry, G.D.; Gazula, D.; Wade, J.K.; Westbergh, P. VCSEL-Based Optical Transceivers for Future Data Center Applications. In Proceedings of the 2018 Optical Fiber Communications Conference and Exposition (OFC), San Diego, CA, USA, 11–15 March 2018; pp. 1–3.
10. Lavrencik, J.; Pavan, S.K.; Thomas, V.A.; Ralph, S.E. Noise in VCSEL-Based Links: Direct Measurement of VCSEL Transverse Mode Correlations and Implications for MPN and RIN. *J. Lightwave Technol.* **2017**, *35*, 698–705. [CrossRef]
11. Liang, C.; Zhang, W.; Ge, L.; He, Z. Mode partition noise mitigation for VCSEL-MMF links by using wavefront shaping technique. *Optics Express* **2018**, *26*, 28641–28650. [CrossRef] [PubMed]

12. Ge, L.; Zhang, W.; Liang, C.; He, Z. Compressed Neural Network Equalization Based on Iterative Pruning Algorithm for 112-Gbps VCSEL-Enabled Optical Interconnects. *J. Lightwave Technol.* **2020**, *38*, 1323–1329. [CrossRef]
13. Yi, L.; Liao, T.; Huang, L.; Xue, L.; Li, P.; Hu, W. Machine Learning for 100 Gb/s/$\lambda$ Passive Optical Network. *J. Lightwave Technol.* **2019**, *37*, 1621–1630. [CrossRef]
14. Tian, F.K.; Zhou, Q.Y.; Yang, C.C. Gaussian mixture model-hidden Markov model based nonlinear equalizer for optical fiber transmission. *Opt. Express* **2020**, *28*, 9728–9737. [CrossRef] [PubMed]
15. Wan, Z.Q.; Yu, Z.M.; Shu, L.; Liang, C.Y.; Zhang, W.J.; Zhang, T.; Li, J.Q.; Xu, K. Volterra equalizer with MLSD for 100 Gb/s PAM4 over up to 160 m OM3-MMF using 20 GHz class 850 nm VCSEL. In Proceedings of the 17th International Conference on Optical Communications and Networks, Zhuhai, China, 16–19 November 2018; Li, Z., Ed.; Spie-Int Soc Optical Engineering: Paris, France, 2019; Volume 11048. [CrossRef]
16. Lin, Y.Y.; Chen, C.J.; Nguyen, H.M.; Chuang, C.Y.; Wei, C.C.; Chen, J.; Shi, J.W. Reduction in Complexity of Volterra Filter by Employing $\ell_0$-Regularization in 112-Gbps PAM-4 VCSEL Optical Interconnect. In Proceedings of the Optical Fiber Communication Conference (OFC), San Diego, CA, USA, 8–12 March 2020; Optical Society of America: Rochester, NY, USA, 2020; p. Th2A.51. [CrossRef]
17. Chuang, C.Y.; Chang, W.F.; Wei, C.C.; Ho, C.J.; Huang, C.Y.; Shi, J.W.; Henrickson, L.; Chen, Y.K.; Chen, J. Sparse Volterra Nonlinear Equalizer by Employing Pruning Algorithm for High-Speed PAM-4 850-nm VCSEL Optical Interconnect. In Proceedings of the 2019 Optical Fiber Communications Conference and Exhibition (OFC), San Diego, CA, USA, 3–7 March 2019; p. 3.
18. Huang, W.; Chang, W.; Wei, C.; Liu, J.; Chen, Y.; Chi, K.; Wang, C.; Shi, J.; Chen, J. 93% Complexity Reduction of Volterra Nonlinear Equalizer by $\ell_1$-Regularization for 112-Gbps PAM-4 850-nm VCSEL Optical Interconnect. In Proceedings of the 2018 Optical Fiber Communications Conference and Exposition (OFC), San Diego, CA, USA, 11–15 March 2018; pp. 1–3.
19. Chuang, C.Y.; Wei, C.C.; Lin, T.C.; Chi, K.L.; Liu, L.C.; Shi, J.W.; Chen, Y.K.; Chen, J. Employing Deep Neural Network for High Speed 4-PAM Optical Interconnect. In Proceedings of the 2017 European Conference on Optical Communication (ECOC), Gothenburg, Sweden, 17–21 September 2017; p. 3. [CrossRef]
20. Zhou, Q.Y.; Yang, C.C.; Liang, A.Z.; Zheng, X.L.; Chen, Z.Y. Low computationally complex recurrent neural network for high speed optical fiber transmission. *Opt. Commun.* **2019**, *441*, 121–126. [CrossRef]
21. Qingyi, Z.; Chuanchuan, Y. AdaNN: Adaptive Neural Network-based Equalizer via Online Semi-supervised Learning. *arXiv* **2019**, arXiv:1907.10258.
22. Zhang, Q.W.; Jiang, Y.T.; Zhou, H.; Deng, C.L.; Duan, S.H.; Wang, Z.C.; Li, Y.C.; Song, Y.X.; Chen, J.; Zhang, J.J.; et al. 137 Gb/s PAM-4 Transmissions at 850 nm over 40 cm Optical Backplane with 25 G Devices with Improved Neural Network-Based Equalization. *Appl. Sci.* **2019**, *9*, 13. [CrossRef]
23. Diamantopoulos, N.P.; Kobayashi, W.; Nishi, H.; Takeda, K.; Kakitsuka, T.; Matsuo, S. 56-Gb/s VSB-PAM-4 over 80-km Using 1550-nm EA-DFB Laser and Reduced-Complexity Nonlinear Equalization. In Proceedings of the 2017 European Conference on Optical Communication (ECOC), Gothenburg, Sweden, 17–21 September 2017; p. 3.
24. Wan, Z.; Li, J.; Shu, L.; Lu, M.; Li, X.; Fu, S.; Xu, K. Nonlinear equalization based on pruned artificial neural networks for 112-Gb/s SSB-PAM4 transmission over 80-km SSMF. *Opt. Express* **2018**, *26*, 10631–10642. [CrossRef] [PubMed]
25. Ge, L.; Zhang, W.; Zhang, Y.; Liang, C.; Du, J.; He, Z. Compressed Nonlinear Equalizers for Optical Interconnects: Efficiency and Stability. In Proceedings of the Optical Fiber Communication Conference (OFC), San Diego, CA, USA, 8–12 March 2020; p. W3D.5. [CrossRef]
26. Stojanovic, N.; Karinou, F.; Qiang, Z.; Prodaniuc, C. Volterra and Wiener Equalizers for Short-Reach 100G PAM-4 Applications. *J. Lightwave Technol.* **2017**, *35*, 4583–4594. [CrossRef]
27. Shu, L.; Li, J.; Wan, Z.; Zhang, W.; Fu, S.; Xu, K. Overestimation Trap of Artificial Neural Network: Learning the Rule of PRBS. In Proceedings of the 2018 European Conference on Optical Communication (ECOC), Rome, Italy, 23–27 September 2018; pp. 1–3.

*sensors*

MDPI

*Article*

# Impact of Equalization-Enhanced Phase Noise on Digital Nonlinearity Compensation in High-Capacity Optical Communication Systems

**Jiazheng Ding [1], Tianhua Xu [1,2,3,*], Cenqin Jin [2], Ziyihui Wang [1], Jian Zhao [1,*] and Tiegen Liu [1]**

1 Key Laboratory of Opto-Electronic Information Technical Science of Ministry of Education, School of Precision Instruments and Opto-Electronics Engineering, Tianjin University, Tianjin 300072, China; whisperil01_@tju.edu.cn (J.D.); zyhwang@tju.edu.cn (Z.W.); tgliu@tju.edu.cn (T.L.)

2 University of Warwick, Coventry CV4 7AL, UK; cenqin.jin@warwick.ac.uk

3 Department of Electronic and Electrical Engineering, University College London, London WC1E 6BT, UK

* Correspondence: tianhua.xu@ieee.org (T.X.); enzhaojian@tju.edu.cn (J.Z.)

Received: 23 June 2020; Accepted: 24 July 2020; Published: 26 July 2020

check for updates

**Abstract:** Equalization-enhanced phase noise (EEPN) can severely degrade the performance of long-haul optical fiber transmission systems. In this paper, the impact of EEPN in Nyquist-spaced dual-polarization quadrature phase shift keying (DP-QPSK), dual-polarization 16-ary quadrature amplitude modulation (DP-16QAM), and DP-64QAM optical transmission systems is investigated considering the use of electrical dispersion compensation (EDC) and multi-channel digital backpropagation (MC-DBP). Our results demonstrate that full-field DBP (FF-DBP) is more susceptible to EEPN compared to single-channel and partial-bandwidth DBP. EEPN-induced distortions become more significant with the increase of the local oscillator (LO) laser linewidth, and this results in degradations in bit-error-rates (BERs), achievable information rates (AIRs), and AIR-distance products in optical communication systems. Transmission systems using higher-order modulation formats can enhance information rates and spectral efficiencies, but will be more seriously degraded by EEPN. It is found that degradations on AIRs, for the investigated FF-DBP schemes, in the DP-QPSK, the DP-16QAM, and the DP-64QAM systems are 0.07 Tbit/s, 0.11 Tbit/s, and 0.57 Tbit/s, respectively, due to the EEPN with an LO laser linewidth of 1 MHz. It is also seen that the selection of a higher-quality LO laser can significantly reduce the bandwidth requirement and the computational complexity in the MC-DBP.

**Keywords:** optical fiber communication; electrical dispersion compensation; multi-channel digital backpropagation; equalization-enhanced phase noise; achievable information rates

## 1. Introduction

With the explosive growth of the 5G and cloud services, the demand for high-capacity optical fiber communication infrastructure has become unprecedentedly urgent. Researchers have utilized dense wavelength division multiplexing (DWDM) and high-order modulation formats to enhance the spectral efficiency (SE) and the capacity of optical transmission systems. Moreover, Nyquist-spaced superchannel transmission systems, composed of closely-spaced WDM channels spaced at, or close to, the symbol rate, have been applied to further increase the spectral efficiencies [1–5]. However, the chromatic dispersion (CD), the polarization mode dispersion (PMD), the laser phase noise (PN), as well as Kerr fiber nonlinearities (NLs) severely degrade the performance of optical communication systems, leading to constraints on achievable information rates (AIRs), a measure of achievable capacity of communication systems [6,7]. Advancements in powerful digital signal processing (DSP) enhanced the reaches of optical communication systems and achieved the compensation of transmission impairments via the use of electronic dispersion compensation (EDC), digital backpropagation (DBP), and carrier phase estimation (CPE) [8–13].

However, in DSP-based coherent systems, the interplay between the laser phase noise and the electronic dispersion compensation can introduce an effect of equalization-enhanced phase noise (EEPN), leading to a significant degradation to the performance of system. A few studies regarding EEPN have been reported for linear optical communication systems, where fiber nonlinearities were neglected. Shieh and Ho have conducted the theoretical derivation and analysis of the generation process of EEPN [14]. Fatadin and Savory have studied the impact of EEPN on the quadrature phase shift keying (QPSK), the 16-ary quadrature amplitude modulation (16-QAM), and the 64-QAM coherent systems using a time-domain dispersion equalizer [15]. Kakkar and Schatz have investigated the effect of the local oscillator (LO) laser linewidth and the symbol rate on the bit-error-rate (BER) of the transmission system [16]. Yoshida and Sugihara simulated the Q-factor of the system under the influence of EEPN by varying the laser linewidth from 100 kHz to 1 MHz [9]. It has been demonstrated that the impact of EEPN scales with the accumulated dispersion, the laser linewidth, the symbol rate, and the modulation format in the communication system.

To compensate for the fiber nonlinearities in optical communication systems, multi-channel digital backpropagation (MC-DBP) has been applied at different bandwidths to mitigate intra-channel and inter-channel fiber nonlinearities, such as self-phase modulation (SPM), cross-phase modulation (XPM), and four-wave mixing (FWM) [13,17–19]. The effect of the number of steps per span, the PMD, the backpropagated signal bandwidth, and the LO phase synchronization have been studied in the optimization of MC-DBP [20–22]. However, the influence of EEPN on the performance of MC-DBP in optical transmission systems was rarely reported and the impact of EEPN has never been quantified [13]. Therefore, it is of great significance to investigate the effect of EEPN on the MC-DBP-based nonlinear compensation in closely-spaced WDM communication systems.

In this paper, the impact of EEPN on the behaviors of EDC and MC-DBP in 9-channel 32-Gbaud dual-polarization (DP) Nyquist-spaced WDM transmission systems has been investigated, where DP-QPSK, DP-16QAM, and DP-64QAM have been applied. The nonlinearity compensation (NLC) performance using the single-channel DBP, the partial-bandwidth DBP, and the full-field DBP (FF-DBP) were numerically simulated with and without the application of EEPN. The EEPN degradations on AIRs in high-speed optical communication systems have also been studied. It is found that EEPN significantly deteriorates the performance of the communication systems, especially in the case of FF-DBP. Systems using higher-order modulation formats can enhance the AIRs and spectral efficiencies, but will be more seriously deteriorated by EEPN. It is also found that the selection of a better-quality LO laser can greatly reduce the bandwidth requirement and the computational complexity in the MC-DBP.

This paper is organized as follows. In Section 2, the numerical transmission system, the principle of EEPN, the mutual information (MI), and the modulation error ratio (MER) are introduced. In Section 3, simulations and results under different scenarios are analyzed. In Section 4, additional insights and discussions are provided. The conclusions are summarized in Section 5.

## 2. Transmission Setup and Principles

### 2.1. Transmission Setup

Figure 1 depicts the 9-channel 32-Gbaud Nyquist-spaced superchannel transmission system using the following modulation formats: DP-QPSK, DP-16QAM, DP-64QAM. At the transmitter (Tx), a 9-line 32-GHz spaced laser comb with a central wavelength of 1550 nm is employed as the phase-locked optical carrier for each sub-channel. Each optical carrier is split into two branches using a polarization beam splitter (PBS) and is then fed into two in-phase and quadrature (I-Q) modulators, respectively. The transmitted data sequences in all channels and polarizations are mutually independent based on the use of pseudo-random binary codes (PRBS) with a pattern length of 216. The Nyquist pulse shaping (NPS) is performed using an ideal root-raised cosine (RRC) filter with a roll-off coefficient of 0.1%.

After the modulation, two orthogonally polarized signal sequences are combined using a polarization beam combiner (PBC) and are then injected into the standard single-mode fiber (SSMF). Numerical simulation of the SSMF is implemented using the split-step Fourier method based on the nonlinear Schrödinger equation (NLSE) with a step size of 0.1 km. Simulation parameters of SSMF include an attenuation coefficient of 0.2 dB/km, CD coefficient of 17 ps/nm/km, and a nonlinear coefficient of 1.2 /w/km, which represents a typical set of SSMF according to the Corning SMF-28 datasheet. An erbium-doped fiber amplifier (EDFA) is used to compensate for the loss of fiber in each span during the transmission. The noise figure of EDFA is 5.0 dB, which is a typical value for standard EDFA [1]. The fiber length of each span is 80 km. At the receiver (Rx), the received signal is mixed with the LO laser to achieve an ideal coherent detection of I-Q components in each polarization state. The signals are sampled using analog-to-digital converters (ADCs) and are then processed by DSP modules which include a dispersion compensation module using EDC, nonlinearity compensation module using MC-DBP, matched filter, CPE, and symbol demodulation. In our simulation, the frequency domain equalizer (FDE) is used for CD compensation [23], and the MC-DBP is used for NLC [24]. The MC-DBP algorithm is implemented based on the reverse split-step Fourier solution of the NLSE and the compensation bandwidth is set as 32 GHz (for 1-channel DBP), 160 GHz (for 5-channel DBP), and 288 GHz (for 9-channel DBP, FF-DBP), respectively. Before the MC-DBP module, an ideal RRC filter with a roll-off of 0.1% is applied to select the compensation bandwidth and to remove the unwanted amplifier spontaneous emission noise (ASE). An ideal CPE is used for the compensation of phase noise, achieved using the conjugate multiplication between the received signal and the extracted intrinsic laser phase noise [13,25]. This is to isolate the impact from EEPN, where no amplitude noise mitigation effect is employed in the CPE module [15,26]. BER and MI estimation are carried out to evaluate the transmission performance. The PMD and the frequency offset are neglected in all simulations.

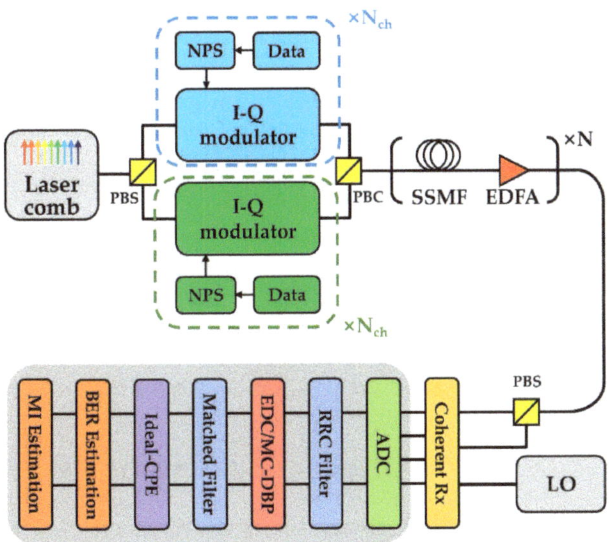


**Figure 1.** Schematic of the 9-channel 32-Gbaud Nyquist-spaced superchannel transmission system. (NPS: Nyquist pulse shaping, PBS: polarization beam splitter, PBC: polarization beam combiner, LO: local oscillator, CPE: carrier phase estimation, MI: mutual information).

### 2.2. *Principle of Equalization Enhanced Phase Noise*

As illustrated in Figure 1, the EDC is used to compensate for dispersion, and the ideal CPE is applied for the CPE in the coherent optical transmission system. The dispersion compensation module (either EDC or MC-DBP) is applied prior to the CPE module [13]. It can be clearly seen that the phase noise from the transmitter passes through the transmission fiber and the dispersion compensation module, so the net dispersion experienced by the phase noise from the transmitter is close to zero. However, the phase noise from the LO laser only goes through the dispersion compensation module. Due to the non-commutability of the phase noise and fiber dispersion, the phase noise of the LO laser will interact with the DSP-based dispersion compensation module. Consequently, the electronic equalizers (EDC or MC-DBP) will enhance the distortions from the LO laser phase noise, and the resulting EEPN will seriously reduce the performance of the large-capacity transmission system.

It has been reported in [11,14,27] that the variance of EEPN has a linear relationship with the accumulated chromatic dispersion, the linewidth of LO laser, and the symbol rate in the system. The noise variance generated by EEPN can be expressed as [14]

$$\sigma^2_{EEPN} = \frac{\pi\lambda^2 D \cdot L \cdot \Delta f_{LO}}{2cT_S} \tag{1}$$

where $\lambda$ is the center wavelength of the carrier laser, $D$ is the dispersion coefficient of SSMF, $L$ is the total length of fiber link, $\Delta f_{LO}$ is the 3-dB linewidth of the LO laser, $c$ is the speed of light in vacuum, and $T_S$ is the symbol period of the transmission system.

Therefore, the total noise variance of coherent transmission system considering the impairment of EEPN is expressed as follows:

$$\sigma^2_{Total} = \sigma^2_{Tx} + \sigma^2_{LO} + \sigma^2_{EEPN} + 2\rho\sigma_{LO}\sigma_{EEPN} \tag{2}$$

$$\sigma^2_{Tx} = 2\pi\Delta f_{Tx} T_S \tag{3}$$

$$\sigma^2_{LO} = 2\pi\Delta f_{LO} T_S \tag{4}$$

where $\sigma^2_{Tx}$ and $\sigma^2_{LO}$ are the variances of the phase noise from the transmitter and receiver, respectively, $\rho$ is the correlation coefficient of the LO phase noise and the EEPN. It has been proved that $\rho \approx 0$ when the length of the optical fiber link exceeds 80 km [11]. Therefore, in a high-speed and long-distance superchannel optical communication system, the total noise variance can be approximately expressed as

$$\sigma^2_{Total} = \sigma^2_{Tx} + \sigma^2_{LO} + \sigma^2_{EEPN} \tag{5}$$

It is noted that when the ideal CPE is applied, the intrinsic phase noise $\sigma^2_{Tx}$ and $\sigma^2_{LO}$ will be fully removed.

### 2.3. Mutual Information and Achievable Information Rate Estimation

Due to the property of the binary symmetric channel, the hard-decision forward error correction (HD-FEC) based code rate of the QPSK and m-QAM signal transmission over an additive white Gaussian noise (AWGN) channel can be expressed as [28,29]

$$R_C = 1 + BER \cdot \log_2(BER) + (1 - BER) \cdot \log_2(1 - BER) \tag{6}$$

In communication systems, mutual information is an indicator of the achievable channel capacity under specific modulation schemes. In dual-polarization coherent optical transmission systems, the $MI$ can be expressed as

$$MI = 2 \cdot \log_2(m) \cdot R_C \tag{7}$$

where $m$ is the order of the modulation format. The *AIR* is also an important indicator, which refers to the net data rate that can be achieved using FEC decoders [30]. In a dual-polarization Nyquist-spaced optical transmission system, AIRs are expressed by MI as

$$AIR = N_{ch} \cdot R_S \cdot MI \tag{8}$$

where $N_{ch}$ is the number of WDM channels in the system, and $R_S$ is the symbol rate of the transmitted signal.

### 2.4. Modulation Error Ratio

The *MER* is also an important indicator to evaluate the quality of a communication system, and it can well reflect the ability of digital receivers in restoring transmitted data [31]. The *MER* is measured using the ratio between the average signal power and the average error power [32], and is defined in *dB* as follows

$$MER(dB) = 10\log_{10}(\frac{P_{signal}}{P_{error}}) \tag{9}$$

where $P_{error}$ is the root mean square (RMS) power of the error vector and $P_{signal}$ is the RMS power of the ideal transmitted signal. It can be found that the MER provides a similar evaluation on the quality of the received signal, compared to the concept of the signal-to-noise-ratio (SNR) [4].

## 3. Simulation Results

In this section, numerical simulations based on the above analyses will be conducted. Firstly, the back-to-back (BtB) performance is investigated to evaluate the benchmark quality of considered transmission systems. Figure 2 shows the BtB performance of the measured BER versus the optical signal-to-noise-ratio (OSNR) for the 9-channel 32-Gbaud Nyquist-spaced DP-QPSK, DP-16QAM, and DP-64QAM systems. It is found that, to achieve a BER target of $10^{-3}$, the required OSNRs in the DP-QPSK, the DP-16QAM, and the DP-64QAM transmission systems are 14.0 dB, 20.6 dB, and 27.0 dB, respectively.

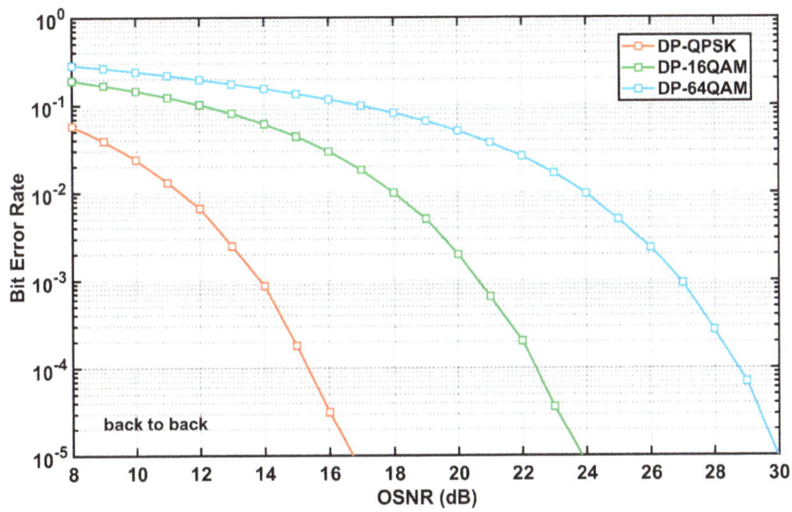


**Figure 2.** Back-to-back (BtB) performance of 9-channel 32-Gbaud Nyquist-spaced dual-polarization quadrature phase shift keying (DP-QPSK), dual-polarization 16-ary quadrature amplitude modulation (DP-16QAM), and dual-polarization 64-ary quadrature amplitude modulation (DP-64QAM) transmission systems.

With and without taking EEPN into account, the BERs and AIRs of the 9-channel Nyquist-spaced transmission systems using EDC and MC-DBP are studied. Modulation formats of DP-QPSK, DP-16QAM, and DP-64QAM were applied. The linewidths of the transmitter laser and the LO laser were both set as 0 Hz to remove the EEPN. To include the EEPN, both the transmitter and LO laser linewidths were set to be equal, and two sets of values were considered: 100 kHz (for common ECL lasers) and 1 MHz (for common DFB lasers), respectively. Single-channel, partial-bandwidth, and full-field DBP were applied to perform nonlinearity compensation in the DSP-based receiver. The case of CD compensation only was also analyzed as reference results. In all following simulations, the MC-DBP was operated with 800 steps per span and a nonlinear coefficient of 1.2 /W/km to ensure its optimal performance.

Simulation results of the 9-channel 32-Gbaud Nyquist-spaced DP-QPSK transmission system with and without the application of EEPN are shown in Figure 3, where the total distance of fiber link is 8000 km. The results are obtained under the use of EDC or MC-DBP with different back-propagated bandwidths. Figure 3a shows the BER versus optical launch power in DP-QPSK transmission system, the system BER using MC-DBP is obviously reduced compared with the EDC-only scheme under different LO linewidths. The compensation effect of FF-DBP is the most obvious because the intra-channel and inter-channel nonlinear interactions in the transmission channels were fully mitigated. When there is no EEPN, the optimum launch powers of −2.0 dBm, −2.0 dBm, −1.0 dBm, and 1.0 dBm with the best BERs of $3.43 \times 10^{-3}$, $2.10 \times 10^{-3}$, $1.09 \times 10^{-3}$, and $1.70 \times 10^{-3}$ are achieved for the scenarios of EDC, single-channel, 5-channel, and full-field DBP, respectively. However, when EEPN is applied with an LO laser linewidth of 100 kHz, the optimum launch powers for EDC, signal-channel, and 5-channel DBP are still −2.0 dBm, −2.0 dBm, and −1.0 dBm, respectively. However, the optimum launch powers for full-field DBP is changed to 0.0 dBm. Corresponding BERs rise to $4.25 \times 10^{-3}$, $2.54 \times 10^{-3}$, $1.45 \times 10^{-3}$, and $1.46 \times 10^{-3}$ compared to the scenario of no EEPN. When the LO laser linewidth rises to 1 MHz, the distortions from EEPN will become greater according to Equation (1). Simulation results in Figure 3a show that the optimum launch powers for EDC, signal-channel, and 5-channel DBP remain constant, respectively, but for FF-DBP, it drops to −1.0 dBm. The system BERs increase further to $1.41 \times 10^{-2}$, $1.07 \times 10^{-2}$, $8.75 \times 10^{-3}$, and $6.95 \times 10^{-3}$ for EDC, signal-channel, 5-channel, and full-field DBP, respectively. It can be clearly seen that the performance of FF-DBP is the most susceptible to EEPN compared to EDC, single-channel, and 5-channel DBP. Due to the EEPN (with an LO laser linewidth of 1 MHz), the best BER increases by two and a half orders of magnitude (from $1.70 \times 10^{-5}$ to $6.95 \times 10^{-3}$) in the case of FF-DBP, while it only increases by less than an order of magnitude (from $3.43 \times 10^{-3}$ to $1.41 \times 10^{-2}$) in the case of EDC.

The relationship between the AIRs and optical launch powers in the 9-channel 32-Gbaud Nyquist-spaced DP-QPSK optical communication system was investigated, as described in Figure 3b. When there is no EEPN, the highest AIRs for the system are 1.11 Tbit/s, 1.13 Tbit/s, 1.14 Tbit/s, and 1.15 Tbit/s for the cases of EDC, single-channel, 5-channel, and full-field DBP, respectively. When the EEPN is applied with an LO linewidth of 100 kHz, it can be seen from Figure 3b that the AIRs of the transmission system are not significantly reduced in all compensation schemes for such an amount of EEPN. However, the degradation on AIRs becomes significant when the LO laser linewidth increases to 1 MHz. The best AIRs now become 1.03 Tbit/s, 1.05 Tbit/s, 1.07 Tbit/s, and 1.08 Tbit/s for the cases of EDC, single-channel, 5-channel, and full-field DBP, respectively. It is noted that, for all launch powers smaller than 0.5 dBm, the AIRs in the cases of single-channel and 5-channel DBP with the LO laser linewidth of 100 kHz are higher than the AIRs in the case of FF-DBP with the LO laser linewidth of 1 MHz. This indicates that a better-quality LO laser, which has a narrower linewidth and leads to a smaller EEPN, can significantly save the bandwidth and the computational effort in the operation of MC-DBP.

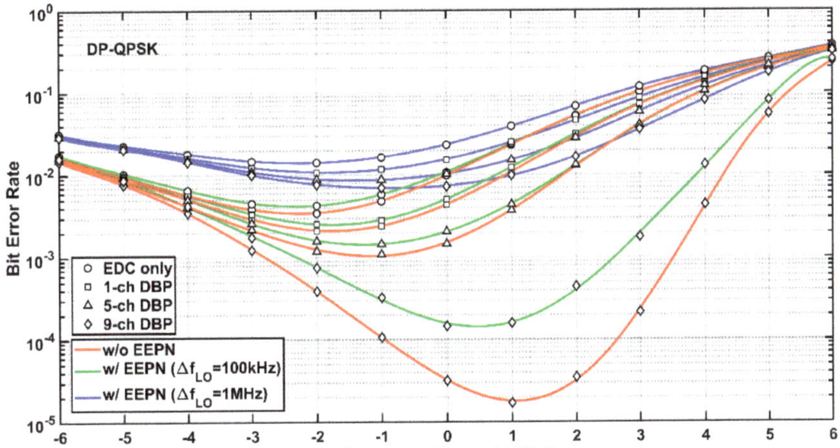


(**a**)

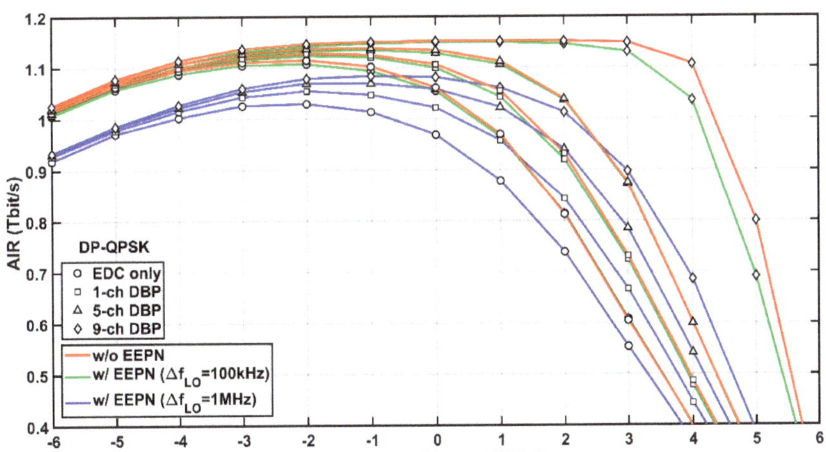


(**b**)

**Figure 3.** Simulation results of the DP-QPSK system with (w/) and without (w/o) equalization-enhanced phase noise (EEPN) at different multi-channel digital backpropagation (MC-DBP) bandwidths. (**a**) Bit-error-rate (BER) versus optical launch power, (**b**) achievable information rate (AIR) versus optical launch power.

Figure 4 shows the simulation results of the 9-channel 32-Gbaud Nyquist-spaced DP-16QAM system at different scenarios of MC-DBP with and without the application of EEPN. The transmission distance is 2000 km. It can be seen from Figure 4a that the optimum launch powers for the use of EDC, single-channel, partial-channel, and full-field DBP are −2.0 dBm, −2.0 dBm, −1.0 dBm, 3.0 dBm, respectively, when there is no EEPN. Corresponding optimal BERs are $6.56 \times 10^{-3}$, $5.11 \times 10^{-3}$, $2.35 \times 10^{-3}$, and $2.36 \times 10^{-3}$, respectively. When the EEPN is applied with an LO laser linewidth of 1 MHz, the optimum launch powers keep the same for the cases of EDC, single-channel, and 5-channel DBP, but decrease to 1.0 dBm for the case of FF-DBP. Meanwhile, the optimal BERs rise to $1.67 \times 10^{-2}$, $1.51 \times 10^{-2}$, $1.51 \times 10^{-2}$, and $5.51 \times 10^{-3}$, respectively. Due to the EEPN with an LO laser linewidth of 1 MHz, the best BER for the FF-DBP increases by two and a half orders of magnitude, similar as the DP-QPSK system. Apparently, the DP-16QAM system is more sensitive to the EEPN than the DP-QPSK system. Because the transmission distance in the DP-16QAM system is only a quarter of that in the DP-QPSK system, this will lead to the reduction of EEPN variance (to a quarter) according to Equation (1).

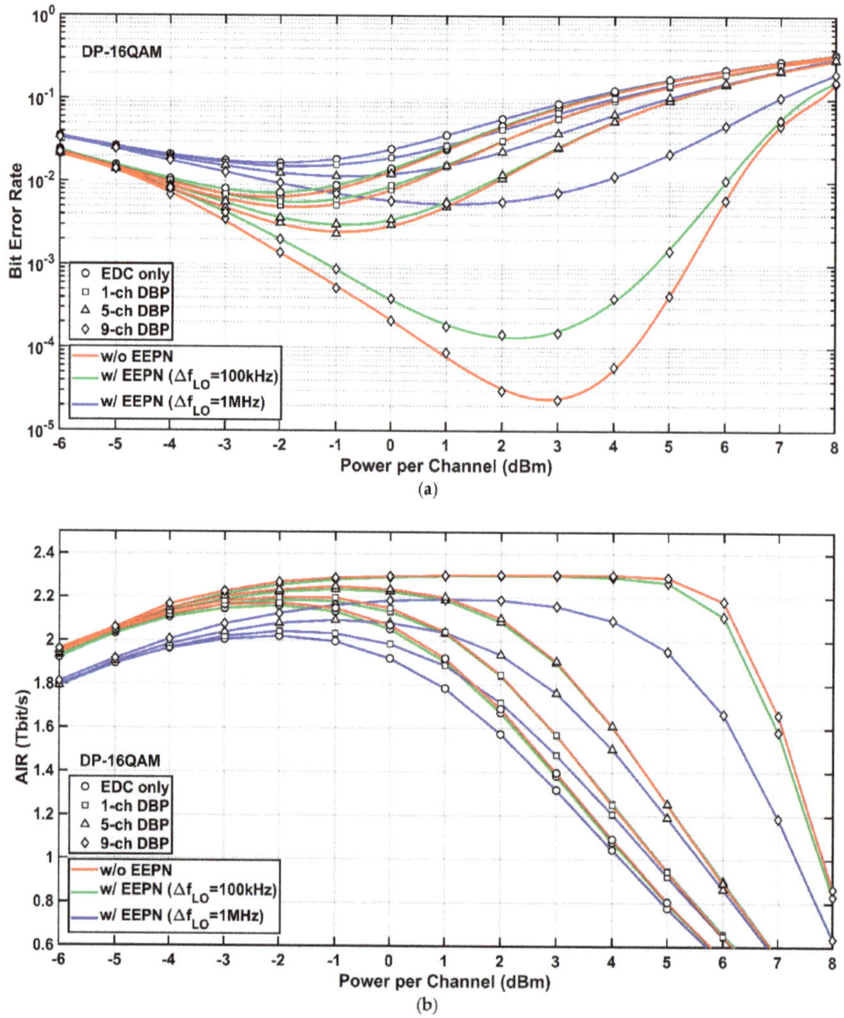


**Figure 4.** Simulation results of the DP-16QAM system with (w/) and without (w/o) EEPN at different MC-DBP bandwidths. (**a**) BER versus optical launch power, (**b**) AIR versus optical launch power.

The AIRs versus optical launch powers in the 9-channel 32-Gbaud Nyquist-spaced DP-16QAM transmission system are shown in Figure 4b. It is found from Figure 4b that, similar to the DP-QPSK system, the EEPN with an LO laser linewidth of 100 kHz poses marginal degradation on AIRs in all EDC and MC-DBP schemes compared to the scenario without any EEPN. However, the EEPN with an LO laser linewidth of 1 MHz will cause a significant degradation to the AIRs. For the FF-DBP, the best AIR drops from 2.30 Tbit/s to 2.19 Tbit/s and the optimum launch power decreases from 3.0 dBm to 1.0 dBm compared to the scenario without any EEPN. It is noted again that, for launch powers smaller than 1.0 dBm, the AIRs in the case of single-channel DBP with the LO laser linewidth of 100 kHz are higher than the AIRs in the case of 5-channel DBP with the LO laser linewidth of 1 MHz, and the AIRs in the case of 5-channel DBP with the LO laser linewidth of 100 kHz are higher than the AIRs in the case of FF-DBP with the LO laser linewidth of 1 MHz. This again demonstrates that the choice of a better-quality LO laser can significantly reduce the bandwidth requirement and the computational complexity in the MC-DBP.

Simulation results of the 9-channel 32-Gbaud Nyquist-spaced DP-64QAM transmission system using EDC and MC-DBP with and without the application of EEPN are illustrated in Figure 5, where the transmission distance is 1200 km. Figure 5a describes that BERs versus optical launch powers in the DP-64QAM superchannel transmission, which shows a similar trend as the BERs the DP-QPSK and the DP-16QAM systems. Due to the EEPN with an LO laser linewidth of 1 MHz, the optimum launch power for the FF-DBP drops from 3.5 dBm to 2.0 dBm, and the best BER increases from $4.63 \times 10^{-3}$ to $3.23 \times 10^{-2}$ compared to the scenario without any EEPN. This is similar as the BER degradation in the DP-QPSK and the DP-16AM system. This also indicates that the DP-64QAM system is more vulnerable to the EEPN than the DP-QPSK and the DP-16QAM systems, since the transmission distance in the DP-64QAM system is only 3/20 of the transmission distance in the DP-QPSK system and 3/5 of the transmission distance in the DP-16QAM system. This will lead to a significant reduction of EEPN variance compared to that in the DP-QPSK and the DP-16QAM systems according to Equation (1). Therefore, the EEPN will pose greater distortions on the systems with higher-order modulation formats, resulting in a stricter requirement on laser linewidths, although they can improve the transmission spectral efficiencies.

Simulation results of AIRs versus optical launch powers for the 9-channel 32-Gbaud Nyquist-spaced DP-64QAM system at different MC-DBP bandwidths, with and without the application of EEPN, are shown in Figure 5b. Again, similar to the DP-QPSK and the DP-16QAM systems, the EEPN with an LO laser linewidth of 100 kHz puts marginal degradation on AIRs in all schemes of EDC and MC-DBP, compared to the scenario without any EEPN. The EEPN with an LO laser linewidth of 1 MHz will pose a significant degradation to the AIRs of the DP-64QAM system. For the FF-DBP, the best AIR decreases from 3.31 Tbit/s to 2.74 Tbit/s. Meanwhile, a similar effect can be found that, for launch powers smaller than 0.5 dBm, the AIRs in the use of single-channel DBP with the LO laser linewidth of 100 kHz are higher than the AIRs in the use of 5-channel DBP with the LO laser linewidth of 1 MHz, and the AIRs in the use of 5-channel DBP with the LO laser linewidth of 100 kHz are higher than the AIRs in the use of FF-DBP with the LO laser linewidth of 1 MHz. This again implies that the use of a better-quality LO laser can significantly reduce the bandwidth requirement and the computational complexity in the MC-DBP. It can also be found in the AIR results that the DP-64QAM system is more susceptible to the EEPN compared to the DP-QPSK and the DP-16QAM systems, since the AIR degradation (~0.57 Tbit/s) for the case of FF-DBP in the DP-64QAM system due to the EEPN with the LO laser linewidth of 1 MHz is more significant than those (~0.11 Tbit/s and ~0.07 Tbit/s, respectively) in the DP-QPSK and the DP-16QAM systems, although the DP-64QAM system has the shortest transmission distance. In addition, the increase in EEPN due to the increase of the LO laser linewidth shows a smaller effect on the AIR degradations in single-channel DBP and partial-bandwidth DBP than in FF-DBP.

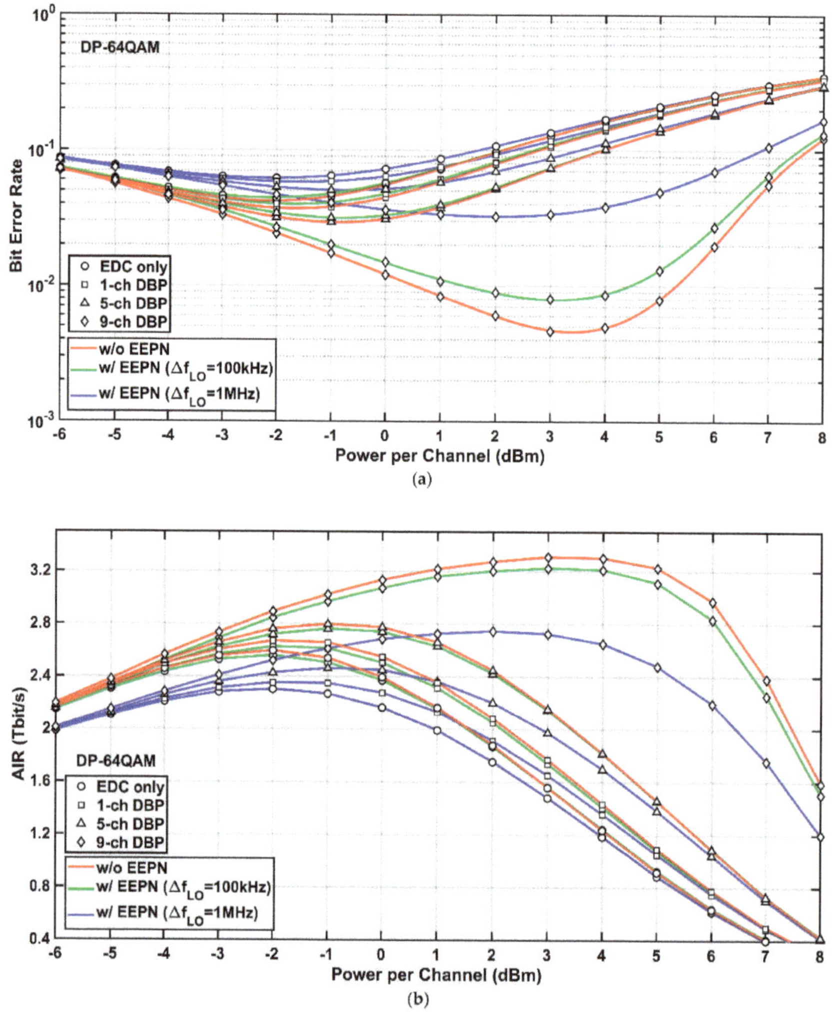


**Figure 5.** Simulation results of the DP-64QAM system with (w/) and without (w/o) EEPN at different MC-DBP bandwidths. (**a**) BER versus optical launch power, (**b**) AIR versus optical launch power.

## 4. Discussions

The MER reflects the ability of digital receivers in restoring transmitted data. Considering the EEPN, the relationship between the MER and the optical launch power in the DP-64QAM communication system using EDC and MC-DBP is investigated in Figure 6, where the transmission distance is 1200 km. It is found that, without EEPN, the maximum MER is 20.81 dB in the case of FF-DBP when the optical launch power is 3 dBm. However, the MER will be reduced to 17.06 dB when the LO laser linewidth is increased to 1 MHz, and the optimal launch power will be reduced to 2 dBm accordingly. It can be found that the EEPN degradations on MERs are also more serious in the case of FF-DBP than those in the scenarios of the EDC, the single-channel, and the partial-bandwidth DBP. This shows a consistent behavior compared to previous analyses.

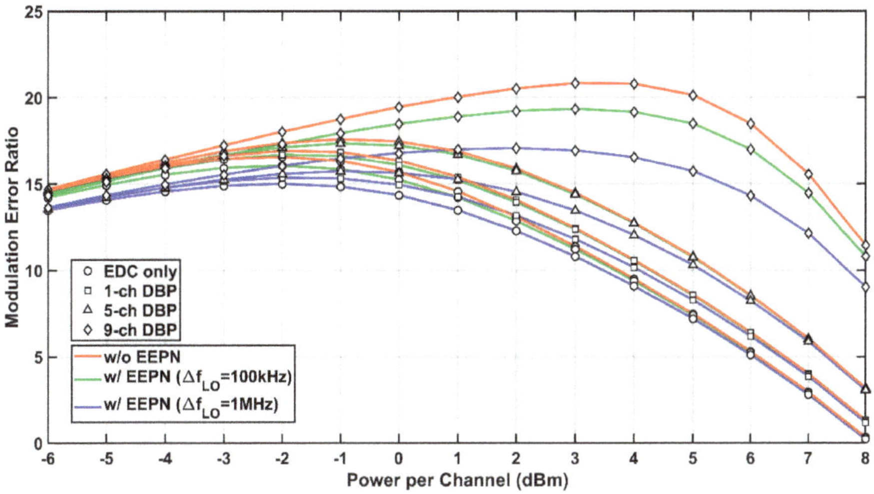


**Figure 6.** Modulation error ratio (MER) versus optical launch power in the DP-64QAM system with (w/) and without (w/o) EEPN at different MC-DBP bandwidths.

To provide a more practical study on the impact of EEPN in optical fiber communication systems, numerical simulations have been carried out based on an 820 km optical fiber field link installed in the eastern coastal area of Sweden. Figure 7 illustrates the block diagram of the 820 km field link, which interconnects seven nodes at Kista, Råsunda, Norrtälje, Östhammar, Gävle, Söderhamn, and Hudiksvall [33]. The transmitter, the receiver, and the DSP setups are the same as those employed in Figure 1. The impact of EEPN on the performance of EDC and MC-DBP in the 820 km transmission system is investigated, as shown in Figure 8, where the modulation format of DP-64QAM is applied. Similar to previous analyses, it can be found that the EEPN also poses significant degradations on the performance of the MC-DBP for such a practical transmission link. For the FF-DBP and the EEPN with an LO laser linewidth of 1 MHz, the optimum launch power drops from 4.0 dBm to 3.0 dBm, and the best BER increases from $4.15 \times 10^{-3}$ to $2.39 \times 10^{-2}$, and the best AIR decreases from 3.32 Tbit/s to 2.89 Tbit/s, compared to the scenario without any EEPN.

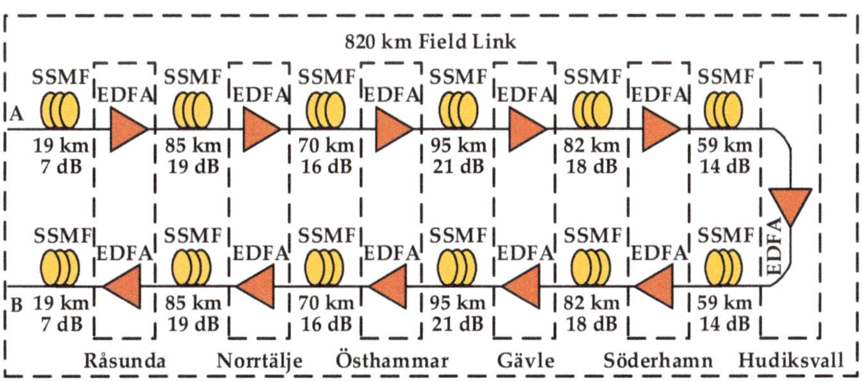


**Figure 7.** Block diagram of 820 km installed fiber field link in the eastern coast of Sweden.

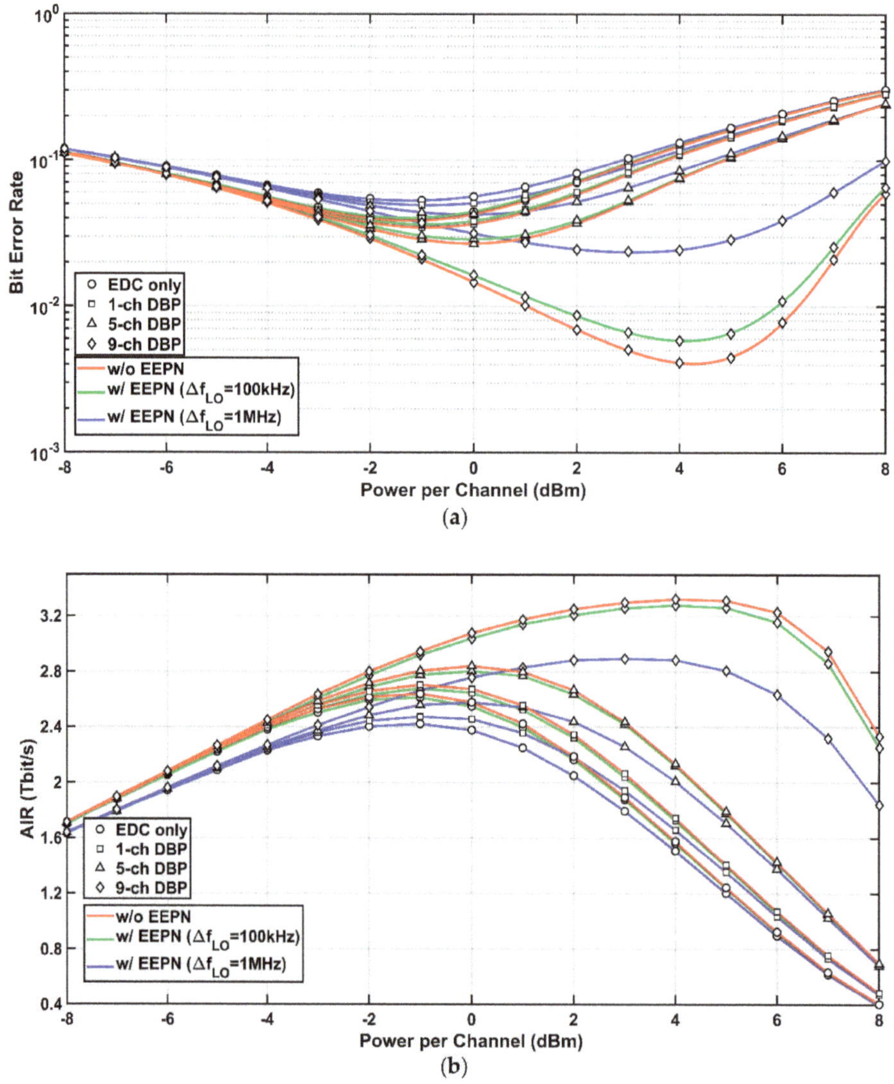


**Figure 8.** Simulation results of the DP-64QAM system with (w/) and without (w/o) EEPN at different MC-DBP bandwidths in the 820 km field link. (**a**) BER versus optical launch power, (**b**) AIR versus optical launch power.

From the above analyses, the intrinsic phase noise from the transmitter and LO lasers in the transmission system was completely compensated by the ideal CPE in the DSP, so the degradations in the system AIRs come from EEPN and the fiber NLs. From the aforementioned results, FF-DBP is the most susceptible to EEPN compared to the single-channel and the partial-bandwidth DBP. This is because the phase change in the DBP (actually the dispersion compensation) filter increases quadratically with the increase of the signal frequency [13,14]. Consequently, EEPN will pose a more serious degradation in the case of FF-DBP. The distortions from EEPN also become more significant for a larger LO linewidth, resulting in a larger increase of BER and a larger reduction in the optimum launch power of FF-DBP. Therefore, in high-capacity optical communication systems, EEPN put strict requirements on the linewidths of the LO lasers, and the restrictions are stronger for the systems using higher-order modulation formats.

The use of higher-order modulation formats can improve the spectral efficiency in optical communication systems. However, it can be seen from the above simulation results that the systems using higher-order modulation formats are more seriously distorted by EEPN. Figure 9 illustrates the maximum AIRs versus transmission distances for the DP-QPSK, the DP-16QAM, the DP-64QAM, and the field-link Nyquist-spaced transmission systems with and without the application of EEPN. Here, the product between AIR and transmission distance $AIR \cdot \sqrt{L}$ is employed as an indicator to evaluate the transmission quality of optical communication systems, the three hyperbolas in Figure 9 are provided using the formula of $AIR \cdot \sqrt{L} = C$ based on the 9-channel 32-Gbaud Nyquist-spaced DP-QPSK system under the transmission distance of 8000 km and three LO laser linewidths (0 Hz, 100 kHz, 1 MHz).

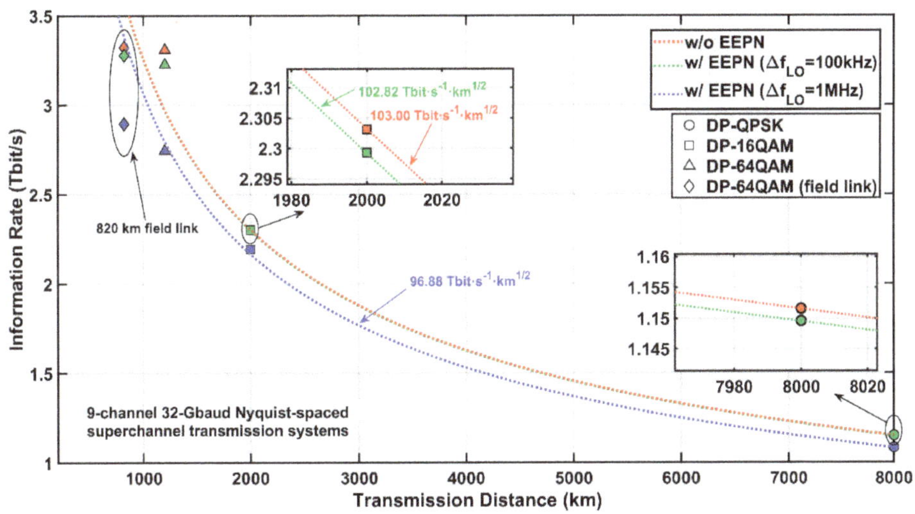


**Figure 9.** Maximum AIRs versus transmission distance for the DP-QPSK, the DP-16QAM, the DP-64QAM, and the 820 km field-link Nyquist-spaced transmission systems with and without EEPN. The three hyperbolas are obtained by $AIR \cdot \sqrt{L} = C$ based on the 9-channel 32-Gbaud Nyquist-spaced DP-QPSK system under the transmission distance of 8000 km and three LO laser linewidths (0 Hz, 100 kHz, 1 MHz).

The AIR-distance product can reach 103.00 Tbit $s^{-1}km^{1/2}$ and 102.82 Tbit $s^{-1}km^{1/2}$ for the DP-QPSK system when the LO linewidths are 0 Hz and 100 kHz, respectively. However, it can only achieve 96.88 Tbit $s^{-1}km^{1/2}$ when the LO linewidth is 1 MHz. It is very interesting that the AIR-distance products of the 9-channel 32-Gbaud Nyquist-spaced DP-16QAM system are basically equal to the AIR-distance products in the DP-QPSK system for all different values of EEPN. For the 9-channel 32-Gbaud Nyquist-spaced DP-64QAM system, the AIR-distance products are 114.63 Tbit $s^{-1}km^{1/2}$ and 111.68 Tbit $s^{-1}km^{1/2}$ for the LO linewidths of 0 Hz and 100 kHz, respectively. This shows a better performance (based on the indicator of the AIR-distance product) compared to the DP-QPSK and the DP-16QAM systems. However, the product drops to 95.05 Tbit $s^{-1}km^{1/2}$ when the LO linewidth rises to 1 MHz, which is similar to the indicators in the DP-QPSK and the DP-16QAM systems. For the 820 km field-link transmission, the system shows a significant performance degradation compared to all above three systems in terms of the AIR-distance product. It can only reach 82.88 Tbit $s^{-1}km^{1/2}$ in the case of FF-BDP when the LO laser linewidth is 1 MHz.

In this research, the impact of EEPN is investigated in optical communication systems based on the use of MC-DBP, which is one of the most promising approaches for nonlinearity compensation in next-generation optical fiber networks, due to its flexible implementation and optimization in the hardware. Meanwhile, the optical phase conjugation (OPC) is another promising technique to compensate for fiber nonlinearities in WDM optical transmission systems [34,35]. In such systems, the phase fluctuation from the high-power pump laser, which is used for realizing the FWM-based OPC, will be transferred into the phase of output signals [13]. Therefore, the EEPN in OPC-based systems arises from the interactions between the phase noise from the high-power pump laser and the fiber dispersion experienced by the phase-conjugated optical signals.

## 5. Conclusions

In this paper, the impact of EEPN on both BER performance and AIRs of Nyquist-spaced superchannel transmission system have been investigated based on the use of EDC and MC-DBP. Different modulation formats including DP-QPSK, DP-16QAM, DP-64QAM have been applied. Distortions from EEPN are more significant in the transmission system when the LO linewidth increases. This results in a reduction in both the AIRs and the AIR-distance products. It is shown that the FF-DBP is more susceptible to EEPN compared to the scenarios of EDC, single-channel, and partial-bandwidth DBP. For the investigated FF-DBP schemes, AIRs are reduced by ~0.07 Tbit/s, ~0.11 Tbit/s, and ~0.57 Tbit/s in the DP-QPSK, the DP-16QAM, and the DP-64QAM systems, respectively, due to the EEPN with an LO laser linewidth of 1 MHz. Systems using higher-order modulation formats will improve the AIRs and spectral efficiencies, but will be more susceptible to EEPN. It is also found that the choice of a better-quality LO laser can significantly reduce the bandwidth requirement and the computational complexity in the MC-DBP. Our research provides a novel insight in the design of long-haul high-capacity optical communication systems using multi-channel nonlinearity compensation techniques.

**Author Contributions:** T.X. and T.L. proposed the concept and initiated the study. J.D., T.X., and C.J. carried out numerical simulations. J.D., T.X., and J.Z. conducted analytical calculations. T.X., J.Z., and T.L. supervised the studies. The paper was written by J.D., T.X., Z.W., J.Z., and T.L. All authors have read and agreed to the published version of the manuscript.

**Funding:** This research received no external funding.

**Conflicts of Interest:** The authors declare no conflict of interest.

## References

1. Agrawal, G.P. *Fiber-Optic Communication Systems*, 4th ed.; John Wiley & Sons Inc.: Hoboken, NJ, USA, 2010.
2. Bayvel, P.; Maher, R.; Xu, T.; Liga, G.; Shevchenko, N.A.; Lavery, D.; Alvarado, A.; Killey, R.I. Maximizing the optical network capacity. *Philos. Trans. R. Soc. A Math. Phys. Eng. Sci.* **2016**, *374*, 20140440. [CrossRef] [PubMed]
3. Ke, J.H.; Gao, Y.; Cartledge, J.C. 400 Gbit/s single-carrier and 1 Tbit/s three-carrier superchannel signals using dual polarization 16-QAM with look-up table correction and optical pulse shaping. *Opt. Express* **2014**, *22*, 71–84. [CrossRef] [PubMed]
4. Xu, T.; Shevchenko, N.A.; Lavery, D.; Semrau, D.; Liga, G.; Alvarado, A.; Killey, R.I.; Bayvel, P. Modulation format dependence of digital nonlinearity compensation performance in optical fibre communication systems. *Opt. Express* **2017**, *25*, 3311. [CrossRef] [PubMed]
5. Essiambre, R.-J.; Tkach, R.W. Capacity Trends and Limits of Optical Communication Networks. *Proc. IEEE* **2012**, *100*, 1035–1055. [CrossRef]
6. Secondini, M.; Forestieri, E.; Prati, G. Achievable Information Rate in Nonlinear WDM Fiber-Optic Systems with Arbitrary Modulation Formats and Dispersion Maps. *J. Light. Technol.* **2013**, *31*, 3839–3852. [CrossRef]
7. Semrau, D.; Xu, T.; Shevchenko, N.A.; Paskov, M.; Alvarado, A.; Killey, R.I.; Bayvel, P. Achievable information rates estimates in optically amplified transmission systems using nonlinearity compensation and probabilistic shaping. *Opt. Lett.* **2016**, *42*, 121. [CrossRef]

8. Zhao, J.; Liu, Y.; Xu, T. Advanced DSP for Coherent Optical Fiber Communication. *Appl. Sci.* **2019**, *9*, 4192. [CrossRef]
9. Yoshida, T.; Sugihara, T.; Uto, K. DSP-based optical modulation technique for long-haul transmission. *Next- Gener. Opt. Commun.* **2015**, *9389*, 93890. [CrossRef]
10. Du, L.B.; Rafique, D.; Napoli, A.; Spinnler, B.; Ellis, A.D.; Kuschnerov, M.; Lowery, A.J. Digital Fiber Nonlinearity Compensation: Toward 1-Tb/s transport. *IEEE Signal Process. Mag.* **2014**, *31*, 46–56. [CrossRef]
11. Xu, T.; Jacobsen, G.; Popov, S.; Li, J.; Friberg, A.T.; Zhang, Y. Analytical estimation of phase noise influence in coherent transmission system with digital dispersion equalization. *Opt. Express* **2011**, *19*, 7756–7768. [CrossRef]
12. Kakkar, A.; Navarro, J.R.; Schatz, R.; Pang, X.; Ozolins, O.; Louchet, H.; Jacobsen, G.; Popov, S. Equalization Enhanced Phase Noise in Coherent Optical Systems with Digital Pre- and Post-Processing. *Photonics* **2016**, *3*, 12. [CrossRef]
13. Xu, T.; Liga, G.; Lavery, D.; Thomsen, B.; Savory, S.; Killey, R.I.; Bayvel, P. Equalization enhanced phase noise in Nyquist-spaced superchannel transmission systems using multi-channel digital back-propagation. *Sci. Rep.* **2015**, *5*, 13990. [CrossRef] [PubMed]
14. Shieh, W.; Ho, K.-P. Equalization-enhanced phase noise for coherent-detection systems using electronic digital signal processing. *Opt. Express* **2008**, *16*, 15718–15727. [CrossRef]
15. Fatadin, I.; Savory, S. Impact of phase to amplitude noise conversion in coherent optical systems with digital dispersion compensation. *Opt. Express* **2010**, *18*, 16273–16278. [CrossRef] [PubMed]
16. Kakkar, A.; Schatz, R.; Pang, X.; Navarro, J.R.; Louchet, H.; Ozolins, O.; Jacobsen, G.; Popov, S. Impact of local oscillator frequency noise on coherent optical systems with electronic dispersion compensation. *Opt. Express* **2015**, *23*, 11221–11226. [CrossRef] [PubMed]
17. Xu, T.; Karanov, B.; Shevchenko, N.A.; Lavery, D.; Liga, G.; Killey, R.I.; Bayvel, P. Digital nonlinearity compensation in high-capacity optical communication systems considering signal spectral broadening effect. *Sci. Rep.* **2017**, *7*, 12986. [CrossRef] [PubMed]
18. Golani, O.; Feder, M.; Shtaif, M. Equalization Methods for Out-of-Band Nonlinearity Mitigation in Fiber-Optic Communications. *Appl. Sci.* **2019**, *9*, 511. [CrossRef]
19. Maher, R.; Xu, T.; Galdino, L.; Sato, M.; Alvarado, A.; Shi, K.; Savory, S.J.; Thomsen, B.; Killey, R.I.; Bayvel, P. Spectrally Shaped DP-16QAM Super-Channel Transmission with Multi-Channel Digital Back-Propagation. *Sci. Rep.* **2015**, *5*, 8214. [CrossRef]
20. Rafique, D.; Ellis, A.D. Impact of signal-ASE four-wave mixing on the effectiveness of digital back-propagation in 112 Gb/s PM-QPSK systems. *Opt. Express* **2011**, *19*, 3449–3454. [CrossRef]
21. Alic, N.; Myslivets, E.; Temprana, E.; Kuo, B.P.-P.; Radic, S. Nonlinearity Cancellation in Fiber Optic Links Based on Frequency Referenced Carriers. *J. Light. Technol.* **2014**, *32*, 2690–2698. [CrossRef]
22. Gao, G.; Chen, X.; Shieh, W. Influence of PMD on fiber nonlinearity compensation using digital back propagation. *Opt. Express* **2012**, *20*, 14406–14418. [CrossRef] [PubMed]
23. Ishihara, K.; Kobayashi, T.; Kudo, R.; Takatori, Y.; Sano, A.; Yamada, E.; Masuda, H.; Miyamoto, Y. Coherent optical transmission with frequency-domain equalization. In Proceedings of the 2008 34th European Conference on Optical Communication, Brussels, Belgium, 21–25 September 2008; Volume 3, pp. 1–2.
24. Ip, E.; Kahn, J.M. Compensation of Dispersion and Nonlinear Impairments Using Digital Backpropagation. *J. Light. Technol.* **2008**, *26*, 3416–3425. [CrossRef]
25. Farhoudi, R.; Ghazisaeidi, A.; Rusch, L.A. Performance of carrier phase recovery for electronically dispersion compensated coherent systems. *Opt. Express* **2012**, *20*, 26568–26582. [CrossRef]
26. Xie, C. Local Oscillator Phase Noise Induced Penalties in Optical Coherent Detection Systems Using Electronic Chromatic Dispersion Compensation. In Proceedings of the Conference on Optical Fiber Communication/International Conference on Integrated Optics and Optical Fiber Communication, San Diego, CA, USA, 24–26 March 2009; pp. 2009–2011.
27. Lau, A.P.T.; Shen, T.S.R.; Shieh, W.; Ho, K.-P. Equalization-enhanced phase noise for 100Gb/s transmission and beyond with coherent detection. *Opt. Express* **2010**, *18*, 17239–17251. [CrossRef]
28. Savory, S. Congestion Aware Routing in Nonlinear Elastic Optical Networks. *IEEE Photon. Technol. Lett.* **2014**, *26*, 1057–1060. [CrossRef]
29. Xu, T.; Xu, T.; Bayvel, P.; Darwazeh, I. Non-Orthogonal Signal Transmission Over Nonlinear Optical Channels. *IEEE Photon. J.* **2019**, *11*, 1–13. [CrossRef]

30. Alvarado, A.; Fehenberger, T.; Chen, B.; Willems, F.M.J. Achievable Information Rates for Fiber Optics: Applications and Computations. *J. Light. Technol.* **2018**, *36*, 424–439. [CrossRef]
31. Mazlumi, F.; Khoshgard, M. Modulation Error Ratio Degradation Due to Error Sources Existing in a DRM Transmitter Using Envelope Elimination and Restoration. In Proceedings of the 2006 European Conference on Wireless Technology, Manchester, UK, 10–12 September 2006. [CrossRef]
32. Hranac, R.; Currivan, B. *Digital Transmission: Carrier-to-Noise Ratio, Signal-To-Noise Ratio, and Modulation Error Ratio*; Cisco Public Inf.: San Jose, CA, USA, 2006; pp. 1–42.
33. Xu, T.; Li, J.; Jacobsen, G.; Popov, S.; Djupsjobacka, A.; Schatz, R.; Zhang, Y.; Bayvel, P. Field trial over 820 km installed SSMF and its potential Terabit/s superchannel application with up to 57.5-Gbaud DP-QPSK transmission. *Opt. Commun.* **2015**, *353*, 133–138. [CrossRef]
34. Hu, H.; Jopson, R.M.; Gnauck, A.H.; Dinu, M.; Chandrasekhar, S.; Liu, X.; Xie, C.; MontoUu, M.; Randel, S.; McKinstrie, C.J. Fiber Nonlinearity Compensation of an 8-channel WDM PDM-QPSK Signal using Multiple Phase Conjugations. In Proceedings of the Conference on Optical Fiber Communication/International Conference on Integrated Optics and Optical Fiber Communication, San Francisco, CA, USA, 11–13 March 2014; p. M3C.2.
35. Phillips, I.D.; Tan, M.; Stephens, M.; McCarthy, M.E.; Giacoumidis, E.; Sygletos, S.; Rosa, P.; Fabbri, S.; Le, S.T.; Kanesan, T.; et al. Exceeding the Nonlinear-Shannon Limit using Raman Laser Based Amplification and Optical Phase Conjugation. In Proceedings of the Optical Fiber Communication Conference, San Francisco, CA, USA, 11–13 March 2014; p. M3C.1.

sensors

MDPI

*Article*

# Time-Domain Blind ICI Compensation in Coherent Optical FBMC/OQAM System

**Binqi Wu [1], Jin Lu [1,*], Mingyi Gao [2], Hongliang Ren [1], Zichun Le [3], Yali Qin [1], Shuqin Guo [1] and Weisheng Hu [4]**

1 College of Information Engineering, Zhejiang University of Technology, Hangzhou 310023, China; wbq@zjut.edu.cn (B.W.); hlren@zjut.edu.cn (H.R.); qyl@zjut.edu.cn (Y.Q.); guosq@zjut.edu.cn (S.G.)
2 School of Electronic and Information Engineering, Soochow University, Suzhou 215006, China; mygao@suda.edu.cn
3 College of Science, Zhejiang University of Technology, Hangzhou 310023, China; lzc@zjut.edu.cn
4 State Key Laboratory of Advanced Optical Communication Systems and Networks, Shanghai Jiao Tong University, Shanghai 200240, China; wshu@sjtu.edu.cn
* Correspondence: lujin@zjut.edu.cn; Tel.: +86-1385-816-9026

Received: 29 August 2020; Accepted: 3 November 2020; Published: 9 November 2020

**Abstract:** A blind discrete-cosine-transform-based phase noise compensation (BD-PNC) is proposed to compensate the inter-carrier-interference (ICI) in the coherent optical offset-quadrature amplitude modulation (OQAM)-based filter-bank multicarrier (CO-FBMC/OQAM) transmission system. Since the phase noise sample can be approximated by an expansion of the discrete cosine transform (DCT) in the time-domain, a time-domain compensation model is built for the transmission system. According to the model, phase noise compensation (PNC) depends only on its DCT coefficients. The common phase error (CPE) compensation is firstly performed for the received signal. After that, a pre-decision is made on a part of compensated signals with low decision error probability, and the pre-decision results are used as the estimated values of transmitted signals to calculate the DCT coefficients. Such a partial pre-decision process reduces not only decision error but also the complexity of the BD-PNC method while keeping almost the same performance as in the case of the pre-decision of all compensated signals. Numerical simulations are performed to evaluate the performance of the proposed scheme for a 30 GBaud CO-FBMC/OQAM system. The simulation results show that its bit error rate (BER) performance is improved by more than one order of magnitude through the mitigation of the ICI in comparison with the traditional blind PNC scheme only aiming for CPE compensation.

**Keywords:** coherent optical communication; offset-quadrature amplitude modulation-based filter-bank multicarrier (FBMC/OQAM); blind phase noise compensation; inter-carrier-interference (ICI); discrete cosine transform (DCT)

## 1. Introduction

A coherent optical offset-quadrature amplitude modulation (OQAM)-based filter-bank multicarrier (CO-FBMC/OQAM) system has recently become a promising candidate for high-speed long-haul optical fiber transmission due to its higher spectral efficiency (SE) and its robustness to channel impairments [1–5]. As a multicarrier modulation (MCM) technique, the FBMC is more robust against pulse dispersion and has less out-of-band leakage than orthogonal frequency division multiplexing (OFDM) [6–14]. Meanwhile, the FBMC/OQAM scheme does not need cyclic prefix and, therefore, its SE can be increased remarkably [15,16].

The performance of optical fiber transmission system can be degraded severely by the phase noise (PN) [17], and the MCM technique is especially vulnerable to PN like OFDM. In CO-FBMC/OQAM systems, the PN generally includes common phase error (CPE) and inter-carrier interference (ICI) [18], similar to that in coherent optical OFDM (CO-OFDM) systems. Nevertheless, the original PN compensation (PNC) algorithms in CO-OFDM cannot be simply transplanted in CO-FBMC/OQAM systems. This is because the OQAM is employed between two immediate adjacent subcarriers, and the orthogonality between subchannels is ensured only in real plane. Due to the nonorthogonality in imaginary plane, the PNC and channel estimation techniques are severely restricted by the so-called intrinsic imaginary interference (IMI) in CO-FBMC/OQAM systems [19–24]. Consequently, many studies on the PNC method focus on the IMI cancellation [25–27]. In [26], a pilot-based PNC method is used to suppress the IMI at the expense of SE, where a pilot structure consists of several additional data symbols surrounding a pilot symbol. Another alternative approach is a coding scheme, and data symbols surrounding a pilot are deliberately coded to eliminate the IMI in that pilot [26,27]. Although the complexity of two mainly pilot-based PNC methods is moderate, their performance is limited by only CPE compensation.

At present, the PNC methods for CO-FBMC/OQAM systems can be classified into two categories: blind and pilot-aided PNC. As mentioned above, to cancel the IMI, the pilot-aided PNC methods have a low complexity and avoid the effect of cycle slips at the cost of the SE loss [25–27]. Compared with the pilot-aided methods, the blind methods can achieve higher SE at the price of higher computational complexity (CC) [28–36]. As a de-facto standard in coherent optical communication systems, the blind phase search (BPS) method can achieve a relatively good PNC effect and a high laser linewidth tolerance. Its main disadvantage is high CC resulting from a large number of test phases. The modified BPS (M-BPS) simplifies the distance calculation in the complex plane as the operations in the real plane [28], and thereby reduces the number of multiplications significantly. However, both blind and pilot-aided methods mainly emphasize on the CPE compensation in frequency domain, and the ICI is not considered to be mitigated and is assumed to be additive noise in these studies. Because the MCM system with a large number of subcarriers can increase its fiber chromatic dispersion (CD) tolerance [6], only CPE compensation is obviously not enough in this case due to serious negative impact of the ICI on the system performance [25–28].

Recently, time-domain ICI compensation has been proposed based on pilot symbols in either wireless or optical FBMC/OQAM systems [18,37]. An orthogonal-basis-expansion (OBE)-based PNC method has been proposed in a polarization-division-multiplexed CO-FBMC/OQAM system [18]. With the aid of pilot symbols, the PN in time domain can be mitigated by estimating the corresponding coefficients of the OBE. In wireless FBMC/OQAM systems, by using the discrete cosine transform (DCT) method, the samples of the PN in time-domain can be described as sums of products of a group of DCT bases and their corresponding coefficients of the DCT [37]. Using pilot symbols, the ICI effect can be tracked by calculating these coefficients of the DCT. It is obvious that the pilot overhead loss is very high for two time-domain pilot-based methods. Several blind ICI compensation schemes have been presented for CO-OFDM systems [38–40], but they could not be used in CO-FBMC/OQAM systems.

In this paper, we propose a blind time-domain ICI compensation scheme for CO-FBMC/OQAM systems by combining the M-BPS method and DCT approximation of PN. The M-BPS method is used to obtain the pre-estimated values of the transmitted signals, with which the DCT coefficients can be obtained from the time-domain compensation model. To improve its performance and reduce its CC, the pre-decision error and the overlapped symbol structure are considered to develop the proposed algorithm. The proposed method is numerically validated in a 30 GBaud back to back (BTB) CO-FBMC/OQAM system. The numerical results show that the proposed algorithm achieves a significantly better performance compared to the M-BPS scheme while having a limited increase in CC.

The remainder of this paper is organized as follows. A theoretical model of CO-FBMC/OQAM system with laser phase noise is derived in Section 2. Section 3 introduces the principle of the proposed PNC method. In Section 4, the overlapped symbol structure in the CO-FBMC/OQAM system is implemented to improve the effectiveness of the algorithm. We numerically investigate the performance of our method and propose the complexity analysis in Section 5. Finally, Section 6 summarizes the paper.

## 2. CO-FBMC/OQAM System with Laser Phase Noise

In this section, the influence of PN is studied in CO-FBMC/OQAM systems, and the analytical expressions of the PN induced interference are derived. As shown in Figure 1, a single polarization CO-FBMC/OQAM BTB system is considered to investigate the PN induced interference and the performance of the proposed PNC algorithm. The single polarization BTB system is used to concentrate on the impact of laser PN in this paper.

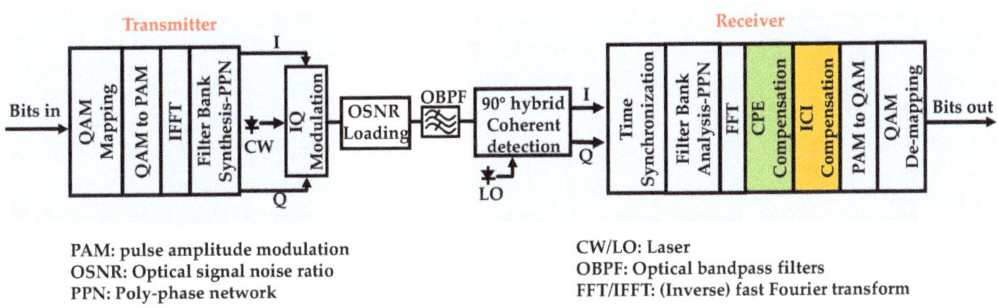


**Figure 1.** Block diagram of CO-FBMC/OQAM BTB system.

At the transmitter, the process of OQAM pre-processing transforms the complex-valued QAM symbols into real-valued symbols, which means that the real-valued symbols are formed with the real and imaginary parts of QAM symbols destaggered by half a QAM symbol period. These real-valued symbols are modulated (i.e., pulse amplitude modulation (PAM)) in such a way that the FBMC system maintains the same data rate as the OFDM system without cyclic-prefix. Then, the PAM symbols are mapped onto a set of orthogonal subcarriers generated by an inverse fast Fourier transform (IFFT). The following operation is a cost-efficient poly-phase network (PPN) for filter bank synthesis implementation. With the digital signal processing (DSP) at the transmitter, its discrete time-domain transmitted signal, $s[k]$, can be expressed as,

$$s[k] = \sum_{n=0}^{N_S-1} \sum_{m=0}^{M-1} a_{n,m} g_{n,m}[k] \tag{1}$$

where $a_{n,m}$ and $g_{n,m}[k]$ are the PAM symbol and the synthesis filter at the $(n,m)$th time-frequency index, respectively; $N_S$ is the total number of CO-FBMC/OQAM blocks; and $M$ is the total number of subcarriers in each CO-FBMC/OQAM block. The $(n,m)$th synthesis filter is described as,

$$g_{n,m}[k] = g[k - n\frac{M}{2}] \cdot e^{j\frac{2\pi}{M}mk} e^{j\psi_{n,m}} \tag{2}$$

$$g[k] = \frac{(4\alpha k/T)\cos[\pi(1+\alpha)k/T] + \sin[\pi(1-\alpha)k/T]}{(\pi k/T)[1-(4\alpha k/T)^2]} \tag{3}$$

where the square root raised cosine function $g[k]$ is employed as the prototype filter with length $L_g$ [41], $\psi_{n,m} = (n+m)/2$ is phase modulation factor, $\alpha$ denotes the roll-off coefficient, and $T$ represents the duration of the QAM complex symbols.

In the BTB CO-FBMC/OQAM system, under the impact of laser PN and additive Gaussian white noise (AWGN), the received signal $r[k]$ can be written as,

$$\begin{aligned} r[k] &= e^{j\varphi[k]}s[k] + w[k] \\ &= e^{j\varphi[k]}\left(\sum_{n=0}^{N_S-1}\sum_{m=0}^{M-1} a_{n,m}g[k-n\tfrac{M}{2}]\cdot e^{j\frac{2\pi}{M}mk}e^{j\frac{\pi(m+n)}{2}}\right) + w[k] \\ &= e^{j\varphi[k]}\left(a_{n_0,m_0}g[k-n_0\tfrac{M}{2}]e^{j\frac{2\pi}{M}m_0k}e^{j\frac{\pi(m_0+n_0)}{2}} + \sum_{(n,m)\neq(n_0,m_0)} a_{n,m}g[k-n\tfrac{M}{2}]e^{j\frac{2\pi}{M}mk}e^{j\frac{\pi(m+n)}{2}}\right) + w[k] \end{aligned} \tag{4}$$

Here, $\varphi[k]$ denotes the PN in the time-domain, and the notation $w[k]$ is the AWGN. The laser PN originates from either the external cavity laser at the transmitter or the local oscillator at the receiver. It is modeled as a random walk Wiener process, and the PN change follows a Gaussian distribution with zero mean and a specific variance $2\pi\cdot\Delta v\cdot T_S$, where $\Delta v$ is the combined laser linewidth, and $T_S$ is the normalized symbol period. Here, $T_S = (1/B_r)/M$, and $B_r$ is the baud rate of QAM symbols. The combined laser linewidth is defined as the sum of linewidths of the transmitter and receiver lasers. For the sake of simplicity, in the BTB system, the transmitter laser has the same linewidth as the received laser, and the combined laser linewidth is twice the linewidth of the single laser.

At the receiver side, the incoming optical wave signal is first coherently detected in a 90° optical hybrid. Subsequently, the obtained digital data are sent to the DSP modules to recover the original data. After a time synchronization is performed perfectly, a time-frequency translation can be completed by a hardware-efficient structure of PPN for filter bank analysis followed by a fast Fourier transform (FFT). Eventually, the complex-valued received demodulated signal at the $(n_0,m_0)$th time-frequency index can be written as,

$$\begin{aligned} D_{n_0,m_0} &= \sum_{k=0}^{L_g-1} r[k]g^*_{n_0,m_0}[k] \\ &= \sum_{k=0}^{L_g-1} r[k]g[k-n_0\tfrac{M}{2}]\cdot e^{-j\frac{2\pi}{M}m_0k}e^{-j\frac{\pi(m_0+n_0)}{2}} \end{aligned} \tag{5}$$

here, the upper script "*" denotes conjugate operation. By substituting Equation (4) into Equation (5), $D_{n_0,m_0}$ can be further expressed as [18],

$$\begin{aligned} D_{n_0,m_0} &= \sum_{k=0}^{L_g-1} r[k]g[k-n_0\tfrac{M}{2}]\cdot e^{-j\frac{2\pi}{M}m_0k}e^{-j\frac{\pi(m_0+n_0)}{2}} \\ &= \eta_{0,0}a_{n_0,m_0} + \sum_{(p,q)\neq(0,0)} a_{n_0+p,m_0+q}\eta_{p,q} + N_{n_0,m_0} \end{aligned} \tag{6}$$

here, $N_{n_0,m_0}$ denotes the filter processed additive noise, and $\eta_{p,q}$ is defined as

$$\begin{aligned} \eta_{p,q} &= \sum_k e^{-j\frac{\pi(p+q)}{2}}e^{-j\frac{2\pi}{M}qk}g[k-n_0\tfrac{M}{2}] \\ &\cdot g^*[k-\tfrac{M}{2}(n_0+p)]\left[\frac{1}{M}\sum_{s=0}^{M-1} e^{-j\frac{2\pi}{M}qs}e^{j\varphi_{n_0}[s]}\right] \end{aligned} \tag{7}$$

at $(p,q) = (0,0)$, $\eta_{0,0}$ is described by the following expression.

$$\begin{aligned} \eta_{0,0} &= \sum_k g\left[k-n_0\tfrac{M}{2}\right]\cdot g^*\left[k-n_0\tfrac{M}{2}\right]\left[\frac{1}{M}\sum_{s=0}^{M-1} e^{j\varphi_{n_0}[s]}\right] \\ &= \frac{1}{M}\sum_{s=0}^{M-1} e^{j\varphi_{n_0}[s]} \end{aligned} \tag{8}$$

In Equation (6), $\eta_{0,0}$ is defined as the CPE for the information symbol $a_{n_0,m_0}$, and it leads to phase rotation of the received constellation diagram. The second item of Equation (6) is defined as the

ICI, which causes divergence of the received constellation diagram. The ICI in CO-FBMC/OQAM systems is a convolution between the coefficient $\eta_{p,q}$ and PAM symbols from different FBMC/OQAM blocks, while the ICI in CO-OFDM systems is a convolution between the ICI coefficients and different sub-carriers in one OFDM symbol. Hence, it is more difficult to compensate the ICI in CO-FBMC/OQAM systems.

In many previous studies, assuming that the PN time-domain samples are the same in one FBMC/OQAM symbol period, the complex-valued received signal $D_{n_0,m_0}$ can be approximately rewritten as [25–29],

$$\begin{aligned} D_{n_0,m_0} &\approx \eta_{0,0} a_{n_0,m_0} + \eta_{0,0} \sum_{k=0}^{L_g-1} \sum_{(p,q)\neq(0,0)} a_{n_0+p,m_0+q} g_{n_0+p,m_0+q}[k] g^*_{n_0,m_0}[k] + N_{n_0,m_0} \\ &= \eta_{0,0}(a_{n_0,m_0} + j u_{n_0,m_0}) + N_{n_0,m_0} \end{aligned} \tag{9}$$

where $u_{n_0,m_0}$ is called the IMI in CO-FBMC/OQAM systems, and it is the interference term which results from neighboring PAM symbols on the interested ($n_0$,$m_0$)th position. Based on the approximation expression above, various studies have focused on the cancellation of the IMI by using pilot symbols [25–27]. However, only CPE compensation effect can be improved through the complete removal of the IMI. Because the ICI effect is the convolution in frequency domain, it is useless to perform the ICI compensation by removing the IMI thoroughly. So in the rest of the paper, the blind ICI compensation is proposed for the CO-FBMC/OQAM systems.

## 3. Principle of the Phase Noise Compensation Method

In this section, a blind PNC scheme is proposed to effectively compensate both CPE and ICI in CO-FBMC/OQAM systems. Figure 2 shows a block diagram of the proposed blind PNC scheme. In this method, a time-domain PNC model is firstly built by using the DCT approximation method, where each PN sample can be expanded as a linear combination of DCT coefficients and a group of DCT bases. To calculate these DCT coefficients in the time-domain model, it is necessary to pre-estimate the transmitted frequency-domain signals. As a result, the CPE is firstly compensated by using the M-BPS algorithm [28], and then the pre-decision process is performed on a part of the compensated signals to obtain the pre-estimated values of the transmitted signals.

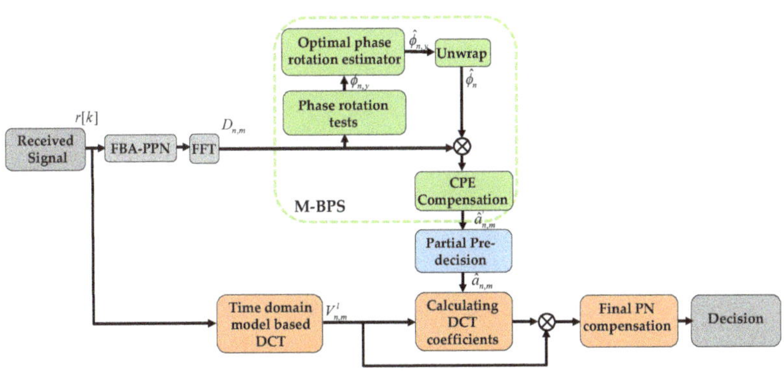


**Figure 2.** Block diagram of the proposed blind phase noise compensation scheme.

After implementing PPN for filter bank analysis and FFT, the M-BPS method is firstly used to compensate the CPE. In the M-BPS method, the signal is firstly rotated by several test phases.

The values of test phases are defined as $\phi_{n,y} = (y/Y) \cdot \pi - \pi/2$, where $y = 1, 2, \ldots, Y$, and $Y$ is the total number of test phases. The rotated version of $D_{n,m}$ can be given by

$$\hat{D}_{n,m,y} = D_{n,m} \cdot e^{-j\phi_{n,y}} \tag{10}$$

The estimated value of the CPE is obtained by minimizing the sum of distances between the hard decision of the received samples after test phase compensation and their projections on the real axis,

$$\hat{\phi}_{n,y} = \underset{\phi_{n,y}}{\operatorname{argmin}} \sum_{m=0}^{M-1} \left| \mathrm{Re}(\hat{D}_{n,m,y}) - DD(\mathrm{Re}(\hat{D}_{n,m,y})) \right| \tag{11}$$

where $\mathrm{Re}(\cdot)$ denotes the real projection operator, and $DD(\cdot)$ is the direct pre-decision operator. Subsequently, the conventional unwrap operation is performed to partly solve the problem of phase ambiguity [42]. Then, the received PAM symbol is obtained as follows:

$$\hat{a}'_{n,m} = \mathrm{Re}(\hat{D}_{n,m}) = \mathrm{Re}(D_{n,m} \cdot e^{-j\hat{\phi}_n}) \tag{12}$$

Next, the estimated values $\hat{a}_{n,m}$ of the transmitted signal are recovered via the pre-decisions upon the signal $\hat{a}'_{n,m}$. Since the performance of the proposed method strongly depends on the pre-decisions, the negative influence of the decision errors should be considered. Here, an effective approach is taken to reduce the decision errors. As shown in Figure 3, taking the 16-OQAM as an example, the constellation points are classified into 4 classes. When the constellation point $\hat{D}_{n,m}$ falls in the border regions between two neighbouring classes, the decision errors often occur. Then, three shadow border regions are regarded as the high decision error probability regions, where the parameter $\delta$ is defined as is the width of a shadow rectangle, and it decides the selected range of the high decision error probability regions. Therefore, if $\hat{D}_{n,m}$ falls in these high decision error probability regions, these signals are thrown away and not used to perform pre-decisions; while $\hat{D}_{n,m}$ does not fall in the shadow regions, a hard pre-decision is made normally. However, if excessive signals are removed from the pre-decision process, this can lead to the performance degradation of the proposed method. Thus, the pre-decision parameter $\delta$ needs to be optimized to achieve good performance.

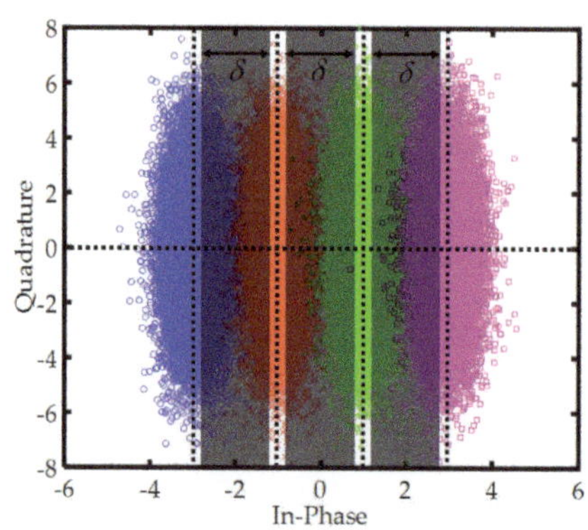


**Figure 3.** Taking the 16-OQAM as an example, the signals with high decision error probability in three shadow regions ($\hat{a}'_{n,m} \in [-2-\delta/2, -2+\delta/2] \cup [-\delta/2, \delta/2] \cup [2-\delta/2, 2+\delta/2]$) not being used to perform the pre-decision.

Secondly, a time-domain PNC model is built to estimate the PN with the aid of DCT. The compensated signal $\hat{r}_n[i]$ can be obtained by multiplying the received signal $r_n[i]$ by the complex conjugate of the estimate of PN,

$$\hat{r}_n[i] = r_n[i]e^{-j\hat{\varphi}_n[i]} \tag{13}$$

where $\varphi_n[i]$ denotes the $i$th time-domain PN sample in the $n$th CO-FBMC/OQAM block. The high frequency components in the PN can be neglected, and the complex conjugate of PN can be expressed as the linear combination of a group of DCT basis and DCT coefficients [37],

$$\boldsymbol{\Phi}_n \approx \boldsymbol{\tau} C_n \tag{14}$$

where $\boldsymbol{\Phi}_n = [e^{-j\hat{\varphi}_n[0]}, e^{-j\hat{\varphi}_n[2]}, \ldots, e^{-j\hat{\varphi}_n[KM-1]}]^T$ and $C_n = [C_n(0), C_n(1), \ldots, C_n(L-1)]^T$ are $L \times 1$ unknown DCT coefficient vector. Here, $[\cdot]^T$ is the transpose operator, and $L$ refers to the length of DCT coefficients. The $L_g \times L$ matrix $\boldsymbol{\tau}$ of DCT basis is given by

$$\boldsymbol{\tau} = \begin{bmatrix} \frac{\sqrt{2}}{2} & \cos\left(\frac{\pi(1+0.5)}{KM}\right) & \cos\left(\frac{2\pi(1+0.5)}{KM}\right) & \cdots & \cos\left(\frac{(L-1)\pi(1+0.5)}{KM}\right) \\ \frac{\sqrt{2}}{2} & \cos\left(\frac{\pi(2+0.5)}{KM}\right) & \cos\left(\frac{2\pi(2+0.5)}{KM}\right) & \cdots & \cos\left(\frac{(L-1)\pi(2+0.5)}{KM}\right) \\ \frac{\sqrt{2}}{2} & \cos\left(\frac{\pi(3+0.5)}{KM}\right) & \cos\left(\frac{2\pi(3+0.5)}{KM}\right) & \cdots & \cos\left(\frac{(L-1)\pi(3+0.5)}{KM}\right) \\ \vdots & \vdots & \vdots & \ddots & \vdots \\ \frac{\sqrt{2}}{2} & \cos\left(\frac{\pi(KM+0.5)}{KM}\right) & \cos\left(\frac{2\pi(KM+0.5)}{KM}\right) & \cdots & \cos\left(\frac{(L-1)\pi(KM+0.5)}{KM}\right) \end{bmatrix} \tag{15}$$

substituting Equation (14) into Equation (13), the compensated temporal signal $\hat{r}_n[i]$ can be rewritten as,

$$\hat{r}_n[i] = \sum_{l=0}^{L-1} r_n[i]\tau_{i,l}C_n(l) \tag{16}$$

hence, the frequency domain compenstated symbol after analysis processing can be expressed as,

$$\begin{aligned} \hat{R}_{n,m} &= \sum_{i=0}^{L_g-1} \hat{r}_n[i]g[i-n\tfrac{M}{2}] \cdot e^{-j\frac{2\pi}{M}mi}e^{-j\frac{\pi(m+n)}{2}} \\ &= \sum_{i=0}^{L_g-1} e^{-j\hat{\varphi}_n[i]}r_n[i]g[i-n\tfrac{M}{2}] \cdot e^{-j\frac{2\pi}{M}mi}e^{-j\frac{\pi(m+n)}{2}} \\ &= \sum_{i=0}^{L_g-1} e^{-j\hat{\varphi}_n[i]}\left(e^{j\varphi_n[i]}s_n[i] + w_n[i]\right)g[i-n\tfrac{M}{2}] \cdot e^{-j\frac{2\pi}{M}mi}e^{-j\frac{\pi(m+n)}{2}} \\ &\approx \underbrace{\sum_{i=0}^{L_g-1} s_n[i]g[i-n\frac{M}{2}] \cdot e^{-j\frac{2\pi}{M}mi}e^{-j\frac{\pi(m+n)}{2}}}_{A_{n,m}} + \xi_{n,m} \end{aligned} \tag{17}$$

where $A_{n,m}$ is the received symbol when the PN is perfectly compensated — i.e., the transmitted PAM symbol $a_{n,m}$ can be recovered by taking real part of $A_{n,m}$ — and $\xi_{n,m}$ is the noise term. By replacing $\hat{r}_n[i]$ from Equation (16) into Equation (17), the symbol $\hat{R}_{n,m}$ is rewritten into

$$\begin{aligned}\hat{R}_{n,m} &= \sum_{i=0}^{L_g-1} e^{-j\varphi_n[i]} r_n[i] g[i-n\tfrac{M}{2}] \cdot e^{-j\frac{2\pi}{M}mi} e^{-j\frac{\pi(m+n)}{2}} \\ &= \sum_{i=0}^{L_g-1}\sum_{l=0}^{L-1} C_n(l)\tau_{i,l} r_n[i] g[i-n\tfrac{M}{2}] \cdot e^{-j\frac{2\pi}{M}mi} e^{-j\frac{\pi(m+n)}{2}} \\ &= \sum_{l=0}^{L-1} C_n(l) \sum_{i=0}^{L_g-1} r_n[i] g[i-n\tfrac{M}{2}]\tau_{i,l} e^{-j\frac{2\pi}{M}mi} e^{-j\frac{\pi(m+n)}{2}} \\ &= \sum_{l=0}^{L-1} C_n(l) V_{n,m}^l\end{aligned} \tag{18}$$

where the symbol $V_{n,m}^l$ satisfies

$$V_{n,m}^l = \sum_{i=0}^{L_g-1} r_n[i] g[i-n\frac{M}{2}]\tau_{i,l} e^{-j\frac{2\pi}{M}mi} e^{-j\frac{\pi(m+n)}{2}} \tag{19}$$

By combining Equation (17) and Equation (18), the received symbol $A_{n,m}$ can be expressed as

$$A_{n,m} \approx \sum_{l=0}^{L-1} C_n(l) V_{n,m}^l - \xi_{n,m} \tag{20}$$

While ignoring the noise term $\xi_{n,m}$, the corresponding estimated vector $\hat{\boldsymbol{A}}_n$ can be approximated as,

$$\hat{\boldsymbol{A}}_n = \mathbf{V}_n \boldsymbol{C}_n \tag{21}$$

where $\hat{\boldsymbol{A}}_n = [\hat{A}_{n,0}, \hat{A}_{n,1}, \ldots, \hat{A}_{n,M-1}]^T$ and $\mathbf{V}_n = [\boldsymbol{V}_n^0, \boldsymbol{V}_n^1, \ldots, \boldsymbol{V}_n^{L-1}]$ $(m \in [0, M-1])$. $\hat{A}_{n,m}$ is the estimated value of $A_{n,m}$, and the $M \times 1$ vector $\boldsymbol{V}_n^l = [V_{n,0}^l, V_{n,1}^l, \ldots, V_{n,M-1}^l]^T$ is a column of the matrix $\mathbf{V}_n$.

Taking the real part of Equation (21) in the left and right sides, the equation is changed into

$$\hat{\boldsymbol{a}}_n = \mathbf{P}_n \boldsymbol{Q}_n \tag{22}$$

where $\hat{\boldsymbol{a}}_n = [\hat{a}_{n,0}, \hat{a}_{n,1}, \ldots, \hat{a}_{n,M-1}]$, $\mathbf{P}_n = [\mathrm{Re}(\mathbf{V}_n) - \mathrm{Im}(\mathbf{V}_n)]$, $\boldsymbol{Q}_n = [\mathrm{Re}(\boldsymbol{C}_n^T)\ \mathrm{Im}(\boldsymbol{C}_n^T)]^T$. $\hat{a}_{n,m}$ is the estimated value of the transmitted PAM symbol, and $\mathrm{Im}(\cdot)$ denotes the imaginary projection operator. As above-mentioned, in order to reduce the decision errors, only partially estimated values of the transmitted signal can be obtained by making a pre-decision on the the constellation points in the low decision error probability regions. When the optimized parameter $\delta$ is selected by balancing performance and complexity of the proposed algorithm, a $Z_n \times 1$ vector is picked out to perform the pre-decision, where $Z_n$ is the total number of the pre-estimated transmitted signals in the $n$th FBMC/OQAM block. The pre-estimated transmitted signal vector can be given by $\mathbf{S}_n\hat{\boldsymbol{a}}_n$, where the permutation matrix $\mathbf{S}_n = [\boldsymbol{\lambda}_{t_1}, \boldsymbol{\lambda}_{t_2}, \ldots, \boldsymbol{\lambda}_{t_{Z_n}}]^T$ is a $Z_n \times M$ matrix, $t_z$ ($z = 1, 2, \ldots, Z_n$) represents the subcarrier index of $z$th pre-estimated transmitted signal, and $\boldsymbol{\lambda}_{t_z}$ is a $M \times 1$ vector $[\underbrace{0, \ldots, 0}_{t_z-1}, 1, \underbrace{0, \ldots, 0}_{M-t_z}]^T$.

Subsequently, Equation (22) can be rewritten as

$$\mathbf{S}_n\hat{\boldsymbol{a}}_n = \mathbf{S}_n\mathbf{P}_n\boldsymbol{Q}_n \tag{23}$$

Finally, the least square (LS) solution of the unknown DCT coefficient vector can be easily found as follows:

$$\boldsymbol{Q}_n = \left((\mathbf{S}_n\mathbf{P}_n)^T(\mathbf{S}_n\mathbf{P}_n)\right)^{-1}(\mathbf{S}_n\mathbf{P}_n)^T(\mathbf{S}_n\hat{\boldsymbol{a}}_n) \tag{24}$$

Once the DCT coefficient vector $\boldsymbol{Q}_n$ is obtained, the PN, including the CPE and ICI, can be compensated via Equation (22).

## 4. Modified Algorithm Based on Overlapped Symbol Structure

Different from OFDM symbols, neighboring temporal FBMC/OQAM blocks are overlapped with each other. Figure 4 shows the time-domain FBMC/OQAM transmitted blocks with overlapped structure. The $i$th sample of $n$th received FBMC/OQAM block can be given by

$$r_n[i] = r[i + (n-1)\frac{M}{2}],\ i \in [1, KM] \tag{25}$$

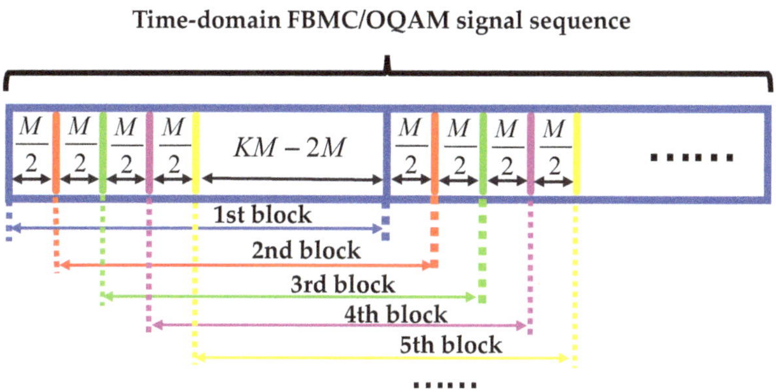


**Figure 4.** Time-domain FBMC/OQAM transmitted blocks with overlapped structure.

Then, the last $KM$-$M/2$ samples of $n$th FBMC/OQAM block are the same as the first $KM$-$M/2$ samples of ($n$ + 1)th FBMC/OQAM block—i.e., $r_n[j + M/2] = r_{n+1}[j],\ j \in [1, KM - M/2]$. As a consequence, the PN sample also satisfies that $\varphi_n[j + M/2] = \varphi_{n+1}[j]$, where $\varphi_{n+1}[j]$ is the $j$th sample of the ($n$ + 1)th PN vector $\varphi_{n+1}$. Let $\boldsymbol{\tau}_{pre}$ and $\boldsymbol{\tau}_{last}$ be the first and the last $KM$-$M/2$ rows of DCT basis $\boldsymbol{\tau}$, repectively, and we can obtain that $\boldsymbol{\tau}_{last}C_n = \boldsymbol{\tau}_{pre}C_{n+1}$. Using LS estimation, the DCT coefficient vector $C_n$ for $n$th CO-FBMC/OQAM block can be obtained as follows,

$$C_n = \left(\boldsymbol{\tau}_{last}^T \boldsymbol{\tau}_{last}\right)^{-1} \boldsymbol{\tau}_{last}^T \boldsymbol{\tau}_{pre} C_{n+1} \tag{26}$$

If the first DCT coefficient vector $C_1$ is known, the other DCT coefficient vectors can be calculated according to the recursive expression in Equation (26). It is obviously unreasonable in such a way to calculate the other DCT coefficient vectors, because the updating between the PN vectors of $\boldsymbol{\varphi}_n$ and $\boldsymbol{\varphi}_{n+1}$ is not completely neglected. Moreover, two truncated matrices of $\boldsymbol{\tau}_{pre}$ and $\boldsymbol{\tau}_{last}$ have very small singular values owing to the circularity of the DCT basis matrix $\boldsymbol{\tau}$. This leads to the significant degradation of the accuracy in the LS estimation. Therefore, an approximate expression between the DCT coefficient vectors of $C_n$ and $C_{n+1}$ is derived to achieve high estimate accuracy.

Although the DCT basis matrix is not an orthogonal matrix in the proposed algorithm, the following condition is still satisfied:

$$\boldsymbol{\tau}^T \boldsymbol{\tau} = b\mathbf{I} \tag{27}$$

where $b$ is a constant, and $\mathbf{I}$ is a $L \times L$ identity matrix. When the difference between the first and the last $M/2$ samples of $\boldsymbol{\varphi}_n$ and $\boldsymbol{\varphi}_{n+1}$ are ignored, $\boldsymbol{\varphi}_{n+1}$ can be approximated as the $M/2$ circular shift of $\boldsymbol{\varphi}_n$. Hence, the terms satisfy that

$$\begin{cases} \boldsymbol{\tau} C_n = \boldsymbol{\varphi}_n \approx \boldsymbol{\mu}_{-1}\boldsymbol{\varphi}_{n+1} = \boldsymbol{\mu}_{-1}\boldsymbol{\tau} C_{n+1} \\ \boldsymbol{\tau} C_{n+1} = \boldsymbol{\varphi}_{n+1} \approx \boldsymbol{\mu}_{+1}\boldsymbol{\varphi}_n = \boldsymbol{\mu}_{+1}\boldsymbol{\tau} C_n \end{cases} \tag{28}$$

where $\mu_{-1}$ and $\mu_{+1}$ are two $L_g \times L_g$ circular shift matrices,

$$\begin{cases} \mu_{-1} = \begin{bmatrix} 0_{\frac{M}{2}\times(L_g-\frac{M}{2})} & \mathbf{I}_{\frac{M}{2}} \\ \mathbf{I}_{L_g-\frac{M}{2}} & 0_{(L_g-\frac{M}{2})\times\frac{M}{2}} \end{bmatrix} \\ \mu_{+1} = \begin{bmatrix} 0_{(L_g-\frac{M}{2})\times\frac{M}{2}} & \mathbf{I}_{L_g-\frac{M}{2}} \\ \mathbf{I}_{\frac{M}{2}} & 0_{\frac{M}{2}\times(L_g-\frac{M}{2})} \end{bmatrix} \end{cases} \tag{29}$$

here, $\mathbf{0}_{m\times n}$ denotes a $m \times n$ zero matrix, and $\mathbf{I}_n$ is a $n \times n$ identity matrix. In combination with Equation (27), Equation (28) can be further expressed as

$$\begin{cases} C_n = \gamma_{-1} C_{n+1} \\ C_{n+1} = \gamma_{+1} C_n \end{cases} \tag{30}$$

$$\begin{cases} \gamma_{-1} = \frac{1}{b}\tau^T \mu_{-1}\tau \\ \gamma_{+1} = \frac{1}{b}\tau^T \mu_{+1}\tau \end{cases} \tag{31}$$

Assuming that three neighboring FBMC/OQAM blocks participate in the estimation process, Equation (21) can be expanded as

$$\begin{bmatrix} \hat{\mathbf{A}}_{n-1} \\ \hat{\mathbf{A}}_n \\ \hat{\mathbf{A}}_{n+1} \end{bmatrix} = \begin{bmatrix} \mathbf{V}_{n-1}\gamma_{-1} \\ \mathbf{V}_n \\ \mathbf{V}_{n+1}\gamma_{+1} \end{bmatrix} C_n \tag{32}$$

The matrices of $\overline{a}_n$ and $\overline{\mathbf{P}}_n$ are separately defined as $\overline{a}_n = [\hat{\mathbf{a}}_{n-1}^T, \hat{\mathbf{a}}_n^T, \hat{\mathbf{a}}_{n+1}^T]^T$ and

$$\overline{\mathbf{P}}_n = \begin{bmatrix} \mathrm{Re}(\mathbf{V}_{n-1}\gamma_{-1}) & -\mathrm{Im}(\mathbf{V}_{n-1}\gamma_{-1}) \\ \mathrm{Re}(\mathbf{V}_n) & -\mathrm{Im}(\mathbf{V}_n) \\ \mathrm{Re}(\mathbf{V}_{n+1}\gamma_{+1}) & -\mathrm{Im}(\mathbf{V}_{n+1}\gamma_{+1}) \end{bmatrix}. \tag{33}$$

Similar to Equations (22)–(24), the LS solution of the unknown DCT coefficient vector can be obtained from the following expression,

$$Q_n = \left(\left(\overline{\mathbf{S}}_n\overline{\mathbf{P}}_n\right)^T\left(\overline{\mathbf{S}}_n\overline{\mathbf{P}}_n\right)\right)^{-1}\left(\overline{\mathbf{S}}_n\overline{\mathbf{P}}_n\right)^T\left(\overline{\mathbf{S}}_n\overline{a}_n\right) \tag{34}$$

where $\overline{\mathbf{S}}_n = [\mathbf{S}_{n-1}^T, \mathbf{S}_n^T, \mathbf{S}_{n+1}^T]^T$.

In the modified proposed algorithm, we throw away the first $M/2$ samples of the $(n-1)$th PN vector $\varphi_{n-1}$ and the last $M/2$ samples of the $(n+1)$th PN vector $\varphi_{n+1}$; in spite of this, the relatively small power is on both ends of impulse response of the prototype filter so that the approximation affects rarely its performance, as the vector $C_n$ is obtained in such a way, and the vectors of $C_{n-1}$ and $C_{n+1}$ are easily calculated using the recurrence relations in Equations (30)–(31). Therefore, only one inverse operation in Equation (34) is required for every three neighboring FBMC/OQAM blocks, and the complexity can be decreased by the simplified operation. The detailed complexity analysis will be described in the next section. The acronyms used in this article are summarized in the following Table 1.

**Table 1.** A summary of abbreviation.

| Abbreviation | Full Name |
|---|---|
| CO-FBMC/OQAM | Coherent optical offset-quadrature amplitude modulation-based filter-bank multicarrier |
| MCM | multicarrier modulation |
| OFDM | orthogonal frequency division multiplexing |
| CO-OFDM | coherent optical OFDM |
| PN | phase noise |
| CPE | common phase error |
| ICI | inter-carrier interference |
| PNC | Phase noise compensation |
| IMI | intrinsic imaginary interference |
| SE | spectral efficiency |
| IFFT | inverse fast Fourier transform |
| FFT | fast Fourier transform |
| CC | computational complexity |
| BPS | blind phase search |
| M-BPS | modified BPS |
| BD-PNC | blind discrete-cosine-transform-based phase noise compensation |
| M-BD-PNC | modified BD-PNC |
| CD | chromatic dispersion |
| OBE | orthogonal-basis-expansion |
| DCT | discrete cosine transform |
| BTB | back to back |
| LS | least square |
| OSNR | optical signal-to-noise ratio |
| HD-FEC | hard-decision forward error correction |
| PPN | poly-phase network |
| DSP | digital signal processing |
| AWGN | additive Gaussian white noise |
| BER | bit error rate |
| $EVM_{RMS}$ | root-mean-square error vector magnitude |

## 5. Simulation Results and Discussion

A 30 GBaud BTB CO-FBMC/OQAM transmission system is simulated to evaluate the performance of the proposed blind PNC scheme, which is built by Optisystem 17.0 and MATLAB. Firstly, a pseudo-random bit sequence with a word length of $2^{17}$ is mapped into m-QAM and the symbol rate is set to 30 GBaud/s. All the complex-valued QAM symbols are converted into the real-valued PAM symbols through QAM to PAM module. The base-band time-domain FBMC/OQAM symbols are generated after all PAM data sequences through IFFT and FBS-PPN. The pulse length of the square root raised cosine filter is chosen as $L_g = 4M$, and its roll-off coefficent $\alpha = 1$.

Then, the base-band FBMC/OQAM signals are modulated onto an optical carrier at 1550 nm using a Mach–Zehnder modulator with a pair of independently controllable branches. At the receiver, the optical signal is coherently detected in a 90° optical hybrid, and the received analog signals are converted into digital signals by analog-to-digital converters with two samples per symbol. Subsequently, the obtained digital symbols are fed into the offline DSP module. With time synchronization being perfectly implemented, all the time-domain symbols are converted to the frequency domain by the PPN for filter bank analysis and FFT. In this paper, perfect channel equalization is considered, and specifically, we suppose that the channel response is flat and time-invariant on each carrier [27]. Next, the blind discrete-cosine-transform-based phase noise compensation (BD-PNC) is implemented, including CPE and ICI compensation. In the related M-BPS algorithm, the total numbers of test phases for 4/16/64-QAM are chosen as 16/32/64, respectively [28]. The original binary bits can be recovered after m-OQAM and m-QAM de-mapping, and the bit error rate (BER) is estimated by comparing them with the transmitted bits.

The effects of two parameters of $\delta$ and $L$ on the performance of the algorithm are investigated in Figure 5a,b. $\delta$ is the width of shadow rectangle in a high decision error probability region, which determines how many signals after CPE compensation join the pre-decision process. As mentioned in Section 3, $\delta = 0$ means that all the data are used to perform the pre-decision while $\delta = 2$ implies that only the leftmost and rightmost data in the I-Q plane participate in the process in Figure 3. The decision error increases with the participation of more data after the CPE compensation. In Figure 5a, $\delta$ is optimized at the length of DCT basis $L = 2$ for the CO-FBMC 4/16/64-QAM BTB system with 512 subcarriers. When the value of $\delta$ is changed from 0 to 1.8, it can be seen that the BER performance does not degrade significantly due to the reduction in decision error probability. Moreover, with the increase in the value of $\delta$, the participation of a low number of data benefits greatly to complexity reduction in the proposed algorithm. Hence, $\delta = 1.8$ is adopted in the remainder of this paper. $L$ denotes the length of the DCT basis. Generally, under a large value of $L$, we can obtain a more accurate approximation of the PN by the DCT in Equation (14). Figure 5b displays the BER performance as a function of $L$ at the linewidth and symbol duration product $\Delta\nu \cdot T_S = 10^{-2}$ with 12 dB optical signal-to-noise ratio (OSNR) and 256/512/1024 subcarriers when using the proposed BD-PNC algorithm and its modified version. The modified algorithm is developed by considering the overlapped symbol structure, which is called the M-BD-PNC scheme. For the BD-PNC, with the increase in $L$, the BER performance makes a small improvement, and is even deteriorated for the 256/512 subcarrier FBMC/OQAM systems. While using the M-BD-PNC method, there is a significant improvement for its BER performance as the value of $L$ is increased. The PN estimation precision depends on the accuracy of the DCT coefficients estimated by the LS method. In Equation (24) or Equation (34), if the matrix $\mathbf{S}_n\mathbf{P}_n$ or $\overline{\mathbf{S}}_n\overline{\mathbf{P}}_n$ does not meet the column full-rank condition, its LS solution could produce a large error. With the increase in the number of subcarriers, more rows in the matrix allow it to meet the column full-rank condition more easily. Hence, when $L$ is increased, the proposed BD-PNC algorithm achieves better performance under a large number of subcarriers; its performance becomes worse in the system with a low number of subcarriers, while in the M-BD-PNC method, $\overline{\mathbf{P}}_n$ is the extended version of $\mathbf{P}_n$ so that it is easier for the matrix $\overline{\mathbf{S}}_n\overline{\mathbf{P}}_n$ to satisfy a column full-rank term. In this case, the PN estimation precision is almost not negatively influenced by the LS method, and a large value of $L$ contributes to the improvement of the PN estimation precision. Therefore, for the M-BD-PNC, the BER performance exhibits a better result at a large $L$. The optimized values of $L$ in the BD-PNC and M-BD-PNC are selected as 2 and 10, respectively.

The CO-FBMC/OQAM system with a large number of subcarriers is robust against fiber CD [12]. However, the PN becomes strong with the increase in the number of subcarriers. It is always a challenge for the existing PNC algorithm to achieve good performance under a large number of subcarriers [18,25–29]. So the performance of the proposed algorithm is evaluated in the BTB system with a large number of subcarriers. In Figure 5c, the influence of the number of subcarriers on the performance of the method is investigated at $L = 2$ and $\delta = 1.8$ using M-BPS and BD-PNC for the CO-FBMC 4/16/64 QAM systems with several different values of $\Delta\nu \cdot T_S$ and OSNR. When the number of subcarriers becomes larger, on one hand, the effect of AWGN is better averaged so that their performance is improved rapidly; on the other hand, the FBMC/OQAM symbol duration increases, and the PN has a larger variance and becomes stronger. As a result, it can be seen that the BER using M-BPS starts to deteriorate at $M = 1024$. Note that the performance using BD-PNC remains almost unchanged when the number of subcarriers is increased from 128 to 1024 for 4/16/64-QAM modulation order. The simulation results prove that its performance is not sensitive to the number of subcarriers in the CO-FBMC/OQAM systems. Under the large number of subcarriers, the proposed algorithm is far superior to the M-BPS. For the 4-QAM modulation at $M = 1024$ with $\Delta\nu \cdot T_S = 5 \times 10^{-2}$ and 11 dB OSNR, the BER performance of BD-PNC is improved by more than one order of magnitude compared to the M-BPS.

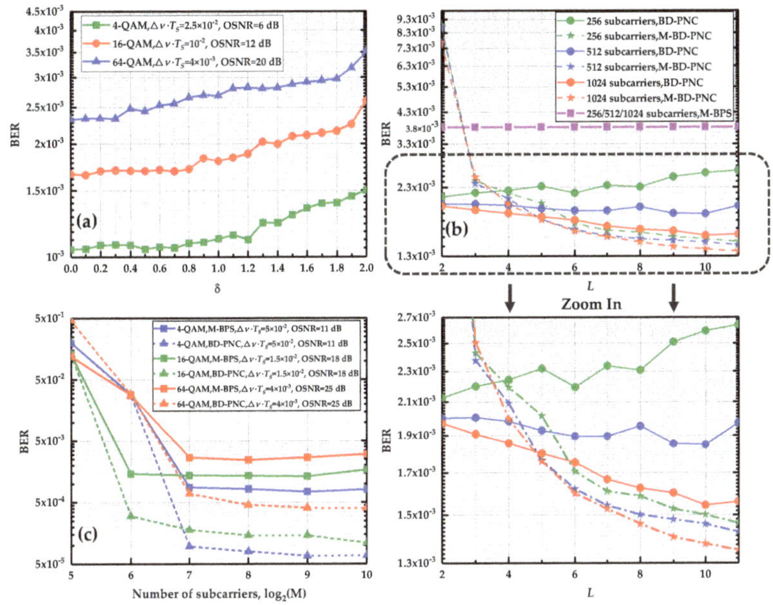


**Figure 5.** (**a**) BER versus the width of shadow rectangle in a high decision error probability region, $\delta$. (**b**) BER versus the length of DCT coefficient, $L$. (**c**) BER versus the number of subcarriers, $M$.

Figure 6a depicts one realization of the real PN and its estimation within six consecutive FBMC/OQAM blocks at the linewidth and symbol duration product $\Delta\nu \cdot T_S = 1.5 \times 10^{-2}$ and an OSNR of 20 dB for 1024 sub-carrier 16-QAM CO-FBMC systems using the BD-PNC and M-BPS. For these time-domain overlapped symbols, the CPE estimated by the M-BPS is only a rough estimate of the PN, which is far from the real PN. The proposed method shows a better phase tracking capability due to the ICI compensation in time-domain. Figure 6b shows the illustration of ICI influence and the necessity of PN estimation in time domain for CO-FBMC/OQAM systems. The constellation diagrams before and after PNC are separately displayed using M-BPS and BD-PNC with an OSNR of 27 dB and combined laser linewidth of 150 kHz. The PN in time domain, including CPE and ICI, is compensated by using BD-PNC, while only CPE is compensated by using M-BPS. It is clear that the PN in time domain is compensated by the proposed BD-PNC method resulting in an improvement of about 125 times in terms of BER over the M-BPS. Moreover, the BD-PNC achieves a root-mean-square error vector magnitude ($EVM_{RMS}$) of 4.7%, which is much lower than the $EVM_{RMS}$ of 7% by the M-BPS.

Supposing that the pre-forward error correction BER should be lower than $3.8 \times 10^{-3}$ for hard-decision forward error correction (HD-FEC) [43], the performance of these PNC algorithms is investigated at the BER target limit. Figure 7a–c show the OSNR penalty as a function of the linewidth and symbol duration product $\Delta\nu \cdot T_S$ using M-BPS and the proposed algorithms for CO-FBMC/OQAM BTB transmission system with 256, 512, and 1024 subcarriers, respectively. The optimal numbers of test phases in the related M-BPS scheme are chosen as 16/32/64 for 4/16/64-QAM modulation order, respectively [28]. When the number of subcarriers is 256 or 512, the length $L$ of DCT basis is set to 2. The length of DCT basis $L$ is selected as 10 with total number of subcarriers $M = 1024$. The parameter $\delta$ is set to 1.8 in this paper. The OSNR penalty is defined as the difference between the required OSNR for a BER of $3.8 \times 10^{-3}$ at a current laser linewidth and zero laser linewidth in CO-FBMC/OQAM systems, where the PNC algorithm is used in the former, and none of the PNC methods is applied in the PN free system in the latter. When no PN exists in the system, the reference OSNRs for 4/16/64-QAM

30 Gbaud signal at the HD-FEC threshold are ~4.4, ~10, and~16.8 dB, respectively. For comparison, an acceptable maximum of 1 dB OSNR penalty is regarded as a threshold value [27].

On the whole, the proposed algorithm provides significantly better performance than M-BPS. In Figure 7a, at 1 dB OSNR penalty threshold, with the number of subcarriers $M = 256$, the M-BPS achieves a $\Delta\nu \cdot T_S$ tolerance of $3.5 \times 10^{-2}$, $7 \times 10^{-3}$, and $10^{-3}$ for 4/16/64-QAM 30 Gbaud signal, respectively. While using the proposed BD-PNC method, under 1 dB OSNR penalty constraint, the maximum $\Delta\nu \cdot T_S$ that can be achieved for 4/16/64-QAM are $5 \times 10^{-2}$, $10^{-2}$, and $1.8 \times 10^{-3}$, respectively. With the increase in the QAM modulation order, the difference between the performance of the two methods is more remarkable. At higher order modulation format, the ICI effect becomes stronger, and the BD-PNC scheme achieves more significant results than M-BPS. In Figure 7b, at $M = 512$, M-BPS and BD-PNC achieve a $\Delta\nu \cdot T_S$ tolerance of $5.6 \times 10^{-4}$ and $1.4 \times 10^{-3}$ for 64-QAM 30 Gbaud signal, respectively.

The studies focus on the investigation of the proposed BD-PNC algorithm's performance with the number of subcarriers $M = 1024$. As a comparison, the performance of the M-BD-PNC algorithm is also evaluated in the 30 GBaud BTB CO-FBMC/OQAM systems with 1024 subcarriers. In Figure 7c, the M-BD-PNC achieves a better $\Delta\nu \cdot T_S$ tolerance than the BD-PNC. This is because the column full-rank condition is easily satisfied in the M-BD-PNC scheme, and its performance benefits from the enhanced estimation accuracy of DCT coefficients by the LS method. Moreover, the M-BD-PNC has a lower CC than the corresponding BD-PNC algorithm, and a detailed analysis of their CC will be described at the end of the section. With a large number of subcarriers, the ICI components in PN have a huge negative impact on the performance. With the ICI effect mitigated, the proposed algorithms offer a significantly better performance compared to the M-BPS under the same QAM modulation order. In particular, the BD-PNC and M-BD-PNC achieve tolerated linewidth and symbol duration products of $2.2 \times 10^{-2}$ and $2.6 \times 10^{-3}$ for 64-QAM 30 Gbaud signal, while the M-BPS can only achieve a $\Delta\nu \cdot T_S$ tolerance of $10^{-3}$.

Finally, we compare the CC of M-BPS, the proposed BD-PNC, and M-BD-PNC algorithms in terms of the number of real multiplications in one FBMC/OQAM block. Since the M-BD-PNC method involves several neighboring FBMC/OQAM blocks, its CC is evaluated as the average CC per block. As shown in Table 2, the CC of the proposed BD-PNC and M-BD-PNC algorithms are compared. According to the block diagram in Figure 2, the CC of the proposed BD-PNC involves the CC of its four DSP steps, including M-BPS, time-domain PNC, calculation of DCT coefficients, and final compensation. The CC of the M-BPS method is given by $O(2MY)$ [28]. The time-domain PNC model is described by Equation (19), and its CC is $O(LM\log_2 M{+}LKM)$. For the BD-PNC, the DCT coefficients are calculated in Equation (24), and its CC is given by $O(8L^3 + 8Z_nL^2 + 2Z_nL)$, where $Z_n$ is the total number of pre-estimated transmitted signals in the $n$th FBMC/OQAM block, while in the M-BD-PNC, the DCT coefficients are obtained by Equations (30) and (34), and real multiplications needed in these equations are $O(8L^3/3 + 8Z_nL^2 + 2ZnL + 2L^2/3)$. For BD-PNC and M-BD-PNC, the final PNC is completed by using Equation (22), and its CC is $O(2ML)$. In order to compare the CC of these algorithms directly, several simulations are carried out to investigate the average value of $Z_n$ at a BER of $3.8 \times 10^{-3}$. In this case, the average value of $Z_n$ is shown in Table 3. Consequently, the required real multiplication numbers per block are counted by type, as shown in Table 4. As a comparison, when all the data after CPE compensation participate in the pre-decision ($\delta = 0$), the required real multiplication numbers of BD-PNC or M-BD-PNC is provided in parentheses. It is observed that the optimization of the parameter $\delta$ in the pre-decision process results in a remarkable decrease in complexity. Furthermore, it is clear that the CC of proposed BD-PNC or M-BD-PNC is not significantly higher than M-BPS, and their CC is less than four times higher than that of M-BPS for 1024 subcarriers 64-QAM 30 Gbaud signals. Nevertheless, the proposed algorithms have a great advantage over M-BPS in terms of the PNC performance. The CC of M-BD-PNC is slightly lower than that of BD-PNC, and its performance is even superior to the BD-PNC algorithm. Therefore, at $M = 1024$, the proposed M-BD-PNC should be the first choice.

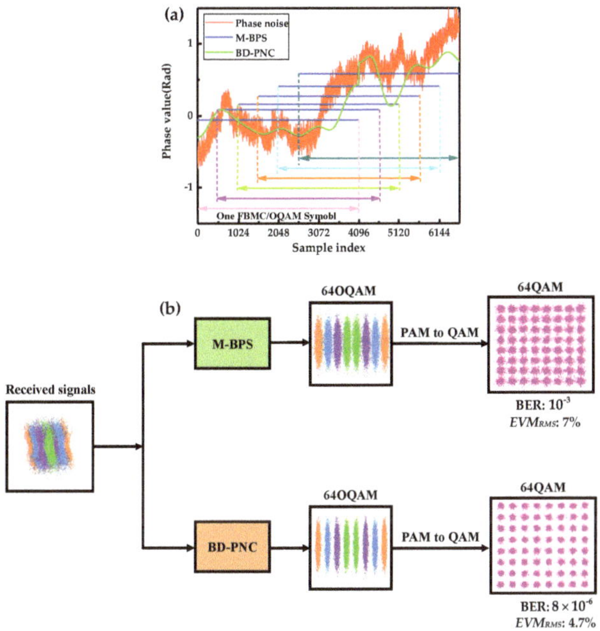


**Figure 6.** (**a**) One realization of the real phase noise and its estimations after the M-BPS and BD-PNC methods at $\Delta\nu \cdot T_S = 1.5 \times 10^{-2}$ and OSNR = 20 dB for 30 Gbaud CO-FBMC 16-QAM BTB transmission systems with 1024 subcarriers. (**b**) Illustration of constellations before and after PNC employing M-BPS and BD-PNC at $\Delta\nu \cdot T_S = 5 \times 10^{-3}$ and OSNR = 27 dB for 30 Gbaud CO-FBMC 64-QAM BTB systems with 1024 subcarriers.

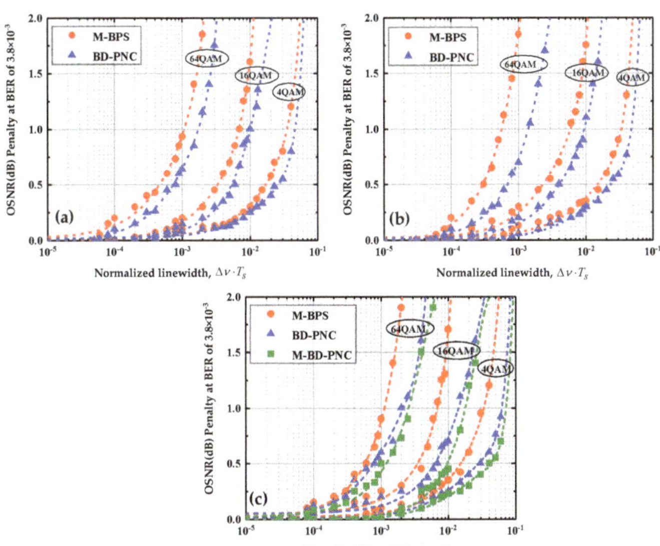


**Figure 7.** OSNR penalty versus $\Delta\nu \cdot T_S$ at a BER of $3.8 \times 10^{-3}$ using M-BPS, the proposed schemes for CO-FBMC 4/16/64-QAM systems with 256 subcarriers (**a**), 512 subcarriers (**b**), and 1024 subcarriers (**c**).

**Table 2.** The CC of the proposed BD-PNC and M-BD-PNC algorithms.

| DSP Step | CC (BD-PNC) | CC (M-BD-PNC) |
|---|---|---|
| CPE pre-compensation (M-BPS) | $O(2MY)$ | $O(2MY)$ |
| Time-domain PNC model | $O(LM\log_2 M + LKM)$ | $O(LM\log_2 M + LKM)$ |
| Calculation of DCT coefficients | $O(8L^3 + 8Z_nL^2 + 2Z_nL)$ | $O(8L^3/3 + 8Z_nL^2 + 2Z_nL + 2L^2/3)$ |
| Final compensation | $O(2ML)$ | $O(2ML)$ |

Note: $M$: number of subcarriers. $Y$: number of test phases. $L$: length of DCT coefficient. $K$: overlap factor. $Z_n$: total number of pre-estimated transmitted signals.

**Table 3.** The average value of $Z_n$.

| | $M$ = 256 | $M$ = 512 | $M$ = 1024 |
|---|---|---|---|
| 4-QAM | 140 | 280 | 570 |
| 16-QAM | 80 | 160 | 320 |
| 64-QAM | 40 | 90 | 180 |

**Table 4.** The required real multiplication numbers per block using several algorithms.

| | M-BPS | BD-PNC | M-BD-PNC |
|---|---|---|---|
| 4-QAM | 8192\|$M$ = 256<br>16384\|$M$ = 512<br>32768\|$M$ = 1024 | 20464 (24640)\|$M$ = 256<br>41888 (50240)\|$M$ = 512<br>672008 (1044288)\|$M$ = 1024 | 666740 (1039000)\|$M$ = 1024 |
| 16-QAM | 16384\|$M$ = 256<br>32768\|$M$ = 512<br>65536\|$M$ = 1024 | 26496 (32832)\|$M$ = 256<br>53952 (66624)\|$M$ = 512<br>499776 (1077056)\|$M$ = 1024 | 494510 (1071800)\|$M$ = 1024 |
| 64-QAM | 32768\|$M$ = 256<br>65536\|$M$ = 512<br>131072\|$M$ = 1024 | 41440 (49216)\|$M$ = 256<br>84200 (99392)\|$M$ = 512<br>450512 (1142592)\|$M$ = 1024 | 445245 (1137300)\|$M$ = 1024 |

## 6. Conclusions

In this paper, we have presented a blind time-domain PNC scheme for the CO-FBMC/OQAM transmission, where the PN sample is approximately expanded by DCT. In the proposed algorithms, the PN in time-domain can be compensated by obtaining their corresponding DCT coefficients. The pre-decision process is firstly performed over the selected signals after CPE compensation by M-BPS, and the decision results are used to calculate these DCT coefficients. This selective decision process reduces the CC largely. Considering the overlapped symbol structure, the modified version of the proposed algorithm is reported with its performance improvement and complexity decrease. Since the proposed algorithms are devoted to removing the ICI effect, they achieve significant performance improvement compared to M-BPS for a simulated 30 GBaud CO-FBMC/OQAM system, especially with a large number of subcarriers and high order QAM modulation format. The algorithm complexity analysis shows that its CC is slightly higher than M-BPS.

**Author Contributions:** All authors contributed extensively to the work presented in this paper. conceptualization, M.G., J.L. and H.R.; methodology, H.R., B.W. and J.L.; software, B.W. and J.L.; validation, J.L., H.R., and B.W.; formal analysis, H.R. and W.H.; investigation, H.R., Z.L., Y.Q. and S.G.; resources, W.H. and H.R.; data curation, H.R., B.W. and J.L.; writing—original draft preparation, B.W.; writing—review and editing, M.G., B.W. and H.R.; visualization, H.R. and B.W.; supervision, H.R.; project administration, H.R.; funding acquisition, W.H. All authors have read and agreed to the published version of the manuscript.

**Funding:** The National Natural Science Foundation of China (NSFC) (62075147,60907032, 61675183 and 61675184); Zhejiang Provincial Natural Science Foundation of China under Grant No.LY20F050009; Open Fund of the State Key Laboratory of Advanced Optical Communication Systems and Networks, China (2020GZKF013). General program of the Education Department of Zhejiang (Y201941823).

**Conflicts of Interest:** The authors declare no conflict of interest.

## References


1. Ding, D.; Sui, X.; Hu, R.; Zhao, C.; Fang, X. K-means-based channel equalization method for optical OFDM/OQAM systems. In Proceedings of the Asia Communications and Photonics Conference (ACPC), Chengdu, China, 2–5 November 2019; pp. 1–3.
2. Yu, Y.; Fang, X.; Mao, J.; Zhang, L.; Gao, X.; Ding, D.; Liu, L. Novel NOMA system for optical communications based on OFDM/OQAM. In Proceedings of the 2019 18th IEEE International Conference on Optical Communications and Networks (ICOCN), Huangshan, China, 5–8 August 2019; pp. 1–3.
3. Suo, Z.; Ding, D.; Yu, Y.; Zhang, L.; Fang, X.; Gao, X. Preamble based frequency domain channel estimation method for optical OFDM/OQAM system. In Proceedings of the 2019 IEEE 11th International Conference on Communication Software and Networks (ICCSN), Chongqing, China, 12–15 June 2019; pp. 13–16.
4. Nguyen, T.-H.; Bramerie, L.; Gay, M.; Kazdoghli-Lagha, M.; Peucheret, C.; Gerzaguet, R.; Gorza, S.P.; Louveaux, J.; Horlin, F. Experimental Demonstration of the Tradeoff Between Chromatic Dispersion and Phase Noise Compensation in Optical FBMC/OQAM Communication Systems. *J. Lightwave Technol.* **2019**, *37*, 4340–4348. [CrossRef]
5. Alaghbari, K.A.; Lim, H.S.; Eltaif, T.A. Compensation of chromatic dispersion and nonlinear phase noise using iterative soft decision feedback equalizer for coherent optical FBMC/OQAM systems. *IEEE J. Lightwave Technol.* **2020**, *38*, 3839–3849. [CrossRef]
6. Huang, Y.-F.; Tsai, C.-T.; Chi, Y.-C.; Huang, D.-W.; Lin, G.-R. Filtered Multicarrier OFDM Encoding on Blue Laser Diode for 14.8-Gbps Seawater Transmission. *J. Lightwave Technol.* **2018**, *36*, 1739–1745. [CrossRef]
7. Oubei, H.M.; Duran, J.R.; Janjua, B.; Wang, H.-Y.; Tsai, C.-T.; Chi, Y.-C.; Ng, T.K.; Kuo, H.C.; He, J.H.; Alouini, M.-S. 4.8 Gbit/s 16-QAM-OFDM transmission based on compact 450-nm laser for underwater wireless optical communication. *Opt. Express* **2015**, *23*, 23302–23309. [CrossRef] [PubMed]
8. Cvijetic, N. OFDM for Next-Generation Optical Access Networks. *J. Lightwave Technol.* **2012**, *30*, 384–398. [CrossRef]
9. Tsonev, D.; Chun, H.; Rajbhandari, S.; McKendry, J.J.D.; Videv, S.; Gu, E.; Haji, M.; Watson, S.; Kelly, A.E.; Faulkner, G.; et al. A 3-Gb/s single-LED OFDM-based wireless VLC link using a gallium nitride $\mu$LED. *IEEE Photon. Technol. Lett.* **2014**, *26*, 637–640. [CrossRef]
10. Dissanayake, S.D.; Armstrong, J. Comparison of ACO-OFDM, DCO-OFDM and ADO-OFDM in IM/DD Systems. *J. Lightwave Technol.* **2013**, *31*, 1063–1072. [CrossRef]
11. Nhan, N.-Q.; Morel, P.; Azou, S.; Morvan, M.; Gravey, P.; Pincemin, E. Sparse Preamble Design for Polarization Division Multiplexed CO-OFDM/OQAM Channel Estimation. *J. Lightwave Technol.* **2018**, *36*, 2737–2745. [CrossRef]
12. Zhao, J.; Townsend, P.D. Dispersion tolerance enhancement using an improved offset-QAM OFDM scheme. *Opt. Express* **2015**, *23*, 17638. [CrossRef]
13. Fang, X.; Xu, Y.; Chen, Z.; Zhang, F. Time-Domain Least Square Channel Estimation for Polarization-Division-Multiplexed CO-OFDM/OQAM Systems. *J. Lightwave Technol.* **2016**, *34*, 891–900. [CrossRef]
14. Farhang-Boroujeny, B. OFDM Versus Filter Bank Multicarrier. *IEEE Signal Process. Mag.* **2011**, *28*, 92–112. [CrossRef]
15. Nissel, R.; Schwarz, S.; Rupp, M. Filter Bank Multicarrier Modulation Schemes for Future Mobile Communications. *IEEE J. Sel. Areas Commun.* **2017**, *35*, 1768–1782. [CrossRef]
16. Kofidis, E.; Katselis, D.; Rontogiannis, A.; Theodoridis, S. Preamble-based channel estimation in OFDM/OQAM systems: A review. *Signal Process.* **2013**, *93*, 2038–2054. [CrossRef]
17. Ding, J.; Xu, T.; Jin, C.; Wang, Z.; Zhao, J.; Liu, T. Impact of Equalization-Enhanced Phase Noise on Digital Nonlinearity Compensation in High-Capacity Optical Communication Systems. *Sensors* **2020**, *20*, 4149. [CrossRef]
18. Fang, X.; Zhang, F. Phase Noise Estimation and Suppression for PDM CO-OFDM/OQAM Systems. *J. Lightwave Technol.* **2017**, *35*, 1837–1846. [CrossRef]
19. Sriyananda, M.G.S.; Rajatheva, N. Analysis of self interference in a basic FBMC system. In Proceedings of the 2013 IEEE 78th Vehicular Technology Conference (VTC Fall), Las Vegas, NV, USA, 2–5 September 2013; pp. 1–5.
20. Ndo, G.; Lin, H.; Siohan, P. FBMC/OQAM equalization: Exploiting the imaginary interference. In Proceedings of the 2012 IEEE 23rd International Symposium on Personal, Indoor and Mobile Radio Communications (PIMRC), Sydney, NSW, Australia, 9–12 September 2012; pp. 2359–2364.

21. Stitz, T.; Ihalainen, T.; Viholainen, A.; Renfors, M. Pilot-Based Synchronization and Equalization in Filter Bank Multicarrier Communications. *EURASIP J. Adv. Signal Process.* **2010**, *2010*, 741429. [CrossRef]
22. Lele, C.; Legouable, R.; Siohan, P. Channel estimation with scattered pilots in OFDM/OQAM. In Proceedings of the 2008 IEEE 9th Workshop on Signal Processing Advances in Wireless Communications, Recife, Brazil, 6–9 July 2008; pp. 286–290.
23. Nissel, R.; Rupp, M. On pilot-symbol aided channel estimation in FBMC-OQAM. In Proceedings of the 2016 IEEE International Conference on Acoustics, Speech and Signal Processing (ICASSP), Shanghai, China, 20–25 March 2016; pp. 3681–3685.
24. Nissel, R.; Rupp, M. Bit error probability for pilot-symbol aided channel estimation in FBMC-OQAM. In Proceedings of the 2016 IEEE International Conference on Communications (ICC), Kuala Lumpur, Malaysia, 22–27 May 2016; pp. 1–6.
25. You, B.; Yang, L.; Luo, F.; Yang, S.; Chen, D.; Ni, Y.; Li, B.; Liu, D. Pilot-based extended Kalman filter for phase noise estimation in CO-FBMC/OQAM systems. *Opt. Commun.* **2019**, *443*, 116–122. [CrossRef]
26. You, B.; Yang, L.; Luo, F.; Fu, S.; Yang, S.; Li, B.; Liu, D. Joint Carrier Frequency Offset and Phase Noise Estimation Based on Pseudo-Pilot in CO-FBMC/OQAM System. *IEEE Photon. J.* **2019**, *11*, 1–11. [CrossRef]
27. Nguyen, T.T.; Le, S.T.; Nissel, R.; Wuilpart, M.; Van Compernolle, L.; Mégret, P. Pseudo-Pilot Coding Based Phase Noise Estimation for Coherent Optical FBMC-OQAM Transmissions. *J. Lightwave Technol.* **2018**, *36*, 2859–2867. [CrossRef]
28. Nguyen, T.-H.; Louveaux, J.; Gorza, S.-P.; Horlin, F. Simple Feedforward Carrier Phase Estimation for Optical FBMC/OQAM Systems. *IEEE Photon. Technol. Lett.* **2016**, *28*, 2823–2826. [CrossRef]
29. Nguyen, T.-H.; Rottenberg, F.; Gorza, S.-P.; Louveaux, J.; Horlin, F. Extended Kalman Filter for Carrier Phase Recovery in Optical Filter Bank Multicarrier Offset QAM Systems. *Opt. Fiber Commun. Conf.* **2017**, *4*, 3. [CrossRef]
30. Tang, H.; Xiang, M.; Fu, S.; Tang, M.; Shum, P.; Liu, D. Feed-forward carrier phase recovery for offset-QAM Nyquist WDM transmission. *Opt. Express* **2015**, *23*, 6215–6227. [CrossRef]
31. Lu, J.; Fu, S.; Tang, H.; Xiang, M.; Tang, M.; Liu, D. Vertical blind phase search for low-complexity carrier phase recovery of offset-QAM Nyquist WDM transmission. *Opt. Commun.* **2017**, *382*, 212–218. [CrossRef]
32. Nguyen, T.H.; Gorza, S.P.; Louveaux, J.; Horlin, F. Low-complexity blind phase search for filter bank multicarrier offset-QAM optical fiber systems. In Proceedings of the Signal Processing in Photonic Communications, Vancouver, Canada, 18–20 July 2016. paper SpW2G.2.
33. Kakkavas, A.; Castañeda, M.; Luo, J.; Laas, T.; Xu, W.; Nossek, J.A. FBMC-OQAM with phase noise: Achievable performance and compensation. In Proceedings of the 2017 IEEE 18th International Workshop on Signal Processing Advances in Wireless Communications (SPAWC), Sapporo, Japan, 3–6 July 2017; pp. 1–5.
34. Doré, J.B.; Berg, V. Blind phase tracking algorithm for FBMC receivers. In Proceedings of the 2015 International Symposium on Wireless Communication Systems (ISWCS), Brussels, Belgium, 25–28 August 2015; pp. 261–265.
35. Rottenberg, F.; Nguyen, T.H.; Gorza, S.P.; Horlin, F.; Louveaux, J. ML and MAP phase noise estimators for optical fiber FBMC-OQAM systems. In Proceedings of the 2017 IEEE International Conference on Communications (ICC), Paris, France, 21–25 May 2017; pp. 1–6.
36. Nguyen, T.-H.; Rottenberg, F.; Gorza, S.-P.; Louveaux, J.; Horlin, F. Efficient Chromatic Dispersion Compensation and Carrier Phase Tracking for Optical Fiber FBMC/OQAM Systems. *J. Lightwave Technol.* **2017**, *35*, 2909–2916. [CrossRef]
37. Ikeuchi, K.; Sakai, M.; Lin, H. Compensation of Phase Noise in OFDM/OQAM Systems. In Proceedings of the 2017 IEEE 86th Vehicular Technology Conference (VTC-Fall), Toronto, ON, Canada, 24–27 September 2017; pp. 1–5.
38. Cao, S.; Kam, P.-Y.; Yu, C. Time-Domain Blind ICI Mitigation for Non-Constant Modulus Format in CO-OFDM. *IEEE Photon. Technol. Lett.* **2013**, *25*, 2490–2493. [CrossRef]
39. Ren, H.; Cai, J.; Ye, X.; Lu, J.; Cao, Q.; Guo, S.; Xue, L.-L.; Qin, Y.; Hu, W. Decision-aided ICI mitigation with time-domain average approximation in CO-OFDM. *Opt. Commun.* **2015**, *347*, 1–7. [CrossRef]
40. Lin, C.-T.; Wei, C.-C.; Chao, M.-I. Phase noise suppression of optical OFDM signals in 60-GHz RoF transmission system. *Opt. Express* **2011**, *19*, 10423–10428. [CrossRef]

41. Bi, M.; Liu, L.; Zhang, L.; Yang, G.; Hu, M.; Li, Q.; Xiao, S.; Hu, W. Low overhead equalization algorithm for simultaneously estimating channel and mitigating intrinsic imaginary interference in IMDD-OQAM-OFDM system. *Opt. Commun.* **2019**, *430*, 256–261. [CrossRef]
42. Zhang, P.; Ren, H.; Gao, M.; Lu, J.; Le, Z.; Qin, Y.; Hu, W. Low-Complexity Blind Carrier Phase Recovery for C-mQAM Coherent Systems. *IEEE Photon. J.* **2018**, *11*, 1–14. [CrossRef]
43. Essiambre, R.-J.; Kramer, G.; Winzer, P.J.; Foschini, G.J.; Goebel, B. Capacity Limits of Optical Fiber Networks. *J. Lightwave Technol.* **2010**, *28*, 662–701. [CrossRef]


**Publisher's Note:** MDPI stays neutral with regard to jurisdictional claims in published maps and institutional affiliations.

sensors

MDPI

Letter

# Optical-Amplifier-Compatible Long-Distance Secure Key Generation Based on Random Phase Fluctuations for WDM Systems

**Ben Wu** [1,*] **and Yue-Kai Huang** [2]

[1] Department of Electrical and Computer Engineering, Rowan University, 201 Mullica Hill Rd., Glassboro, NJ 08028, USA

[2] NEC Laboratories America, Inc., Princeton, NJ 08540, USA; kai@nec-labs.com

* Correspondence: wub@rowan.edu

Received: 23 September 2020; Accepted: 3 November 2020; Published: 5 November 2020

**Abstract:** We proposed and experimentally demonstrated a secure key generation and distribution system that is compatible with optical amplifiers and standard wavelength-division multiplexing (WDM) transmission systems. The key is generated from the phase fluctuations induced by environmental instabilities. The key generation system is tested in a 240 km bidirectional fiber-pair link with multiple optical amplifiers. To demonstrate the compatibility with WDM systems, 38 WDM channels are transmitted together with the key distribution channel. The secret key is protected against eavesdropping and coherence detection attack by the wide-band property of the signal carrier and the fast-changing rate of the phase fluctuations.

**Keywords:** fiber optics communications; optical security and encryption; phase fluctuations

## 1. Introduction

Cryptography and data encryption are the fundamental elements of network security [1]. The effectiveness of the data encryption relies on the scheme to generate and distribute the key securely. If the key is attacked and exposed to an eavesdropper, the encrypted signal can be easily decrypted. Traditional key generation schemes are based on software algorithms. For example, the widely used Rivest–Shamir–Adleman (RSA) cryptosystem is based on the difficulty to factor the product of two large prime numbers [2]. The security of the system relies on computational resources available for the eavesdropper, while the increasing power of quantum computing raises a potential challenge to the software-based key generation schemes [3–6].

The physical layer encryption and key generation scheme provides an alternative solution to the network security [7–9]. One of the most widely studied physical layer key generation schemes is the quantum key distribution (QKD) [10–12]. The security of a QKD system relies on the fact that without knowing the eigenvalues of the quantum states, the eavesdropper cannot measure the signal, which is the quantum state of a single photon. Compared with traditional software-based key distribution methods, the physical layer key generation methods take advantages of the physical properties of the transmission media. If these physical properties are not properly recorded in real time, the data are lost permanently. The physical property for QKD is the quantum state of a single photon. If the eavesdropper does not know the eigenvalues of the quantum states, the original signals are lost at the time as the signals are being received by the eavesdropper. In this case, even the eavesdropper has unlimited computational power; the physical properties of the transmission media are not fully digitized for post-processing and signal recovery.

In this paper, we experimentally demonstrate a key generation and distribution system that is compatible with both optical amplifiers for long-range transmission and standard wavelength-division multiplexing (WDM) systems. The keys are generated from the random fluctuations of phase in fiber interferometers [13]. Environmental instabilities, including temperature changes and mechanical vibrations, induce random refractive index changes in fiber interferometers, which are deployed at each one of the communicating pair with long optical delays, and are then converted to random phase signals at both ends to generate digital keys. As a physical layer key generation scheme, the proposed system generates keys from phase randomness and if the analog phase signals are not recorded properly, the keys are lost and cannot be recovered.

The key generation system is demonstrated in a 240 km long fiber link with three optical amplifiers, and the key distribution channel cotransmits with 38 WDM communication channels. The proposed system is compatible with WDM channels and can be implemented over the existing fiber-optic networks based on three facts: (1) Standard optical fiber and fiber amplifiers are used to transmit signals for the key generation system. The key generation channel shares the C-band optical amplifiers with the other 38 WDM channels to compensate the attenuation from standard single mode fiber (SSMD). (2) The key generation system uses the bandpass filter with the same power extinction ratio as the standard WDM filters to download the key generation channel from the networks. (3) Since the power level of the key generation channel is comparable to the power level of the other WDM channels, photodetectors for standard WDM channels are directly used to receive signals of the key generation channel, and the key generation channel is installed with low cost. The proposed system can be implemented by plugging in a pair of key generators to the existing fiber networks for the two users (Alice and Bob) that require synchronized keys.

The security of the proposed system is based on the random phase fluctuations, and such phase fluctuations always exist in fiber interferometers. A pair of matched optical delay lengths is introduced to the two users that share keys. The matched optical delay lengths are in km range and coherence length of the broadband source is in mm range. If the eavesdropper tries to measure the phase fluctuations, another optical delay length has to be applied. The additional optical delay length, which is also in km range, introduces new phase fluctuations, so the eavesdropper cannot accurately measure the keys.

## 2. Experimental Setup

Figure 1 shows the experimental setup of the key generation system, which consists of two Mach–Zehnder (MZ) delayed interferometers. The synchronized key signals are the results of the interference from interferometers at both ends. Both Alice and Bob send and receive signals with phases affected by the local temperature changes and mechanical vibrations. Since the signals in both directions go through the same path, the phase fluctuations and the interference results at the MZI outputs are the same at both ends, making key sharing possible. Broadband sources are used as the signal carriers to prevent the eavesdropper from measuring the phase directly with coherent detection. In the experiment, both Alice and Bob use a filtered amplified spontaneous emission (ASE) with bandwidth 150 GHz and center wavelength 1547.6 nm as the light source. Since the broadband source has short coherence length, which is 2 mm, the optical delays placed in Alice's (L1) and Bob's (L2) interferometers have to be matched within the 2 mm range to recover the phase information for key generation [14]. The optical delays used in the proposed system are typically in the range of tens of kilometers. Without knowing the delays in the interferometers, it is extremely difficult for the eavesdropper to scan the optical delays and find the coherence length in mm resolution. The delays (L1 and L2 in Figure 1) range from 20 m to 26 km in the experiment, which means the eavesdropper has to find a 2 mm range in up to a 26 km range in order to find the matching condition. More importantly, it should be noted that by building another fiber interferometer for eavesdropping, new phase randomness is created in addition to the transmitted phase information, thus reducing the accuracy of the eavesdropper's measurement of the phase information.

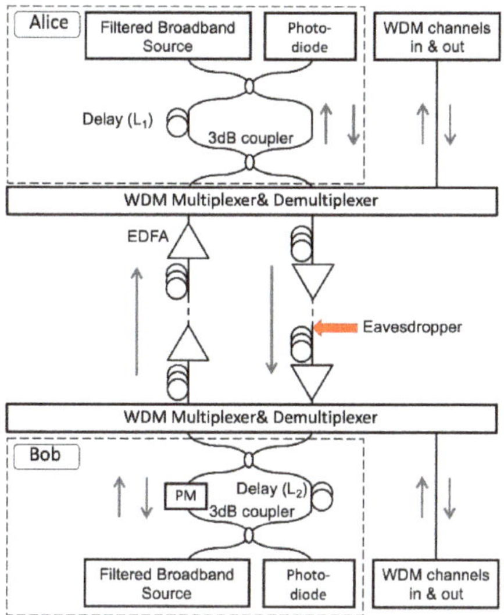


**Figure 1.** Experimental setup. (EDFA: erbium-doped fiber amplifier; WDM: wavelength-division multiplexing; PM: phase modulator).

The selection of 150 GHz bandwidth is a balanced solution that considers both the security of the system and the compatibility to WDM channels. The coherence length is inversely proportional to the bandwidth. Using the entire C-band for key generation enables the coherence length to be shorter than 2 mm, which creates a challenge for the eavesdropper to find the coherence length and the matching condition. However, it does not allow the existence of the other WDM channels. With 150 GHz bandwidth, the key generation channel can coexist with the other WDM channels in C-band. The selection of 26 km delay length not only provides large key space that challenges the eavesdropper to find the matching condition, but also generates fast-changing phase fluctuations that smear out the spectrum signatures that indicate the delay length. The experimental results for the spectrum signatures with different delay lengths are discussed in Section 3.2.

## 3. Results and Analysis

### 3.1. Key Generation without Phase Filter

We demonstrated the transmission of the key distribution channel with 38 WDM channels using standard C-band transmission window. The key generation system is tested in both back-to-back transmission and a 240 km link. The 240 km bidirectional fiber-pair link includes three 80 km standard single mode fiber (SSMF) spans, each with erbium-doped fiber amplifier (EDFA) to compensate the average fiber loss of 17.5 dB. The 38 dummy WDM channels were emulated by carving another ASE source using a flexible-band wavelength selective switch (WSS) at 100 GHz spacing (Figure 2). The spectrum shows that the power spectral density (PDF) of the key distribution channel is comparable to the PDF of the other WDM channels, so the key generation channel can be downloaded from the network by using a bandpass filter with the same power extinction ratio as the standard WDM filters.

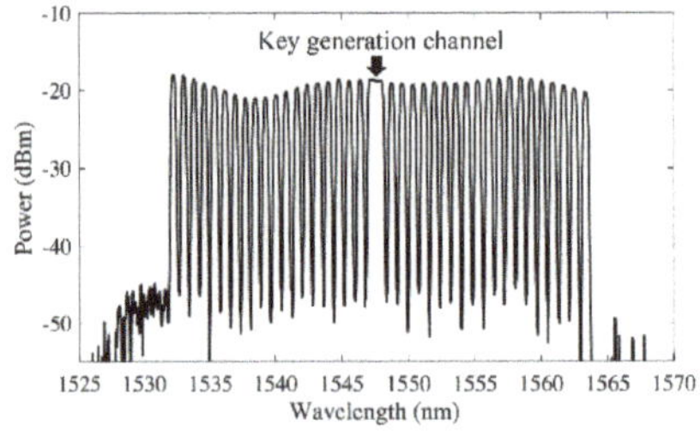


**Figure 2.** Spectrum of the key generation channel and the dummy WDM channels.

When 26 km optical delays are applied by both Alice's and Bob's MZIs, the two interference results by direct photo-detection match each other in back-to-back testing (Figure 3a,b). The phase randomness is generated from the 26 km optical delays deployed by both Alice and Bob. The spectrum analysis further proves the matched signal. Figure 3c is the spectrum of signal received by Alice, and shows that the signal exhibits frequency content up to 2 kHz. Figure 3d is the differential spectrum between the received signals, and is obtained by the Fourier transform of Alice's signal minus Bob's signal. The comparison between Alice's spectrum (Figure 3c) and the differential spectrum (Figure 3d) shows that Alice's spectrum is significantly larger than the differential spectrum up to 1 kHz, which further proves that Alice's signal matches with Bob's signal up to 1 kHz.

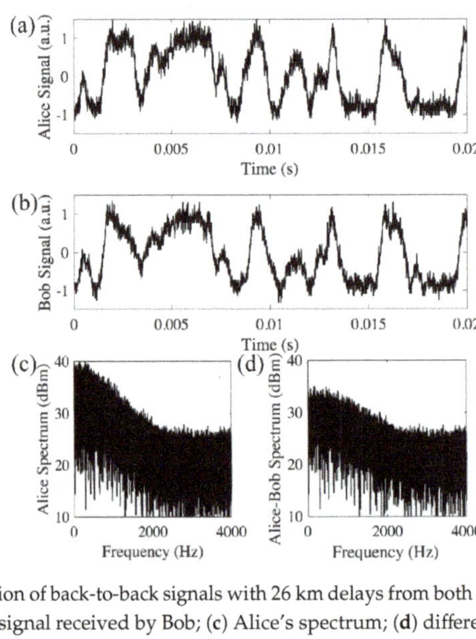


**Figure 3.** Direct detection of back-to-back signals with 26 km delays from both Alice and Bob. (**a**) Signal received by Alice; (**b**) signal received by Bob; (**c**) Alice's spectrum; (**d**) differential spectrum between Alice and Bob has a much smaller amplitude than Alice's spectrum (a.u.: arbitrary unit).

Figure 3 shows that Alice's signal matches with Bob's signal in the back-to-back transmission. After 240 km transmission, the interference results from Alice and Bob do not match (Figure 4a,b). This is because other than the rate of phase change induced by the environmental instability, the time of flight for the signal to travel between Alice and Bob also needs to be considered. In the experiment, the average time for the phase changes induced by the 26 km long MZI is in the order of 1 ms (Figure 3), while it takes the signal 1.2 ms to travel 240 km in the fiber. Thus, the signals in the two opposite directions will experience different phase changes. Figure 4 shows that the average phase changes induced by both the 26 km long delay of the MZI and the 240 km fiber are much faster than the phase changes in the back-to-back transmission system (Figure 3). The spectrum of the received signal has frequency contents above 10 kHz (Figure 4c). The comparison between spectrum received by one receiver (Alice in Figure 4c) and the differential spectrum (Figure 4d) shows whether the received signals are matched at different frequencies. Figure 3c, d shows that in a back-to-back link, the received signals are matched from direct current (DC) to 1 kHz, and in a logarithmic scale, most of the signal power lies within the range of DC to 1 kHz. In the 240 km link, Alice's spectrum (Figure 4c) and the differential spectrum (Figure 4d) have the same amplitude from DC to 20 kHz, which means the received signal does not match in any frequency components.

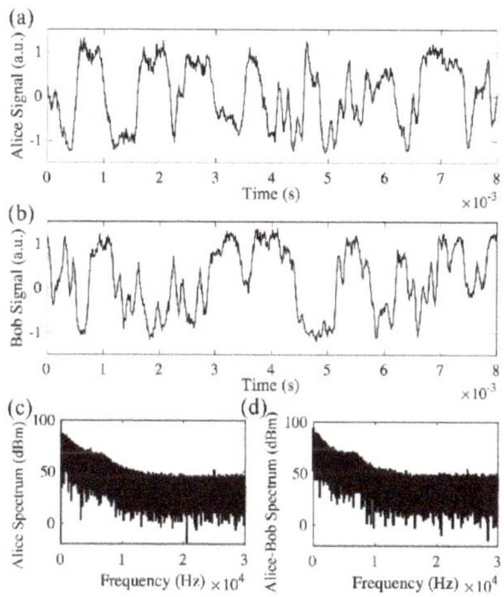


**Figure 4.** Direct detection of the signal after 240 km transmission with 26 km delays from both Alice and Bob. (**a**) Signal received by Alice; (**b**) signal received by Bob; (**c**) Alice's spectrum; (**d**) differential spectrum between Alice and Bob has the same amplitude as Alice's spectrum.

There are two possible solutions to match the signals and synchronize the key: (1) The phase fluctuation speed can be reduced by using delay lengths less than 26 km, so the signals transmitted in both of the directions experience the same change. (2) A low-pass filter can be applied to remove the fast-changing components of the phase changes. Section 3.2 shows that the first solution undermines the security of the system. If the phase fluctuation speed is reduced, an eavesdropper is able to measure the delay length by the spectral pattern of the received signal. Section 3.3 demonstrates that with the second solution, the system generates matched keys in a secure way.

### 3.2. Security against Man-In-The-Middle Attack

The speed of phase fluctuations can be reduced by reducing the optical delay lengths in MZIs, since the changing speed of the phase is proportional to the optical delay length in the interferometer [15,16]. However, the slower phase changing speed poses a potential security weakness to the system. Eavesdroppers can use coherent detection to measure the location of the spectral fringes within the key signal spectrum. With the spectral fringes, it is possible to measure the optical delay lengths applied by Alice and Bob [17–19]. The challenge for an eavesdropper is that enough frequency resolution is required to observe the spectral fringes, and the frequency resolution is determined by the length of temporal data collected by the eavesdropper. With longer optical delay in MZ, a higher frequency resolution is required, and at the same time the fast phase changes will essentially smear out measurement results.

To determine the appropriate delay length that can defend the spectral-fringe-based attack, we emulate the eavesdropper (red arrow in Figure 1) and perform the attacks to the systems with different delay lengths. An ultra-stable fiber laser with 400 Hz linewidth is used as the local oscillator (LO) for the coherent detection. The Fourier transforms of the coherently detected signals show that when 20 m and 2.3 km delays are used (Figure 5a,b), the fringes in the spectrum can be clearly identified to track the phase fluctuation information within the interferometer. As for the 26 km delay, the fringes are smeared out by the fast-changing phase due to the long measurement time frame needed to achieve the frequency resolution required (Figure 5c). The experimental results show that the delay length of the fiber interferometer has to be longer than a few kilometers in order to protect against coherent detection attack.

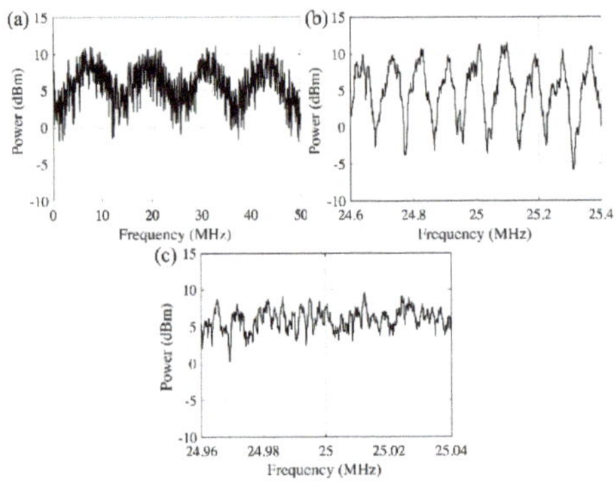


**Figure 5.** Spectrum of the signal from fiber interferometer with different optical delay length. (**a**) 20 m delay; (**b**) 2.3 km delay; (**c**) 26 km delay.

### 3.3. Key Generation with Phase Filter

Since longer delay length is needed to protect the signal from eavesdropping, to match the interference results and enable the key exchange between Alice and Bob, the only solution is to remove the fast-changing component of the phase fluctuations so the slow-changing component of the phase fluctuations can be used for key generation. However, the low-pass filter cannot be directly applied on the interference results, which is the cosine term of the phase, because cosine is not a linear function for Fourier transform. Figure 4c, d compares the Fourier transforms of Alice's direct detected signal and the differential spectrum between Alice's and Bob's. The results show that Alice's spectrum has the same order of magnitude as the differential spectrum from DC to 10 kHz, meaning that the interference results do not match in either low- or high-frequency ranges.

In order to detect the actual interference phase results, we use a phase modulator that switches between 0° and 90° at a rate of 100 kHz, which is more than 10 times faster than the phase changing rate (Figure 6a,b). Since the phase modulator switches at least one order of magnitude faster than the random phase signal, we assume that a neighboring 0° and 90° phase shift from the phase modulator is applied on the same random phase. In this case, both sine and cosine of the random phase are measured to obtain the random phase value. After digitization, a low-pass filter with 3 dB cutoff frequency at 50 Hz is applied to the measured phase signal digitally to remove the mismatched high-frequency phase components. Figure 7a, b shows the matched results from the cosine functions of the two filtered phase signals. The signals are transmitted in the 240 km bidirectional link. The self-correlation and cross-correlation of Alice's and Bob's filtered signals are plotted in Figure 7c. The self-correlations of Alice's and Bob's individual signals show the true temporal randomness of the generated key. The cross-correlation between Alice's and Bob's signals shows the matching performance of the key exchange with correlation peak of 0.8. The comparison between Alice's spectrum (Figure 7d) and the differential spectrum between Alice and Bob (Figure 7e) shows that the differential spectrum has a much smaller amplitude than Alice's spectrum. The differential spectrum further proves that with the 50 Hz low-pass filter applied on the phase, Alice's signal matches with Bob's signal.

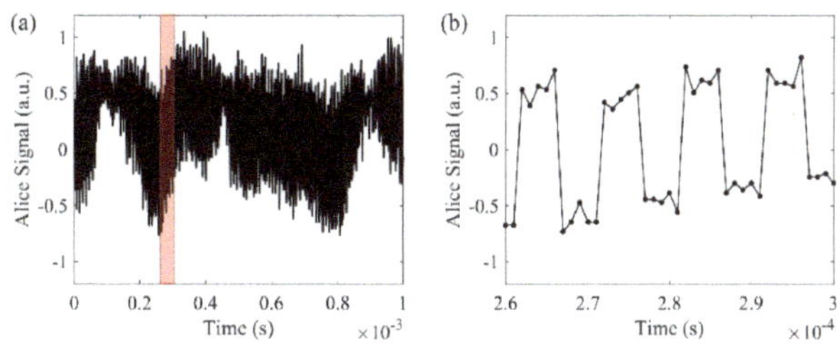


**Figure 6.** (**a**,**b**) are signals with phase modulator; (**b**) is the enlarged view of the red region in (**a**).

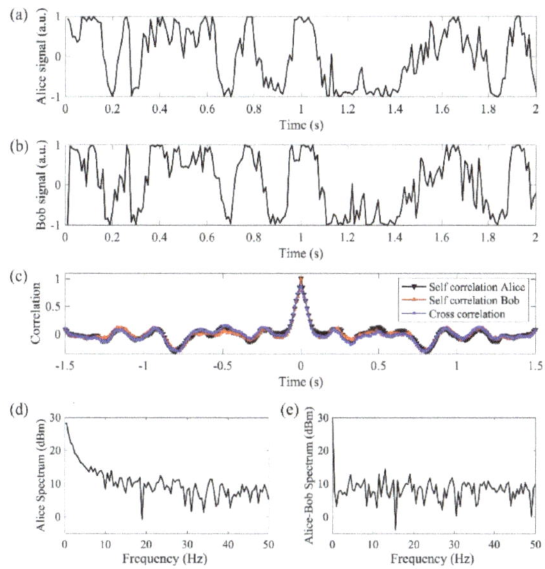


**Figure 7.** (**a**,**b**) are the received signals from Alice and Bob with high-frequency phase fluctuations filtered; (**c**) shows the self-correlations and cross-correlation of the signals; (**d**) Alice's spectrum; (**e**) differential spectrum between Alice and Bob has a much smaller amplitude than Alice's spectrum.

To extract binary keys from the analog signals and reduce key errors, a buffer region is set in between the high and low levels, and the sampled points within the buffer region are dropped. The analog signals used to extract keys are sine and cosine functions of the recovered and filtered phases, which range from −1 to 1. The boundary of the buffer region is $-v_o$ and $v_o$ ($0 < v_o < 1$). If the analog signal is higher than the buffer region, a bit 1 is generated as the key. If the analog signal is lower than the buffer region, a bit 0 is generated as the key. The buffer region is optimized to minimize the dropped samples and key error rate (Table 1). Both sine and cosine functions of the measured phase are used to generate binary keys. The key generation achieves a rate of 90 bps and an error rate of $5 \times 10^{-3}$, which can be corrected by hard decision FEC technique [20,21] and key reconciliation [22]. To ensure true randomness of the generated key, the 90 bps is further reduced to 20 bps based on the width of the correlation peak in Figure 7c. The width of correlation peak is 100 ms, which means two bits with intervals equal to or larger than 100 ms are not correlated. By using the cosine of the phase, a bit rate of 10 bps is obtained. Since sine and cosine functions are orthogonal, another 10 bps is obtained with the sine function.

**Table 1.** Ratios of dropped/accepted sample and key error.

| Description | Percentage |
|---|---|
| Dropped samples | 54% |
| Accepted samples | 46% |
| Key error | 0.5% |
| Correct key | 99.5% |

By using standard optical amplifiers in WDM networks, the transmission distance can go beyond 240 km. The maximum transmission distance depends on two factors: (1) Key rate: To remove the unsynchronized phase signal, the 3 dB cutoff frequency of the low-pass phase filter for the 240 km link is 50 Hz. With the transmission distance longer than 240 km, the cutoff frequency is inversely proportional to the transmission distance. (2) Signal-to-noise ratio (SNR): The signal carrier of the key generation channel is wide-band ASE noise. The SNR of the key distribution channel is different from the SNR of a standard communication channel that uses a laser as the signal carrier [23]. For the standard communication channel, the ASE noise only exists in the denominator of the SNR. For the key distribution channel, the power spectral density of the ASE noise affects both the numerator and denominator of the SNR, which means the key distribution channel has shorter maximum transmission distance than the standard communication channel. The theorical model and experimental results of the maximum transmission distance for a communication channel that uses ASE noise as the signal carrier is analyzed in [23].

### 3.4. Analysis of Information Leakage

Section 3.2 has demonstrated that if the optical delay is longer than a few kilometers, the eavesdropper is not able to measure the spectral fringes and thus cannot measure the optical delay by coherence detection. This section shows that the system is protected from the information leakage, and even if the eavesdropper knows the right optical delay length, he or she cannot measure the phase information from Alice and Bob. The analysis in this section is based on the assumption that the eavesdropper uses brute force attack to scan the 26 km fiber with high resolution and finds the 2 mm coherence length. Under such assumption, the eavesdropper uses the same optical delay as the authorized users to measure the phase information from Alice ($\varphi_A$) or Bob ($\varphi_B$). The 26 km optical delay applied by the eavesdropper generates additional phase randomness ($\varphi_E$). Although the eavesdropper can use a feedback system to control the phase randomness, the feedback system applies to the sum of the phase change $\varphi_A + \varphi_E$, which means the eavesdropper can only measure $\varphi_A + \varphi_E$ or $\varphi_B + \varphi_E$ and cannot accurately measure either $\varphi_A$ or $\varphi_B$.

### 3.5. Description of the Protocol

In this section, the key distribution protocol is summarized step by step based on the experimental setup and results. The protocol includes the following five steps: (1) Alice and Bob match optical delays with preshared length. Based on the secure analysis in Section 3.2, the matched delay lengths range from a few kilometers to less than 100 km. The accuracy of the matched delay lengths is within 2 mm, which is the coherence length of the light source. (2) The delay lengths generate random phase differences between two light paths at the interferometers on both Alice's ($\varphi_A = \varphi_{A1} - \varphi_{A2}$) and Bob's ($\varphi_B = \varphi_{B1} - \varphi_{B2}$) sides, where $\varphi$ represents phase, A and B represent Alice and Bob, respectively, and 1 and 2 represent two paths in the fiber interferometers. In both of the transmission directions, from Alice to Bob and from Bob to Alice, the signal carriers experience the same phase change ($\varphi_{A1} - \varphi_{A2} + \varphi_{B1} - \varphi_{B2}$), which generates synchronized keys. (3) The phase changes $\varphi_{A1} - \varphi_{A2}$ and $\varphi_{B1} - \varphi_{B2}$ are time-dependent, and have frequency components from DC to 10 kHz (Figure 4). It takes 1.2 ms for the signal carriers to transmit through the 240 km fiber link, and if either $\varphi_{A1} - \varphi_{A2}$ or $\varphi_{B1} - \varphi_{B2}$ changes during the 1.2 ms, Alice and Bob receive mismatched signals. Therefore, low-pass digital filters are used at both Alice's and Bob's sides to remove the fast-changing frequency components of the phase changes $\varphi_{A1} - \varphi_{A2}$ and $\varphi_{B1} - \varphi_{B2}$., and only the slow-changing frequency components are used to generate the keys. (4) A phase modulator that switches between 0° and 90° is applied at the receiver (Figure 1), so both sine and cosine of the phase changes are received, and the phase information can be recovered. (5) Alice and Bob both generate binary digital keys based on the synchronized analog signals. A buffer region is set in between the high and low levels. Sampled signals above and below the buffer region generate digital keys of 1 and 0, and the sampled signals within the buffer region are dropped.

## 4. Conclusions

We propose and experimentally demonstrate a key distribution system that is compatible with optical amplifiers for long-distance communication and standard WDM systems. The keys are generated and digitized from the random phase fluctuations caused by environmental instabilities in fiber interferometers. A bidirectional transmission link with multiple EDFAs and 240 km length of optical fiber is demonstrated. The signal carrier of the key distribution channel is wide-band ASE noise from EDFA and shares C-band transmission spectrum with 38 neighboring WDM channels. The wide-band ASE noise and the fast-changing phase protect the optical delay in the key distribution system from being measured by an eavesdropper with coherent detection attack.

**Author Contributions:** B.W. and Y.-K.H. carried out the experiment and performed signal processing. The paper was written by B.W. and edited by Y.-K.H. All authors have read and agreed to the published version of the manuscript.

**Funding:** This research received no external funding.

**Acknowledgments:** The authors would like to gratefully thank Fatih Yaman for valuable discussions.

**Conflicts of Interest:** The authors declare no conflict of interest.

## References

1. Stinson, D.R. *Cryptography Theory and Practice*, 4th ed.; Chapman & Hall: London, UK, 2019; pp. 1–13.
2. Rivest, R.L.; Shamir, A.; Adleman, L. A method for obtaining digital signatures and public-key cryptosystems. *Commun. ACM* **1978**, *21*, 120–126. [CrossRef]
3. Shor, P.W. Polynomial-time algorithms for prime factorization and discrete logarithms on a quantum computer. *SIAM J. Compt.* **1997**, *26*, 1484–1509. [CrossRef]
4. Steane, A. Quantum computing. *Rep. Prog. Phys.* **1998**, *61*, 117–173. [CrossRef]
5. Mavroeidis, V.; Vishi, K.; Zych, M.D.; Jøsang, A. The impact of quantum computing on present cryptography. *Int. J. Adv. Comp. Sci. Appl.* **2018**, *9*, 1–10. [CrossRef]
6. Mosca, M. Cybersecurity in an era with quantum computers: Will we be ready? *IEEE Secur. Priv.* **2018**, *16*, 38–41. [CrossRef]
7. Kapov, N.S.; Furdek, M.; Zsigmond, S.; Wosinska, L. Physical-layer security in evolving optical networks. *IEEE Commun. Mag.* **2016**, *54*, 110–117. [CrossRef]
8. Wand, T.L.; Gariano, J.A.; Djordjevic, I.B. Employing Bessel-Gaussian beams to improve physical-layer security in free-space optical communications. *IEEE Photonics J.* **2018**, *10*, 7907113.
9. Tanizawa, K.; Futami, F. Single-channel 48-Gbit/s DP PSK Y-00 quantum stream cipher transmission over 400-and 800-km SSMF. *Opt. Express* **2019**, *27*, 25357–25363. [CrossRef] [PubMed]
10. Lo, H.K.; Curty, M.; Tamaki, K. Secure quantum key distribution. *Nat. Photonics* **2014**, *8*, 596–604. [CrossRef]
11. Shor, P.W.; Preskill, J. Simple proof of security of the BB84 quantum key distribution protocol. *Phys. Rev. Lett.* **2000**, *85*, 441–444. [CrossRef] [PubMed]
12. Grosshans, F.; Assche, G.V.; Wenger, J.; Brouri, R.; Cerf, N.J.; Grangier, P. Quantum key distribution using gaussian-modulated coherent states. *Nature* **2003**, *421*, 238–241. [CrossRef] [PubMed]
13. Kravtsov, K.; Wang, Z.; Trappe, W.; Prucnal, P.R. Physical layer secret key generation for fiber-optical networks. *Opt. Express* **2013**, *21*, 23756–23771. [CrossRef] [PubMed]
14. Wu, B.; Tait, A.N.; Chang, M.P.; Prucnal, P.R. WDM optical steganography based on amplified spontaneous emission noise. *Opt. Lett.* **2014**, *39*, 5925–5928. [CrossRef] [PubMed]

15. Minar, J.; Riedmatten, H.; Simon, C.; Zbinden, H.; Gisin, N. Phase-noise measurements in long-fiber interferometers for quantum-repeater applications. *Phys. Rev. A* **2008**, *77*, 052325. [CrossRef]
16. Wanser, K.H. Fundamental phase noise limit in optical fibers due to temperature fluctuations. *Electron. Lett.* **1992**, *28*, 53–54. [CrossRef]
17. Wu, B.; Huang, Y.K.; Zhang, S.; Shastri, B.J.; Prucnal, P.R. Long range secure key distribution over multiple amplified fiber spans based on environmental instabilities. In Proceedings of the CLEO Conference, OSA, San Jose, CA, USA, 5–10 June 2016.
18. Liao, Y.; Wang, J.; Wang, S.; Yang, H.; Wang, X. Spectral characteristics of the microfiber MZ interferometer with a knot resonator. *Opt. Commun.* **2017**, *389*, 253–357. [CrossRef]
19. He, W.; Zhu, L.; Dong, M.; Lou, X.; Luo, F. Wavelength-switchable and stable-ring-cavity, erbium-doped fiber laser based on Mach–Zehnder interferometer and tunable filter. *Laser Phys.* **2018**, *28*, 045104. [CrossRef]
20. 100G CI-BCH-3 eFEC Technology. Available online: www.microsemi.com (accessed on 5 September 2020).
21. Bocherer, G.; Schulte, P.; Steiner, F. Probabilistic shaping and forward error correction for fiber-optic communication systems. *J. Lightwave Technol.* **2019**, *37*, 230–244. [CrossRef]
22. Brassard, G.; Salvail, L. Secret-Key Reconciliation by Public Discussion. In Proceedings of the Workshop on the Theory and Application of Cryptographic Techniques EUROCRYPT 1993, Lofthus, Norway, 23–27 May 1993; pp. 410–423.
23. Wu, B.; Shastri, B.J.; Prucnal, P.R. System performance measurement and analysis of optical steganography based on noise. *IEEE Photonic Technol. Lett.* **2014**, *26*, 1920–2923. [CrossRef]


**Publisher's Note:** MDPI stays neutral with regard to jurisdictional claims in published maps and institutional affiliations.

sensors

MDPI

Letter

# MongoDB Database as Storage for GPON Frames

**Martin Holik *, Tomas Horvath, Vaclav Oujezsky, Petr Munster, Adrian Tomasov and Sobeslav Valach**

Department of Telecommunication, Brno University of Technology, Technicka 12, 616 00 Brno, Czech Republic; horvath@feec.vutbr.cz (T.H.); oujezsky@feec.vutbr.cz (V.O.); munster@feec.vutbr.cz (P.M.); xtomas32@stud.feec.vutbr.cz (A.T.); valach@feec.vutbr.cz (S.V.)

* Correspondence: xholik11@stud.feec.vutbr.cz; Tel.: +420-541-146-923

Received: 17 September 2020; Accepted: 29 October 2020; Published: 30 October 2020

**Abstract:** This work is focused on creating an open-source software-based solution for monitoring traffic transmitted through gigabit passive optical network. In this case, the data are captured by the field-programmable gate array (FPGA) card and reassembled using parsing software from a passive optical network built on the International Telecommunication Unit telecommunication section (ITU-T) G.984 gigabit-capable passive optical network GPON recommendation. Then, the captured frames are converted by suitable software into GPON frames, which will be further processed for analysis. Due to the high transfer rate of GPON recommendations, the work describes the issue of writing to the Mongo database system. In order to achieve the best possible results and minimal loss of transmitted frames, a series of tests were performed. The proposed test scenarios are based on different database writing approaches and are implemented in the Python and C# programming languages. Based on our results, it has been shown that the high processing speed is too high for Python processing. Critical operations must be implemented in the C# programming language. Due to rapid application development, Python can only be used for noncritical time-consuming data processing operations.

**Keywords:** analysis; FPGA; GPON; MongoDB; storing

## 1. Introduction

Optical networks are divided into active optical networks (AONs) and passive optical networks (PONs). Active networks require some of the network elements to be powered and are used primarily in transport core networks. On the contrary, the elements of the passive network do not require any power supply and, thanks to their low purchase price, are used primarily as a so-called last mile solution to connect end users. They are also common in networks combining metallic and optical lines, which are created by the gradual replacement of metallic lines [1–5].

Currently, the most widespread International Telecommunication Union telecommunication section (ITU-T) recommendations include Gigabit-capable passive optical networks (GPON) and 10-Gigabit-capable passive optical network (XG-PON), which are based on time division multiplex (TDM) [6,7]. The size of the frames is therefore not exactly given but is limited by time periods of 125 μs. In the ascending direction, the concept of time division into 125 μs sections is retained, but it is supplemented by regular time ticks/marks, which are used for synchronization. To enable bidirectional communication, different wavelengths are used for uplink and downlink. For GPON, a signal with a wavelength of 1490 nm is used for the downlink and 1310 nm for the uplink and 1577 nm in the downlink and 1270 nm in the uplink direction for XG-PON [2–4,6,8–11].

The problem of data traffic analysis lies in the large flow of data that is transmitted through passive optical networks. This is an area of big data processing, where it is impossible to analyze communication in the usual way. Although there are tools for data analysis in a commercial environment, the problem of these tools is associated not only with a problematic and very expensive licensing policy but also with closed program code and the inability to use the interface for self-testing. Today's analyzers are difficult to reach for small Internet services providers (ISPs) with hundreds of customers. Our solution is focused on providing a modular solution that should be distributed under open-source licenses. Our proposed solution is built with field-programmable gate array (FPGA) card and appropriate control software that will be installed on the operating system.

The main contribution of the article is the experimental verification of the write speed of traffic transmitted in the downstream direction of the passive optical network. This work is focused on traffic from the GPON network, but later it is planned to create a version for the analysis of XG-PON networks. After the successful storage of all important data, the data will be further processed in order to detect hidden security threats using artificial intelligence.

The rest of this paper is structured as follows. Section 2 provides an overview of the related works. Section 3 describes the necessary prerequisites that are required for testing including a data structure built on Mongo database. Section 4 is focused on the description of test scenarios. The individual scenarios are divided according to the way in which GTC frames are written to the repository and according to the programming languages used. Section 5 provides a discussion of the results. Finally, Section 6 concludes the paper.

## 2. Related Works

In recent years, several PON-related studies have been published. The authors in [12] describe a novel method for low-latency 10-Gigabit-capable symmetric passive optical network (XGS-PON) mobile front-haul for small cell in cloud radio access network (C-RAN) based on traffic estimation. They proposed the adaptive-Learning dynamic bandwidth allocation (DBA), which reduces jitter in optical distribution network (ODN), packet loss ratio, delay, and increasing utilization performance. The article [13] deals with the impact of report message scheduling (RMS) in 1/10 G passive optical networks. The researchers reduce the idle time of channel during a reporting optical network unit (ONU) buffer occupation for data transmission. The work [14] also deals with a dynamic bandwidth allocation algorithm for long-reach systems. The primary purpose of this research is to reduce inefficiency occurs in the DBA upstream protocols because of the large propagation time between optical line termination (OLT) and ONUs. This algorithm is independent of the SI and uses real data queuing information without prediction. In [15] the authors present a novel opportunity for data transmission in GPON with sensing data encapsulation. The sensory system uses different wavelengths; the results are transmitted in GPON encapsulation method (GEM) frames.

Another research field in passive optical network is their software defined controlling. In general, software defined networks (SDNs) are a very popular and promising solution in metro and core networks; the software controlling can be applied in access networks [16–23]. A big network with multiple OLTs will require a complex support of technician; however, the SDN controlling with one central point may help to make a management easier. Multi-OLT networks have to control DBA algorithms to ensure the quality of service (QoS) [24–26].

A data analysis is not a trivial task in passive optical networks defined by ITU recommendations. First of all, all data are encapsulated into GPON transmission convergence (GTCs). The structure of GTC is completely different in comparison with Ethernet frame; however, Ethernet frames are encapsulated into them. In recently years, many publications have been published about simulations of data traffic in passive optical networks [27–29]. Note that each simulation tool supports physical or above layers; this means that the physical parameters will be evaluated or data transmission will occur separately. Nowadays, the available tools for data analysis in passive optical networks are in stock but with one main limitation—postprocessing data representation [30,31].

Analyzing the transmission convergence layer of passive optical networks is not a new problem. Several patents in the field of passive optical network analysis have been published in the past.

Still, current developments in technology and world affairs place greater emphasis on detecting unwanted processes that could cause the leakage of sensitive information. The work [32] discloses a method of reducing a PON network's energy consumption by monitoring the physical layer and the GTC layer. In the work [33], the authors proposed a GPON network extender built on an FPGA array. For this extender to work properly, adding a sync bit sequence to the optical layer during coarse wavelength division multiplexing (CWDM) transmission was necessary. The designed extender with OLT unit allows them to connect up to 512 client devices but changes the activation process of ONU units according to ITU-T recommendation [3] were necessary. The article [34] also deals with the issue of security using FPGA arrays. The authors created an advanced encryption standard (AES) encryption algorithm for hardware transmission encryption using the FPGA card. As part of the Fabulous EU Strep project, the authors dealt with the simulation of traffic transmission and its digital processing using algorithms in FPGA arrays [35].

Previously mentioned works, focused on FPGA card data processing, are primarily focused on increasing the quality of operation of passive optical networks or security from the perspective of third parties, for example, against eavesdropping. The focus of our research is to analyze the communication itself against unwanted processes that are not defined in the ITU-T recommendations of GPON networks. Currently, the market offers many devices of different manufacturers that are not compatible with each other due to various implementations of optional values or specification in their recommendations. Mutual incompatibility and the use of one manufacturer's equipment on a single network could be a hidden threat to sensitive data leakage. This research should make it possible to verify that the data transmitted follow the ITU-T recommendations.

## 3. Prerequisites

Our test scenario was designed with regard to the requirements for capturing traffic from passive optical networks. The entire circuit diagram of the topology of the currently used optical network is shown in Figure 1. The FPGA module developed by our partner is connected to the passive optical network, which captures the transmitted bits in the downlink direction, i.e., from the OLT unit to the ONU unit [36]. The captured data are reassembled into higher layer frames and stored in a folded form. Depending on the recommendation used, it is necessary to perform the optimal design of the data structure for the maximum speed of writing data to the storage. The second, but less critical, operation is the optimization of the output for future reading and subsequent data processing.

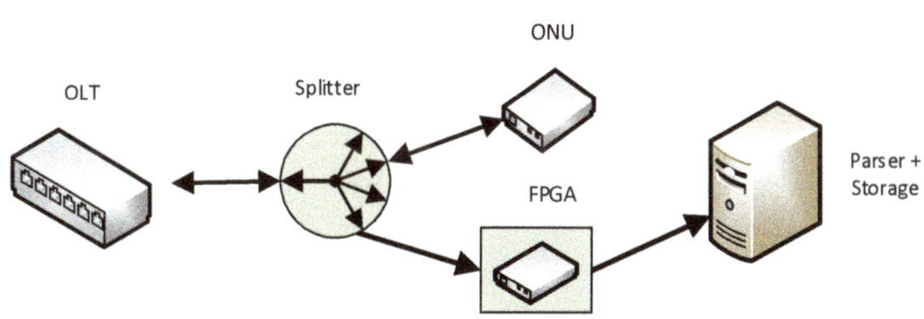


**Figure 1.** Scheme of our measurement topology.

The FPGA card (named Cecilia) is specially designed for processing and transferring data streams from the GPON and Ethernet interfaces up to 100 Gbit data rates. The card is equipped by a middle class Xilinx's Kintex UltraScale + FPGA—XCKU11/15P-2FFVE1517. The FPGA provides 20 pairs of gigabit transceiver Y (GTY) transceiver operating up to 28.21 Gbit/s and 32 pairs of gigabit transceiver H (GTH) transceiver with maximum achievable bit rate 16.375 Gbit/s [37]. Figure 2 depicts a block drawing showing card configuration and Figure 3 shows the main peripherals.

This parsing software utilizes one GTH transceiver connected to the small form-factor pluggable (SFP+) cages. One cage is used for upstream and the second one for downstream monitoring and stream capturing. The captured data are transferred directly into a host computer main memory by peripheral component interconnect express (PCIe) interface. Due to a low rate stream, the PCIe can be configured as x4 interface to save FPGA resources and lower power consumption.

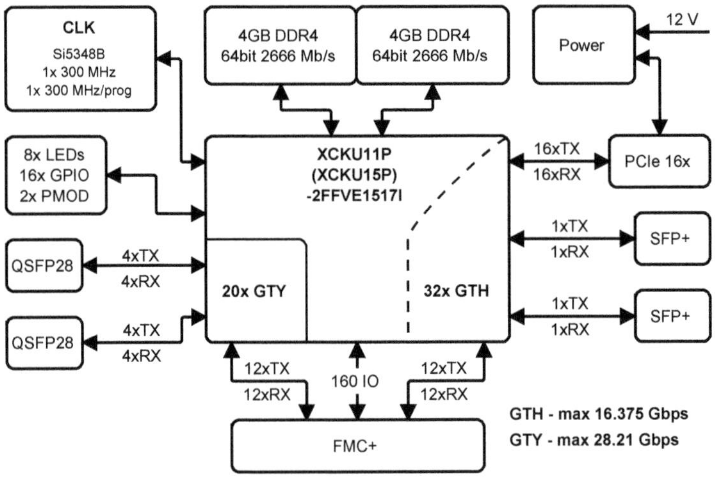


**Figure 2.** The block diagram of developed field-programmable gate array (FPGA) card.

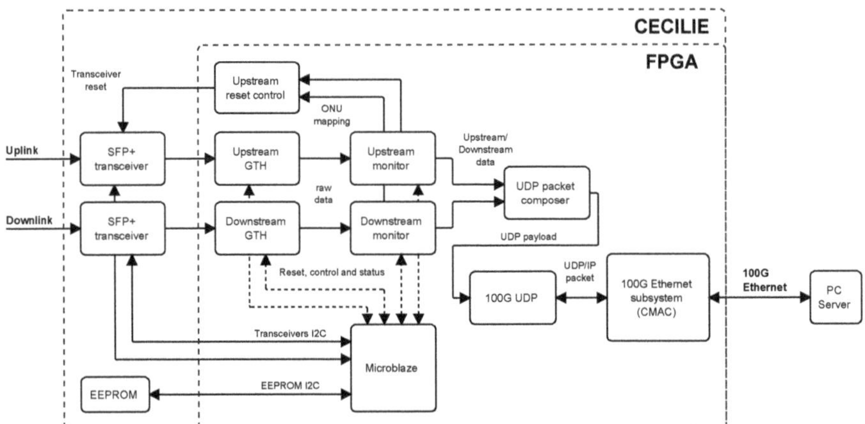


**Figure 3.** The internal configuration of Cecilia card.

A firmware implementation is divided into three blocks. The first block captures data from downstream and upstream links and provides raw data to the monitoring block handling ONU's mapping and upstream reset generation. The downstream capture unit oversamples input stream by ratio of five and digitally recovers incoming data.

The time stamped data are sent through the PCIe subsystem to the computer host main memory. The logic utilization is 9.5% of the XCKU11P device without any special optimization technique. The design includes debug support features consuming more than 50% of overall application.

In general, this type of application can be processed by application-specific integrated circuits (ASICs) or Network processors. The FPGA is more flexible and gives us a high level of freedom in development of new monitoring techniques and ideas.

The Cecilia card allows two connection modes. The first mode preverification (PV) can be used as a standalone solution and it is connected to the server that processes the data via a 10 gigabit Ethernet port. The second option is to use the card as a server solution when inserted into the PCIe slot of the server on which the control software is running. During our tests of the FPGA card, a stand-alone connection was used. The whole concept of data capture consists of four layers. The first and, at the same time, the lowest layer is the FPGA card itself, which sends via the Ethernet interface to the server, where they are captured by the listening module. The intercepted data is processed at the application module level by parsing software written in the C# programming language, allowing data to be stored in a repository for GPON frames. The stored data is ready for processing or can be viewed using a web interface written in the Python programming language [36,38,39]. Our solution does not include a step to create COM-callable wrapper and call using Python's component object model/object linking and embedding (COM/OLE) from the complexity perspective.

During the tests, a passive optical network based on the ITU-T G.984 GPON recommendation was used, which offers a downstream transmission speed of 2.488 Gb/s [8]. The total number of GEM frames per second is very difficult to estimate because it is a problem to estimate the correct number of frames with relevant data. However, it is relatively easy to determine the lower limit. A GEM frame can carry a maximum of 4095 bytes of data, and therefore a minimum of about 76,000 GEM frames per second can be expected. Assuming that one GEM frame carries half the maximum size, the storage must handle twice, that is, 152 thousand GEM frames per second. All transmitted GEM frames are encapsulated as user data into 8000 GPON frames.

## 4. Testing Scenarios

The first data storage tests were performed using Microsoft structured query language (MS SQL). MS SQL is a database system designed for a large volume of data, and a suitable design can divide the data into tables to avoid duplicate records [38–40]. Better results were obtained using the MongoDB document database system. This system's advantages include native data storage in JavaScript object notation (JSON) format, which is more advantageous not only for easier and faster writing to the database but also for the reading speed, which is often higher than the MS SQL database. The second advantage of using a Mongo database was eliminating problematic data conversion, which caused up to 90 percent loss of reshaped frames in the previous case.

The Mongo database is supported by many programming languages, including Python and C#, with which we performed write testing. The MongoRepository class has implemented methods for testing all tested scenarios, as shown in the unified modeling language (UML) diagram in Figure 4. This class is used as a general model and for GEM frames, GPON frames, bandwidth maps (BWmaps), GemRepository, GtcRepository, and BwMapRepository. Each class represents one collection in the Mongo database. The data were not intentionally divided into smaller collections, not only due to the possible slowdown in the writing speed, but also the complexity of the logic that would put the data back into the final frames. The UML diagram is only a general design. The individual naming of methods in the UML diagram is adapted to the naming conventions of the programming language used.

Individual frames are stored as Mongo database documents and as a whole form a collection. Information about the smallest unit of the GPON frame is stored in individual documents, i.e., a GPON header without a BWmap field, a BWmap field and all GEM frames transmitted in the body of the GPON frame. The data is stored in binary javascript object notation (BSON) format, which is convenient for further processing. The proposed structure in the Mongo database allows to store data in inconsistent or nonstandard form against relational databases, which can be especially useful when storing undefined traffic compared to the G.984 recommendation.

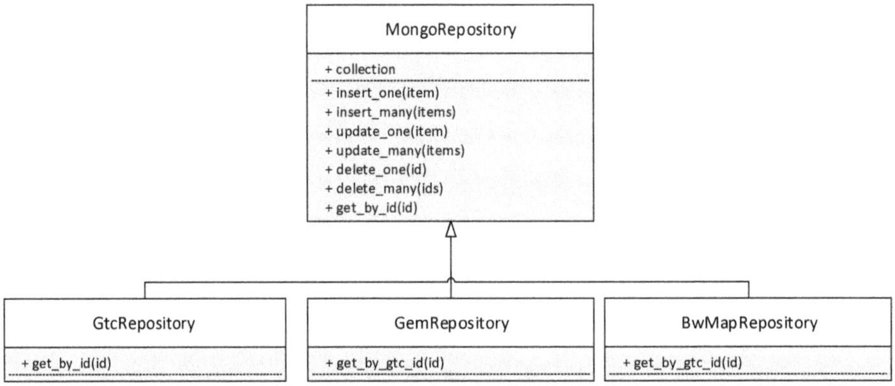


**Figure 4.** UML diagram—MongoRepository.

To achieve the most realistic simulation of real operation, each collection (GTC, GEM, BWmap) in the database was written for each scenario. GTC documents and bandwidth maps are connected by a direct link (1:1), while multiple links (1:N) are proposed between GTC and GEM documents. From the above, it is clear that the GEM of documents will be many times more and therefore the speed of writing will be limited by writing GEM documents.

The test data set is the same for both languages and is based on the data actually captured by the FPGA card. In order to make the operation uniform during the test and to be able to realistically compare the results between the individual test scenarios, one GPON frame is selected from the real operation. This GPON framework is periodically generated by a script in a given language and sent for storing in the database.

The test scenarios were run on a computer station with the same configuration as shown in Table 1.

**Table 1.** Hardware used for testing.

| Component | Caption | Key Parameters |
|---|---|---|
| CPU | AMD Ryzen 7 2700X | 3.2 GHz, 8 physical cores |
| RAM | HyperX Predator 4 × 8 GB | 32 GB DDR4 3333 MHz |
| HDD | M.2 SSD WD SN500 | Writing 1450 MB/s |

### 4.1. Serial Data Writing

Serial writing method is a way in which individual data are written sequentially. Due to the proposed structure of the Mongo database, this is a way in which the GPON header of the frame is written first, including both payload length indicator (Plend) fields but without the BWmap field. After writing the header, the BWmap field is written and finally the document containing the GEM frames.

### 4.2. Mass (Batch) Data Writing

There are two ways to test batch enrollment. As in the previous case, the GPON header of the frame without the BWmap field is written first. The remaining parts of the GPON frame are stored in a field in the writing application and linked to the document key of the specific GPON frame, which was obtained after writing the specific GPON frame header to the Mongo database. Two batches are stored in the repository, the first with BWmap fields and the second with individual GEM frames.

The second way of writing is very similar. The principle of writing is the same, differing only in the amount of data that is written in the batch. In contrast to the previous method, where all BWmap fields and GEM frames were written at once for the whole batch of GPON frames, the write is always related to one GPON frame.

### *4.3. Asynchronous Writing Method*

The asynchronous writing method is based on writing after the previous method, i.e., after GTC frames. In this test, the emphasis is on parallelism. Unlike the previous methods, each write waited for its result, while the asynchronous write waited for the result at the end of the test. Better results should be obtained by using asynchronous code. The principle of writing is shown in Figure 5.

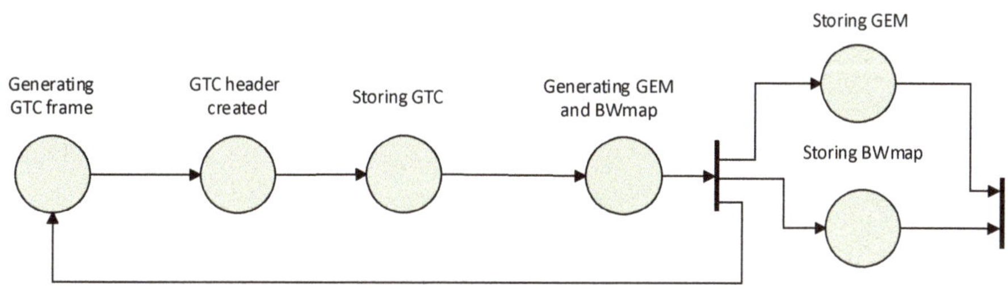


**Figure 5.** Asynchronous writing method.

### *4.4. Asynchronous Bulk Write with Parallel Generation*

This test scenario is again a modification of the previous scenario. The only difference is when BWmap fields and GEM frames are written. The previous scenario saved the GTC header, when the unique key of the document was returned, then the GEM header of the frame and the BWmap field were generated, which were then stored asynchronously. In this scenario, parallelization is already achieved when saving the GPON frame. While waiting for the document's GTC key to be obtained, a BWmap and GEM frame field is generated. While the key is being added to the second generated data, the process of generating another GPON frame is already started. Detailed process flow is shown in Figure 6.

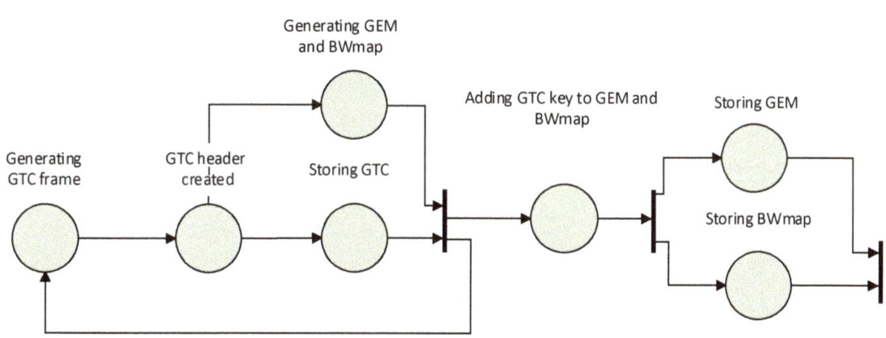


**Figure 6.** Asynchronous bulk write with parallel generation.

## 5. Results

Serial write speeds for Python ranged from approximately 2100 to 2900 GEM documents per second. In most cases, C# achieved faster write speeds ranging from 1500 to 4100 documents per second. The dependence of the write speed on the number of GEM documents can be seen in Figure 7. Serial writing does not seem appropriate, because its speed is many times less than the minimum required.

The bulk write speed ranged from 6000 to 32,000 GEM headers per second for Python and from approximately 8500 to 97,000 GEM headers per second for C#. As can be seen from Figure 7, the bulk writing of all GEM documents is faster for smaller numbers of GEMs. At more than 250 GEM, the performance difference begins to decrease and is almost zero for 800 elements.

The bulk write speed ranged from 6000 to 32000 GEM headers per second for Python and from approximately 10000 to 325000 GEM headers per second for C#. Asynchronous storage in Python provided almost no acceleration compared to bulk writing. Performance testing in C# shows that asynchronous access is not suitable for small numbers of GEM headers up to 50, where the write speed is slower than for bulk write. However, for larger quantities, there is a significant increase in performance, with both the minimum and expected write speeds achieved in both scenarios. An hybrid solution (precompiled C# package) might solve the issue.

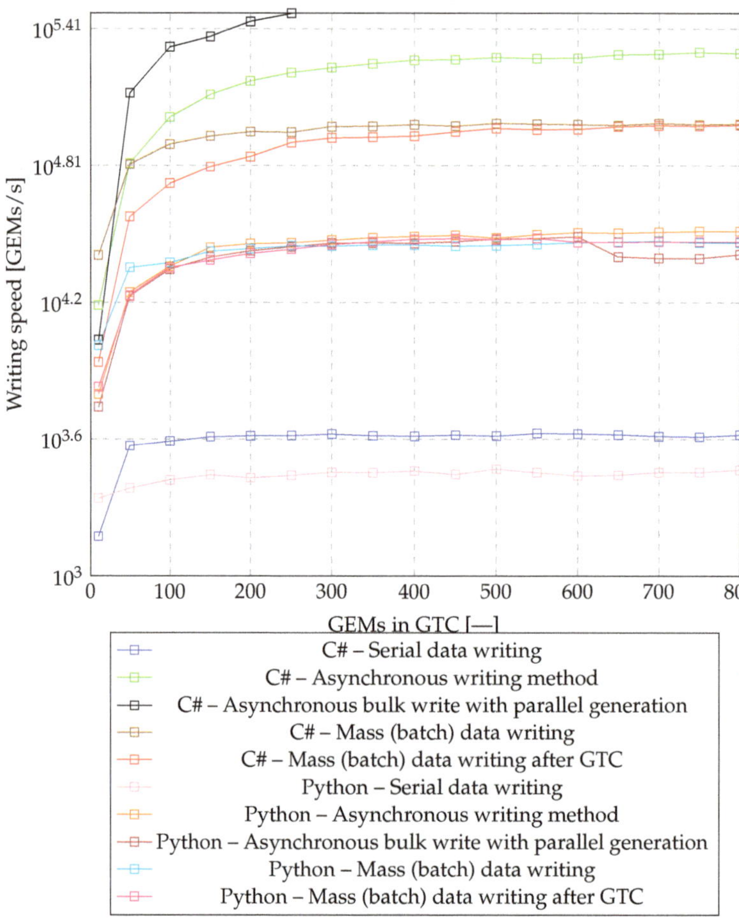


**Figure 7.** Graph of dependence of writing speed based on the used writing technique.

The results obtained, on which the above graphs are based, are also introduced in Table 2. The complete testing differed based on the change in the number of GEM frames encapsulated in GPON frames. During testing, results were measured for 10, 50, 100, 150, 200, 250, 300, 350, 400, 450, 500, 550, 600, 650, 700, 750, 800 GEM frames encapsulated in GPON frames. Due to the high number of measured values, only a subset of them was listed in the table.

Table 2. Reached results from writing tests.

| GEMs in GTCs | 10 | 100 | 200 | 300 | 400 | 500 | 600 | 700 | 800 |
|---|---|---|---|---|---|---|---|---|---|
| C# – serial | 1488 | 3903 | 4131 | 4206 | 4114 | 4121 | 4211 | 4102 | 4163 |
| C# – batch all | 25,707 | 79,302 | 90,050 | 94,817 | 96,642 | 97,867 | 96,790 | 97,943 | 97,312 |
| C# – batch GTC | 15,456 | 104,167 | 149,925 | 171,821 | 185,529 | 190,549 | 189,095 | 196,574 | 198,265 |
| C# – asynchronous GTC | 8681 | 53,135 | 69,881 | 84,507 | 86,003 | 92,920 | 92,061 | 95,720 | 96,165 |
| C# – parallel generating | 10,905 | 212,314 | 275,482 | 313,152 | 325,468 | 314,861 | 313,972 | 316,170 | 312,744 |
| Python – serial | 2182 | 2632 | 2693 | 2842 | 2883 | 2940 | 2747 | 2844 | 2919 |
| Python – batch all | 10,300 | 23,856 | 27,466 | 28,198 | 28,557 | 28,341 | 29,254 | 29,617 | 29,168 |
| Python – batch GTC | 6751 | 22,655 | 26,188 | 28,655 | 30,161 | 30,075 | 29,299 | 29,435 | 29,439 |
| Python – asynchronous GTC | 6278 | 22,915 | 28,893 | 29,943 | 31,052 | 30,612 | 32,316 | 32,466 | 32,784 |
| Python – parallel generating | 5520 | 22,301 | 26,893 | 29,071 | 29,022 | 30,257 | 31,028 | 24,946 | 25,959 |

## 6. Discussion

The results of this work are based on previous experiments with the MS SQL database, described in the publications [38–40]. The main requirement for future work was to optimize the writing speed as much as possible and also to enable easy data acquisition, which may not comply with G.984 recommendations. Based on these requirements, the Mongo database system was selected and a series of experiments to write to the database were performed. Because these are high data rates, we consider working with data to be a big data area. The experiments performed were to show whether it is possible to reach the minimum writing speed with the selected area. Two programming languages were used during the experiments. The Python programming language was chosen as the environment for rapid application development and the language that will be used for future processing, for example using analysis in TensorFlow. The second programming language that is used to implement test scenarios is C#. The parsing software, described in publication [38], is implemented in this programming language.

After initial examination of the writing possibilities, it was quite clear that the serial writing would be completely unsuitable for the described bit rate, but it was mentioned for interest. The aim of the established tests was to parallelize the writing of data to the database as much as possible and thus achieve maximum writing speed. From the achieved values in Table 2, it is clear that the Python programming language is very slow and therefore unsuitable for very fast operations. During tests with the C# programming language, it has been shown that sufficient speed can be achieved to write data fast enough with zero GPON frame rate.

The proposed experiments aimed to show the possibility of writing data from passive optical networks based on the G.984 GPON standard, which are currently still used in Europe. Due to newer types of passive optical networks (e.g., XG-PON), these tests can be used for testing if the input GPON frames are converted to XG-PON frames. By converting to a newer type of passive optical networks, the speed limits of both the programming language and the data storage itself will probably be reached and it will be necessary to adjust the whole principle of frame storage.

## 7. Conclusions

This paper summarizes the issue of recording traffic from a passive optical network to a repository built on the popular Mongo document database. The main goal of this work was to select a suitable way for data to be stored in the repository. Based on the performed testing, it is possible to use the Mongo database to store such a large amount of data, but even so, values at the limit of the required standards are achieved. If we wanted to apply current methods for storing traffic from faster networks (such as XG-PON), it would be necessary to use other technologies such as Apache Kafka. The stored data can also be processed by the tested Python language, because there is no longer a speed-limiting element. The data stored in the repository will be used for the following analysis of the operation, for example using artificial intelligence.

Future work will be focused on the analysis of stored data, design of methods for traffic analysis and modification of the presented solution for use in passive optical networks with higher data transmission. In addition to the ability to use other technology, processing speed can be increased by adjustment of the computer configuration. Multiprocessing might be used to increase the parallelization of computational operations to achieve better results.

The second option is to increase the number of solid-state drives (SSDs) and create a raid array. Another thread of this work could be focused on the data preprocessing performance directly on the FPGA card. Which could, after appropriate consideration, filter, ignore, or delete unnecessary data. For example, the same repeated physical synchronization (Psync) fields can be ignored. They indicate only the beginning of the GPON frame.

The last option to achieve better write results, which can also be a big challenge, is to send data directly to the Mongo database from the FPGA card. Artificial intelligence neural networks appear to be suitable methods for analysis.

**Author Contributions:** Conceptualization, M.H. and V.O.; Data curation M.H., T.H., and P.M.; Funding acquisition T.H. and V.O.; Investigation M.H., T.H., V.O., and P.M.; Methodology, M.H., T.H., V.O., P.M., and S.V. Software, M.H. and V.O. Validation, T.H. and P.M.; Resources T.H. and V.O.; Software M.H., T.H., P.M., and A.T.; Validation M.H., T.H., V.O., P.M., A.T., and S.V.; Writing—original draft M.H., T.H., V.O., P.M., A.T., and S.V.; Writing—review and editing M.H., T.H., V.O., P.M., A.T., and S.V. All authors have read and agreed to the published version of the manuscript.

**Funding:** The research described in this paper was financed by a grant from the Ministry of the Interior of the Czech Republic, Program of Security Research, VI20192022135, PID VI3VS/746 for "Deep hardware detection of network traffic of next generation passive optical network in critical infrastructures".

**Conflicts of Interest:** The authors declare no conflict of interest.

## Abbreviations

The following abbreviations are used in this manuscript:

| | |
|---|---|
| AES | Advanced encryption standard |
| AON | Active optical network |
| ASIC | Application-specific integrated circuit |
| BSON | Binary JSON |
| BWmap | Bandwidth map |
| C-RAN | Cloud radio access network |
| COM | Component object model |
| CPU | Central processing unit |
| CWDM | Coarse wavelength division multiplexing |
| DBA | Dynamic bandwidth algorithm |
| FPGA | Field-programmable gate array |
| GEM | GPON encapsulation method |
| GPON | Gigabit-capable passive optical network |
| GTC | GPON Transmission Convergence |
| GTH | Gigabit transceiver H |
| GTY | Gigabit transceiver Y |
| HDD | Hard disk drive |
| ISP | Internet services provider |
| ITU-T | International Telecommunication Unit telecommunication section |
| JSON | JavaScript object notation |
| MS SQL | Microsoft structured query language |
| ODN | Optical distribution network |
| OLE | Object linking and embedding |
| ONU | Optical network unit |
| OLT | Optical line termination |
| PCIe | Peripheral Component Interconnect Express |

| | |
|---|---|
| Plend | Payload length indicator |
| PON | Passive optical network |
| Psync | Physical synchronization |
| PV | Preverification |
| QoS | Quality of Service |
| RAM | Random access memory |
| RMS | Report message scheduling |
| SDN | Software defined network |
| SFP | Small form-factor pluggable |
| SSD | Solid-state drive |
| TDM | Time division multiplex |
| UML | Unified modeling language |
| XG-PON | 10Gigabit-capable passive optical network |
| XGS-PON | 10-Gigabit-capable symmetric passive optical network |

## References


1. Singh, J.; Garg, A.K. Optimal solutions of integrated optical and wireless applications using GPON-RoF technologies. In Proceedings of the 3rd International Conference on Electronics, Communication and Aerospace Technology (ICECA), Coimbatore, India, 12–14 June 2019; pp. 526–531.
2. Hood, D.; Trojer, E. *Gigabit-Capable Passive Optical Networks*; Wiley: Hoboken, NJ, USA, 2011.
3. International Telecommunication Union. *G.984.3 Gigabit-Capable Passive Optical Networks (G-PON): Transmission Convergence Layer Specification*, 1st ed.; International Telecommunication Union: Geneva, Switzerland, 2014.
4. Hantoro, G.D.; Wibisono, G. GPON performance analysis for 5G backhaul solutions. In Proceedings of the 2018 IEEE Region 10 Conference, Jeju Island, Korea, 28–31 October 2018; pp. 1544–1547.
5. Zin, A.M.; Idrus, S.M.; Ismail, N.A.; Ramli, A.; Butt, R.A. Energy efficient performance evaluation of XG-PON for sustainable green communication infrastructure. In Proceedings of the 2018 Progress in Electromagnetics Research Symposium (PIERS-Toyama), Toyama, Japan, 1–4 August 2018; pp. 950–955.
6. International Telecommunication Union. *G.984.2 Gigabit-Capable Passive Optical Networks (G-PON): Physical Media Dependent (PMD) Layer Specification*, 1st ed.; International Telecommunication Union: Geneva, Switzerland, 2003
7. International Telecommunication Union. *G.987.1: 10-Gigabit-Capable Passive Optical Networks (XG-PON): General Requirements* , 1st ed.; International Telecommunication Union: Geneva, Switzerland, 2016
8. Cale, I.; Salihovic, A.; Ivekovic, M. Gigabit passive optical network—GPON. In Proceedings of the 2007 29th International Conference on Information Technology Interfaces, Cavtat, Croatia, 25–28 June 2007; pp. 679–684.
9. International Telecommunication Union. *G.984.1: Gigabit-Capable Passive Optical Networks (GPON): General Characteristics*, 1st ed.; International Telecommunication Union: Geneva, Switzerland, 2008
10. Menoutis, G.; Foteas, A.; Liakopoulos, N.; Georgis, G.; Reisis, D.; Synnefakis, G. A configurable transmitter architecture organization for XG-PON OLT/ONU/ONT network elements. In Proceedings of the 2015 IEEE International Conference on Electronics, Circuits, and Systems (ICECS), Cairo, Egypt, 6–9 December 2015; pp. 673–676.
11. Pachnicke, S.; Eiselt, M.H.; Grobe, K.; Elbers, J. The frontiers of optical access networks. In Proceedings of the 2015 International Conference on Optical Network Design and Modeling (ONDM), Pisa, Italy, 11–14 May 2015; pp. 12–15.
12. Mikaeil, A.; Hu, W.; Hussain, S.; Sultan, A. Traffic-Estimation-Based Low-Latency XGS-PON Mobile Front-Haul for Small-Cell C-RAN Based on an Adaptive Learning Neural Network. *Appl. Sci.* **2018**, *8*, 1–15, doi:10.3390/app8071097.
13. Mercian, A.; McGarry, M.P.; Reisslein, M. Impact of report message scheduling (RMS) in 1G/10G EPON and GPON. *Opt. Switch. Netw.* **2014**, *12*, 1–13, doi:10.1016/j.osn.2013.11.004.
14. Sales, V.; Segarra, J.; Prat, J. An efficient dynamic bandwidth allocation for GPON long-reach extension systems. *Opt. Switch. Netw.* **2014**, *14*, 69–77, doi:10.1016/j.osn.2014.01.009.

15. Wang, Y.; Zhu, Z.; Wang, L.; Bai, J. A novel proposal of GPON-oriented fiber grating sensing data digitalization system for remote sensing network. *Opt. Commun.* **2016**, *366*, 1–7, doi:10.1016/j.optcom.2015.12.029.
16. Das, S.; Ruffini, M. A Variable Rate Fronthaul Scheme for Cloud Radio Access Networks. *J. Light. Technol.* **2019**, *37*, 3153–3165, doi:10.1109/JLT.2019.2912127.
17. Talli, G.; Slyne, F.; Porto, S.; Carey, D.; Brandonisio, N.; Naughton, A.; Ossieur, P.; McGettrick, S.; Blumm, C.; Ruffini, M.; et al. SDN Enabled Dynamically Reconfigurable High Capacity Optical Access Architecture for Converged Services. *J. Light. Technol.* **2017**, *35*, 550–560, doi:10.1109/JLT.2016.2604864.
18. Kosmatos, E.; Uzunidis, D.; Matrakidis, C.; Stavdas, A.; Horlitz, S.; Pfeiffer, T.; Lord, A. Building a Truly Dynamic Filterless Metro Network by Reusing a Commercial PON's Data-Plane and a Novel SDN-Enabled Control-Plane. *J. Light. Technol.* **2019**, *37*, 6033–6039, doi:10.1109/JLT.2019.2945410.
19. Yeh, C.H.; Chow, C.W.; Yang, M.H.; Hsu, D.Z. A Flexible and Reliable 40-Gb/s OFDM Downstream TWDM-PON Architecture. *IEEE Photonics J.* **2015**, *7*, 1–9, doi:10.1109/JPHOT.2015.2504970.
20. Pakpahan, A.F.; Hwang, I.S.; Nikoukar, A. OLT Energy Savings via Software-Defined Dynamic Resource Provisioning in TWDM-PONs. *J. Opt. Commun. Netw.* **2017**, *9*, 1019–1029, doi:10.1364/JOCN.9.001019.
21. McGettrick, S.; Slyne, F.; Kitsuwan, N.; Payne, D.B.; Ruffini, M. Experimental End-to-End Demonstration of Shared N. *J. Light. Technol.* **2016**, *34*, 4205–4213, doi:10.1109/JLT.2016.2593661.
22. Yin, S.; Shen, T.S.; Bi, Y.; Jin, J.; Oyama, T.; Kazovsky, L.G. A Novel Quasi-Passive, Software-Defined, and Energy Efficient Optical Access Network for Adaptive Intra-PON Flow Transmission. *J. Light. Technol.* **2015**, *33*, 4536–4546, doi:10.1109/JLT.2015.2477036.
23. Quadri, C.; Premoli, M.; Ceselli, A.; Gaito, S.; Rossi, G.P. Optimal Assignment Plan in Sliced Backhaul Networks. *IEEE Access* **2020**, *8*, 68983–69002, doi:10.1109/ACCESS.2020.2986535.
24. Mustak, M.S.; Hossen, M.; Saha, S. Weight-based bandwidth allocation algorithm for improving the QoSs of Multi-OLT PON in downstream direction. In Proceedings of the 5th International Conference on Advances in Electrical Engineering (ICAEE), Dhaka, Bangladesh, 26–28 September 2019; pp. 663–667.
25. Hossen, M.; Hanawa, M. Dynamic Bandwidth Allocation Algorithm With Proper Guard Time Management Over Multi-OLT PON-Based Hybrid FTTH and Wireless Sensor Networks. *J. Opt. Commun. Netw.* **2013**, *5*, 802–812, doi:10.1364/JOCN.5.000802.
26. Liu, Z.; Gan, C.; Xie, W.; Yan, Y.; Qiao, H. Algorithm of both release and allocation bandwidth for downstream channel in multi-OLT PON. *IET Commun.* **2018**, *12*, 824–831, doi:10.1049/iet-com.2017.0876.
27. Peng, Z.; Radcliffe, P. Modeling and simulation of Ethernet Passive Optical Network (EPON) experiment platform based on OPNET Modeler. In Proceedings of the IEEE 3rd International Conference on Communication Software and Networks, Xi'an, China, 27–29 May 2011; pp. 99–104, doi:10.1109/ICCSN.2011.6013671.
28. Wu, X.; Brown, K.; Sreenan, C.; Alvarez, P.; Ruffini, M.; Marchetti, N.; Payne, D.; Doyle, L. An XG-PON module for the NS-3 network simulator. In Proceedings of the Sixth International Conference on Simulation Tools and Techniques, Cannes, France, 5–7 March 2013; pp. 195–202, doi:10.4108/icst.simutools.2013.251605.
29. Nakayama, Y.; Yasunaga, R. ITU TWDM-PON module for ns-3. *Wirel. Netw.* **2020**, *1*, 1–12, doi:10.1007/s11276-019-02236-8.
30. Horvath, T.; Munster, P.; Jurcik, M.; Koci, L.; Filka, M. Timing measurement and simulation of activation process in GPON networks. *Opt. Appl.* **2015**, *45*, 1–14, doi:10.5277/oa150403.
31. Horvath, T.; Krkos, R.; Dubravec, L. Deep data analysis in gigabit passive optical networks. *Opt. Appl.* **2017**, *47*, 157–170, doi:10.5277/oa170114.
32. Meng, L.; Peng, H.; Zeng, J. Hardware Platform System of GPON ONU System Designed Based on FPGA. CN101365250A, 8 December 2010.
33. Doo, K.-H.; Lee, S.-S.; Kim, W.-W. Design of a retimed long-reach GPON Extender using FPGA. In Proceedings of the Digest of the 9th International Conference on Optical Internet (COIN 2010), Jeju, Korea, 11–14 July 2010; pp. 1–3.
34. Vinh, T.Q.; Park, J.-H.; Kim, Y.-C.; Kim, K.-O. An FPGA implementation of 30Gbps security module for GPON systems. In Proceedings of the 8th IEEE International Conference on Computer and Information Technology, Sydney, Australia, 8–11 July 2008; pp. 868–872.
35. Straullu, S.; Savio, P.; Nespola, A.; Chang, J.; Ferrero, V.; Gaudino, R.; Abrate, S. Demonstration of upstream WDM+FDMA PON and real time implementation on an FPGA platform. In Proceedings of the 2015

European Conference on Optical Communication (ECOC), Valencia, Spain, 27 September–1 October 2015; pp. 1–3.
36. Oujezsky, V.; Horvath, T.; Jurcik, M.; Skorpil, V.; Holik, M.; Kvas, M. Fpga network card and system for gpon frames analysis at optical layer. In Proceedings of the 2019 42nd International Conference on Telecommunications and Signal Processing (TSP), Budapest, Hungary, 1–3 July 2019; pp. 19–23.
37. DFC Design. lCecilie—xPON module. https://www.dfcdesign.cz/en/cecilie-xpon-module (accessed on 14 October 2020).
38. Jurcik, M.; Horvath, T.; Oujezsky, V.; Skorpil, V.; Holik, M. GPON parser for database analysis. In Proceedings of the 2019 42nd International Conference on Telecommunications and Signal Processing (TSP), Budapest, Hungary, 1–3 July 2019; pp. 347–350, doi:10.1109/TSP.2019.8768849.
39. Horvath, T.; Jurcik, M.; Oujezsky, V.; Skorpil, V. GPON analyzer—Frame parser module. In Proceedings of the 2019 42nd International Conference on Telecommunications and Signal Processing (TSP), Budapest, Hungary, 1–3 July 2019; pp. 748–752, doi:10.1109/TSP.2019.8768882.
40. Holik, M.; Horvath, T.; Oujezsky, V. Application for GPON Frame Analysis. *Electronics* **2019**, *8*, 11, doi:10.3390/electronics8060700.

**Publisher's Note :** MDPI stays neutral with regard to jurisdictional claims in published maps and institutional affiliations.

sensors

MDPI

Article

# A QoE-Aware Energy Supply Scheme over a FiWi Access Network in the 5G Era

**Chao He** [1,2,3] **and Ruyan Wang** [1,2,3,*]

1 School of Communication and Information Engineering, Chongqing University of Posts and Telecommunications, Chongqing 400065, China; d170101004@stu.cqupt.edu.cn
2 Chongqing Key Laboratory of Optical Communication and Networks, Chongqing 400065, China
3 Chongqing Key Laboratory of Ubiquitous Sensing and Networking, Chongqing 400065, China
* Correspondence: wangry@cqupt.edu.cn

Received: 3 June 2020; Accepted: 5 July 2020; Published: 7 July 2020

check for updates

**Abstract:** Integrated fiber-wireless (FiWi) should be regarded as a promising access network architecture in future 5G networks, and beyond; this due to its seamless combination of flexibility, ubiquity, mobility of the wireless mesh network (WMN) frontend with a large capacity, high bandwidth, strong robustness in time, and a wavelength-division multiplexed passive optical network (TWDM-PON) backhaul. However, the key issue in both traditional human-to-human (H2H) traffic and emerging Tactile Internet is the energy conservation network operation. Therefore, a power-saving method should be instrumental in the wireless retransmission-enabled architecture design. Toward this end, this paper firstly proposes a novel energy-supply paradigm of the FiWi converged network infrastructure, i.e., the emerging power over fiber (PoF) technology instead of an external power supply. Then, the existing time-division multiplexing access (TDMA) scheme and PoF technology are leveraged to carry out joint dynamic bandwidth allocation (DBA) and provide enough power for the sleep schedule in each integrated optical network unit mesh portal point (ONU-MPP) branch. Additionally, the correlation between the transmitted optical power of the optical line terminal (OLT) and the quality of experience (QoE) guarantee caused by multiple hops in the wireless frontend is taken into consideration in detail. The research results prove that the envisioned paradigm can significantly reduce the energy consumption of the whole FiWi system while satisfying the average delay constraints, thus providing enough survivability for multimode optical fiber.

**Keywords:** FiWi access network; energy efficiency; power over fiber; TWDM-PON; delay analysis

## 1. Introduction

Ever-increasing green communication may still become one of the major concentrations for information and communication technology (ICT) development and evolution [1,2]. Meanwhile, the access networks of the ICT sector consume the greatest part of the whole communication architecture because of increasing network device and traffic volumes. International Mobile Telecommunications-2020 (IMT-2020) firstly proposed a 5G flower consisting of nine key performance indicators [3]. Importantly, except for cost-effectiveness, the other eight key performance indicators accepted by the International Telecommunication Union Radio Communications Department (ITU-R) are the user experience rate, spectral efficiency, mobility, delay, connection density, network energy efficiency, traffic density, and peak data rate. Thereinto, network energy efficiency can be further extended to the three typical application scenarios [1–3].

When it comes to the 5G-XHaul network architecture, the seamless convergence of the fiber-wireless (FiWi) broadband access network is a cutting-edge candidate paradigm, which provided network operators with the integrated advantages of a high reliability, large capacity, and low loss of the optical backhaul subnetwork with ubiquity, flexibility, and mobility of the wireless/cellular frontend subnetwork, as well as further provided both fixed subscribers and mobile users with an extensive broadband data transmission service [4–6]. Just for the fronthaul systems of the 5G and beyond networks, the seamless combination of a wireless access signal in the mm Wave bands and fiber-optic is proposed due to strong demand for a lower latency, high data rate, and ultra-density. The high performance of the radio access signal and optical fronthaul system is evaluated by optical self-heterodyne and subcarrier multiplexing intermediate frequency, respectively [7]. To compensate for the path loss in the mmWave frequency bands, except for beamforming and massive multiple-input multiple-output (MIMO), a self-steering phased array beamformer for the FiWi mobile fronthaul was first demonstrated without any external tuning management [8]. Van et al. evaluated the possibilities, challenges, and guidelines of power-saving mechanisms (PSM) for the Internet of Thing (IoT) application over a FiWi access network through adapting the optical network unit (ONU) sleep mode and discontinuous reception (DRX) mechanism for IoT devices [9]. Furthermore, multiple time-driven smart grid sensors report their monitored data to the energy control system utilizing the time-division multiplexing access (TDMA) mechanism. In addition, the enhanced smart grid based on a dependable FiWi access network can be deployed to further shed some light on big data acquisition, smart metering based on multistage stochastic programming, probabilistic availability quantification, as well as the total cost of ownership (TCO) and risk [10]. The most critical issues for FiWi networks, e.g., propagation delay diversity, quality of service (QoS) provisioning, survivability and reliability, prediction and estimation in the propagation delays, dynamic bandwidth allocation (DBA), network architecture optimization, as well as scheduling strategies, have received a great deal of attention, especially power-saving methods [11]. It is worth mentioning that FiWi access networks have been recently receiving significant attention in IoTs [9], smart grids [10], smart cities [12], as well as Tactile Internet [13].

Generally, the energy efficiency issue is one of the major challenges for an integrated FiWi access network, which can allow operators to decrease their OPEX and extend the battery life of mobile devices when considering the QoS constraint. A promising solution is converting the network components with a lower load and leisure into the sleep state, which focuses either on wireless/cellular fronthaul or optical access network (OAN) backhaul [9,14–16]. More specifically, for the wireless/cellular fronthaul segment, the wireless station (STA) adapts PSM, and the mesh points (MPs) are transformed into a doze in the IEEE 802.11 family in terms of WiFi coverage. Besides, user equipment (UE) makes the utmost of the timeout-driven DRX/discontinuous transmission (DTX) mechanisms in 4G-and-beyond-enabled cellular coverage. Meanwhile, for the fiber backhaul segment, either ONU sleep or the combination of optical line terminal (OLT) sleep with ONU sleep is separately leveraged in time-division multiplexed passive optical networks (TDM-PONs) and time and wavelength division multiplexed PONs (TWDM-PONs). It is important to note, however, that the existing literature that particularly concentrates on QoE-aware energy schemes over FiWi access networks covers the following several segments: (1) network topology design, by optimizing the number of network device deployment and periodically implementing network topology reconfiguration [14]; (2) energy-efficient DBA, by integrating the power scheduling of wireless frontend and fiber backhaul [15,17,18]; (3) dynamic adaptive mechanism, by adapting the power state of ONU according to the dynamic traffic profile [16,19,20]; (4) BSs energy consumption minimization with UE connection constraints, at the expense of the successful connection number of end equipment [21,22]; and (5) service class resource management-based PSM, worldwide interoperability for microwave access (WiMAX) and long-term evolution-advanced (LTE-A) possessing five and eight service classes, respectively, and PON consisting of three service classes via incorporating service class differentiation into the PSM of green wireless optical broadband access networks (WOBAN) [23].

Recently, in the power-efficient cloud radio access network (CRAN) operation, given the fact that the ONUs' sleep phase in the envisioned remote radio heads (RRHs) was fragmented into several timeslots, the electrical power spent by both the ONU sleep and active states was only supplied by the OLT through the power over fiber (PoF) technology, so that mobile network operators can allow ONUs to efficiently operate without external power supplement [24–26]. However, the vision of providing external power to all ONU-MPPS will involve high costs, especially if the ONU-MPP over FiWi architecture is deployed where external power is not available, such as in remote mountainous areas, mobile communications in the military, etc. For this reason, PON utilizing PoF is considered as a key power supply technology. The combination of OLT sleep with ONU sleep was explained in the proposed FiWi network; however, the network equipment responsible for the wireless frontend and optical backhaul was not involved in the PoF-empowered power saving. In this paper, we take into consideration not only the converged ONU mesh portal points (ONU-MPPs) sleep scheduling scheme in the polling cycle time, but also the optical subscriber unit (OSU) for the PoF technology and data communication between the OLT and integrated ONU-MPPs over dedicated wavelengths.

More specifically, the main contributions and highlights of this paper are organized as follows:

- We propose a FiWi access network architecture based on a seamless integration paradigm of a multi-hop wireless mesh network (WMN) frontend and TWDM-PON backhaul. Then, to ensure the quality of service (QoS) characteristics and PSM formulation, we leverage the future-proof time-division multiplexed access (TDMA) scheme to jointly synchronize the DBA process. Meanwhile, we analyze the energy harvesting and conversion in the ONU module in detail.
- We apply the service-oriented DBA scheme to quantify timeslot allocation occupied by the active and sleep state of the ONU-MPP. In PoF-enabled energy supply, we investigate the energy-aware transmitted optical power scheme to achieve the minimum energy consumption, which can be adjusted to the optimum level.
- Given the minimized energy consumption and acceptable data communication delay, we derivate the mean tolerant end-to-end traffic delay over the envisioned FiWi access network, and the bridge of the correlation function between the QoE value and transmitted optical power.

The logical architecture of this paper is organized as follows. Section 2 reviews the related work of the traditional energy conservation for the FiWi access network. The provision of communication services to all integrated ONU-MPPs via PoF technology, instead of an external power supply, over the FiWi access network is envisioned in Section 3 from the viewpoint of network architecture. Section 4 elaborates on the mathematical model of energy consumption minimization, including joint bandwidth allocation and PSM problem formulation, PoF-enabled harvested power, and the correlation between multi-hops and QoE value. Under the human-to-human (H2H) traffic delay constraint, numerical analysis and discussions are evaluated in Section 5. Finally, Section 6 provides the conclusions.

## 2. Related Work of Energy Conservation over FiWi Network

The growing traffic volume requested by the end-users renders the energy consumption increase of the access network. Either the low-loaded or the leisure ONU-MPPs in the backhaul OAN was transformed into the sleep state as much as possible so that the hosting residual traffic can be rerouted to the other active ONU-MPPs to further improve the energy efficiency (EE) and network resource utilization [16,20]. However, the energy consumption of the smart FiWi access networks can be susceptible to the length of the ONU-MPP sleep state period in the polling cycle time. Togashi et al. demonstrated that, the longer the ONU sleep phase, the longer the traffic delay and the higher energy consumption [27]. Toward this end, minimizing the energy consumption of the FiWi candidate was created with only ONU-MPP sleep in mind, and the network components of the wireless frontend were neglected.

It is challenging to ensure that joint wireless (e.g., multi MP radio interfaces) and optical (e.g., ONU) power state scheduling is leveraged to carry out the energy-saving design. The optimal sleep state scheduling was achieved by leveraging the optical and radio frequency paradigm to reduce delays and energy consumption, and a two-step QoS-aware energy-efficient FiWi scenario serving as a benchmark was proposed [19]. A combination of the ONU sleep method with the radios off has been proposed to facilitate energy conservation, QoS provisioning, and efficient traffic rerouting, employing bonding wireless and optical power-state scheduling [28]. If the power-saving state scheduling in the two subnetworks are not synchronized with each other, all the FiWi networks would cause more energy consumption and additional delays. To address this issue, a cooperative ONU sleep mechanism was proposed to reduce the power consumption by dynamically integrating ONU sleep with the STA PSM control mechanism [29]. A comprehensive power-saving model of STA, MP, and ONU was proposed to not only synchronize the energy conservation scheduling strategy of the two subnetworks by leveraging the TDMA technology, but also to mitigate the end-to-end delay by taking advantage of an M/G/1 queuing model [16]. Importantly, both the QoS guaranteeing or energy-saving mode was mismatched in both segments, and the traffic transmission delay was increasing. The energy-saving modes in both coverage-centric WLAN and capacity-centric PON were referred to as STA PSM and ONU sleep, while the QoS policies were hybrid coordination function-controlled channel access (HCCA) and DBA, respectively. The smart integration of ONU sleep with STA PSM and DBA with HCCA was proposed to decrease the transmission delay [30].

It is worth mentioning that the Tactile Internet has recently attracted extensive research efforts due to common features of very little delays, an ultra-high reliability, H2H/machine-to-machine (M2M) coexistence, data-centric technology, as well as security. Van et al. has envisioned an M2M communication architecture over the emerging FiWi enhanced LTE networks, in which the M2M devices' DRX mechanism in the cellular frontend and ONUs' power-saving mode in the optical backhaul were devised to enhance energy saving and decrease packet delay comprehensively; accordingly, the semi-Markov process and M/G/1 queue paradigm were modeled analytically, respectively [13].

The first three sections take into consideration energy consumption minimization from the perspective of the IoT or H2H traffic over the FiWi access network. Conversely, when it comes to the FiWi-enhanced LTE-A HetNets, to achieve power consumption minimization, BSs should be switched into the sleep state as many times as possible. On the other hand, to acquire stable service maximization, BSs should be switched into the active state as many times as possible and consume more energy inevitably. There must be an optimal number of active BSs between power consumption minimization and stable service maximization. Toward this end, the power consumption minimization problem under stable service constraints was studied over the proposed architecture through a heuristic greedy algorithm, brute force algorithm, and snowball rolling algorithm [21,22].

On account of achieving energy saving, existing approaches have started to be implemented in the deployment of network devices, front-end backhaul cooperative energy-saving, ONU power state scheduling, reduction user access ratio, accessibility, traffic distribution, and service class differences. Nevertheless, research attention in the area of PoF technology over the promising FiWi access network is still in its infancy, in particular when considering energy consumption minimization.

## 3. PoF-Enabled Energy Supply Paradigm over FiWi Access Network

The envisioned network architecture as shown in Figure 1 covers the core network (CN), bearer network, and access network, which is characterized by FiWi broadband access networks integrating a WMN frontend subnetwork adapting multi-hop with a TWDM-PON backhaul subnetwork exploiting PoF technology, as well as an optical backbone network crossing between the CN and central office (CO). Considering the PoF technology in the FiWi access network, a passive splitter in the RN is applied to provide multiple ONU-MPPs with a communication service via a single optical fiber cable. Meanwhile, an OLT associated with multiple ONU-MPPs supplies energy to each ONU-MPP with the help of the optical-fiber cable [28,31]. On the wireless network frontend, which consists of wireless access and a wireless backhaul, various kinds of new emerging terminal devices, e.g., smartphones, wearing devices, wireless sensors, unmanned aerial vehicles, connected vehicles, power grid stations, and so forth, access the integrated ONU-MPPs through an uplink PS-poll frame and downlink Beacon frame, and the ONU-MPP further allocates the requested upstream (US) subslot to each STA in the next polling cycle time by employing TDMA fashion in each ONU-MPP branch. More specially, a subset of MPs close to the terminal devices are viewed as mesh access points (MAPs), which provide the associated STAs with an access service. A subset of the MPs adjacent to the ONUs are identified as MPPs, which allocate the required bandwidth to its accommodated MPs and portion transmission subslot. Besides, other residual MPs act as relay nodes and forward the data traffic between the MAPs and ONU-MPPs. It is foreseen that the mobile access frontend in terms of the WMN paradigm provides ubiquitous connectivity for numerous smart devices and enhances access capacity.

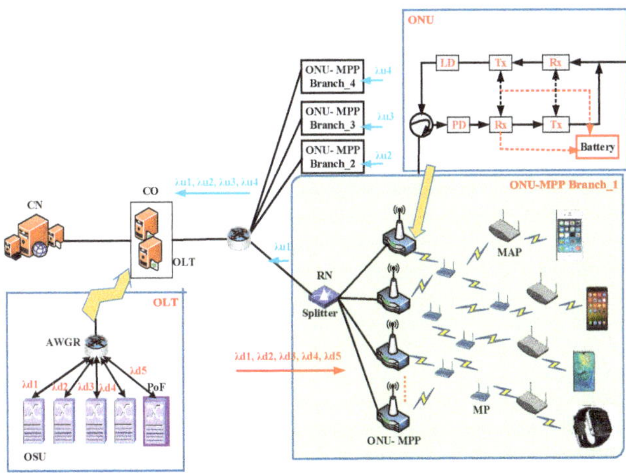


**Figure 1.** FiWi architecture illustration of the time and wavelength division multiplexed passive optical network (TWDM-PON) backhaul exploiting the power over fiber (PoF) and wireless mesh network (WMN) frontend.

Because of the subset of integrated ONU-MPPs in the envisioned FiWi broadband access network, consisting of an ONU module, the battery-powered module in a continuous active state, and the MPP module, which is an important network component, it exchanges data communication services between the WMN frontend and TWDM-PON optical fiber backhaul. In other words, the ONU-MPP receives the *PS-poll frames* from all the STAs and transmit the *Report frames* to the OLT in the uplink direction, while receiving the *Gate frames* from the OLT and then transmitting the *Beacon frames* to each STA in the downlink direction. In addition, the converged ONU-MPPs can flexibly switch between the sleep phase and active phase according to the service-oriented associated STAs.

On the optical backhaul segment, the OLTs located in the CO manage the resource allocation of the integrated ONU-MPPs via wavelength-division multiplexing (WDM); broadcast the data communication traffic stemming from the CO from the OLT to multiple integrated ONU-MPPs; and provide integrated ONU-MPPs with power supply via PoF technology, which is capable of converting an optical signal, e.g., receiving unnecessary data that is not meant for the accommodated ONU-MPPs, into electrical power consumed by the integrated ONU-MPPs. Here, the transmitted optical power of the OLT is restrained by the CO due to the CO controlling both wireless components and optical components in the envisioned FiWi access networks. In the downlink direction, the four OSUs used for data communication and the one OSU applied to the PoF technology are firstly converted into an optical signal in an arrayed waveguide grating router (AWGR), then transmitted over the multimode fiber-enabled optical distribution networks between the OLT and the ONU-MPP. Usually, the primary remote node (RN) contains a splitting ratio of 1:4, and the secondary RN can adopt a typical splitting ratio of 1:64.

## 4. Problem Formulation of Energy Consumption Minimization

In this section, the energy consumption of all the integrated ONU-MPPs is provided by PoF technology instead of external power supply, which is represented by high costs and operational spending. In addition, a joint bandwidth allocation and sleep scheduling scheme of integrated ONU-MPPs is introduced, which is crucial to improving energy efficiency. Finally, the energy supply required by the integrated ONU-MPPs scheduling strategy is taken into consideration via PoF-enabled harvested power technology.

### 4.1. *Joint Bandwidth Allocation and Power-Saving Method*

Unlike the fact that some remote radio heads (RRHs), consisting of an ONU module, battery module, and antenna module, can switch from the active state into sleeping mode only if the residual energy of the battery module in the RRH is below a certain level [25], without taking into consideration the real-time traffic requirements of the aware ONU-MPP sleep scheduling strategy. In this paper, however, each integrated ONU-MPP allocates a subslot to its associated wireless STAs in order to achieve a bandwidth allocation and energy-saving mode scheduling as shown in Figure 2. Here, in order to notably achieve energy conservation, it is clear from Figure 2 that in a PON polling cycle time $T_c$, an OLT can provide polling service for $N$ ONU-MPPs. Any integrated ONU-MPP only operates in a certain timeslot $T_{sl}$, and then sleeps in other timeslots $(N-1)T_{sl}$, and its timeline consists of a data interval, reservation interval, and vacation interval in the uplink, receiving data internally in the downlink. Besides, an ONU-MMP associates $M$ STAs, where an STA transmits data to the associated ONU-MPP in a small timeslot, and then enters into a sleep state while receiving data from an ONU-MPP in a certain timeslot $T_{sl}$. In other words, the duration length of the active state and sleep state is defined as $T_a^i$, $T_a^i = T_{sl}$, where $i$ is the index of the integrated ONU-MPP, i.e., $i \in [1, 2, \ldots, N]$ and $T_s^i$, $T_s^i = (N-1)T_{sl}$, respectively. The set of integrated ONU-MPP branches and that of each ONU-MPP accommodates STAs are denoted by $O = \{o_1, \ldots, o_i, \ldots, o_N\}$ and $S = \left\{s_{i,1}, \ldots, s_{i,j}, \ldots, s_{i,M}\right\}$, $j \in [1, 2, \ldots, M]$, respectively. In addition, the polling cycle time $T_c$ is denoted by the summation of both the ONU-MPPs' allocated timeslot $T_{sl}$ and wake-up overhead time $T^{oh}_{onu\text{-}mpp}$, which ensures all the ONU-MPP components can enter the sleep phase in every polling cycle time. Meanwhile, the achieved bandwidth $B_i$ for the converged ONU-MPP $o_i$ is represented by the total bandwidth $B$ provided by the OLT and the total number of ONU-MPP accommodated STAs, i.e., $B_i = \frac{B}{N}$.

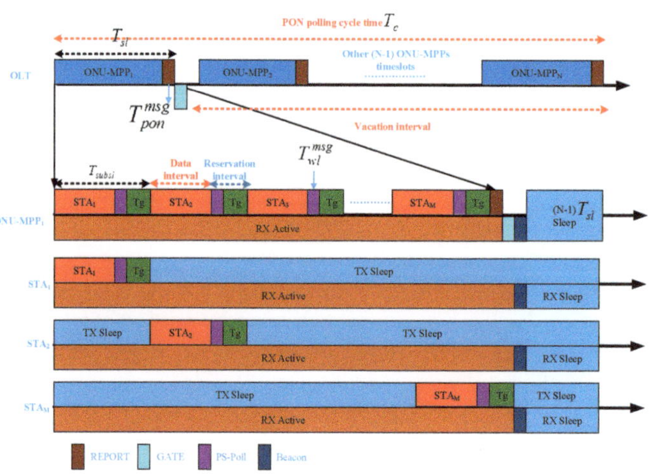


**Figure 2.** Power-saving illustration of the FiWi access network.

In each uplink timeslot $T_{sl}$ allocated by the OLT, the STA $s_{i,j}$ successively transmits an uplink data interval, a *PS-poll frame*, and a guard frame, which is used to mitigate the interference between the two adjacent STAs, whereas always keeping the TX sleep state in the residual timeslot of the allocated timeslot $T_{sl}$. Therefore, each STA must possess an active phase of $T_{subsl}$ and a sleep phase of $T_{sl} - T_{subsl}$. Meanwhile, all the STAs still receive the downlink data traffic from the associated ONU-MPP in the allocated timeslot $T_{sl}$. More specifically, due to the PoF technology-enabled power supply, the energy consumption of the envisioned FiWi access network is calculated for an ONU-MPP component as follows:

$$E_c = p^a_{onu\text{-}mpp}\left(T_{sl} + T^{oh}_{onu\text{-}mpp}\right) + p^s_{onu\text{-}mpp}\left(T_c - T_{sl} - T^{oh}_{onu\text{-}mpp}\right) \tag{1}$$

where $p^a_{onu\text{-}mpp}$ and $p^s_{onu\text{-}mpp}$ denote the power consumed by the active and sleep mode of each converged ONU-MPP, respectively. For the sake of simplification, we can assume that the energy consumption in the wakeup process is the same as that in the active phase. However, the total energy consumption of a traditional ONU-MPP in a polling cycle time can be calculated as

$$E = T_c \cdot p^a_{onu-mpp} \tag{2}$$

In other words, the accommodated ONU-MPP is always in an active state. When the energy-saving scheme is taken into consideration, the energy efficiency in the considered FiWi access networks is verified as follows:

$$\eta = \frac{E - E_c}{E} \tag{3}$$

In this regard, after substituting the polling cycle time and Equation (2) into Equation (3), Equation (3) can be further rewritten as $\eta = \frac{N-1}{N} \cdot \frac{p^a_{onu\text{-}mpp} - p^S_{onu\text{-}mpp}}{p^a_{onu\text{-}mpp}}$. It is worth mentioning that the energy efficiency $\eta$ is denoted as a function of both the total number of converged ONU-MPPs and the power consumption in a different state mode for the sleep scheduling paradigm and the harvested power strategy. Give the potential advantages of the PoF technology, the effect of the transmitted optical power on the energy harvest is remarkable due to the optical power that can be transformed into electrical power in the ONU module.

### 4.2. PoF Technology-Enabled Power Supply

In our envisioned FiWi broadband access network, the crucial CO contains multiple OLTs represented by the set $L \in \{l_1, l_2, \ldots, l_k, \ldots, l_L\}$. Similarly, four OSUs are used for data communication, but the remaining one is PoF technology via a dedicated optical wavelength and accommodates four integrated ONU-MPP branches. In the TWDM-PON utilizing PoF technology, the transmitted optical power originating from the OSUs in OLT is susceptible to optical fiber power loss, which is the result of the fiber attenuation factor of both the feeder fiber and distribution fiber, segregation ratio of the passive splitter in the RN, and photoelectric conversion efficiency at the photodiode (PD) component in the ONU module. Therefore, the electrical power of the ONU-MPP $o_i$ receiving from the OLT $l_k$ is attenuated, and further can be expressed as follows:

$$P^{rx}_{l_k,o_i} = \frac{P^{tx}_{l_k} \cdot \Gamma\left(d_{l_k,o_i}\right) \cdot \zeta}{N} \tag{4}$$

where $P^{tx}_{l_k}$ and $P^{rx}_{l_k,o_i}$ are the transmitted optical power of the OLT $l_k$ and the received power of the electrical signal for the ONU-MPP $o_i$ addressed to the OLT $l_k$, respectively. Furthermore, $\zeta$, $d_{l_k,o_i}$, and $N$ are defined as the photoelectric conversion efficiency of the PD, the link transmission distance between the OLT $l_k$ and the ONU-MPP $o_i$, and the segregation ratio of the RN, respectively. Given the effect of link transmission distance and fiber attenuation factor on the optical fiber power loss factor $\Gamma\left(d_{l_k,o_i}\right)$, $\Gamma\left(d_{l_k,o_i}\right) = 10^{\left(-d_{l_k,o_i}\right) \cdot \alpha / 10}$ explains the mutual relationship in detail, in which $\alpha$ is the fiber attenuation factor in dB/km.

The receiving component Rx of the battery module can convert the transmitting unnecessary electrical signal after the PD component into available electrical power. In addition, the battery module of the integrated ONU-MPP can always store the harvested electrical power during the sleep phase or active phase. The whole evolution process is referred to as PoF technology. Note, however, that the harvested electrical power of the battery module in the ONU-MPP $o_i$ both in the sleep and active mode is differentiated, and can be calculated as Equations (5) and (6), respectively.

$$HP^{s}_{l_k,o_i} = P^{rx}_{l_k,o_i} \tag{5}$$

$$HP^{a}_{l_k,o_i} = P^{rx}_{l_k,o_i}\left(1 - \frac{\beta_{l_k,o_i}}{N}\right) \tag{6}$$

where $\beta_{l_k,o_i}$ is the link resource utilization between the OLT $l_k$ and the ONU-MPP $o_i$. Therefore, the harvested electrical energy of the ONU-MPP $o_i$ in a polling cycle time can be derived as follows:

$$HE_{l_k,o_i} = HP^{a}_{l_k,o_i}\left(T_{sl} + T^{oh}_{onu\text{-}mpp}\right) + HP^{s}_{l_k,o_i}\left(T_c - T_{sl} - T^{oh}_{onu\text{-}mpp}\right) \tag{7}$$

It can be anticipated that the effect of the transmitted optical power $P^{tx}_{l_k}$ on the harvested electrical energy is remarkable. In order to facilitate the energy-efficient operation of the envisioned FiWi access network, it is required along with the strict constraint, i.e., $HE_{l_k,o_i} \geq E_c$.

## 4.3. *Correlation between Multi-Hops and QoE Value*

As shown in Section 4.1 and Figure 2, any integrated ONU-MPP polls the data interval $X_u$, the reservation interval $V_u$ on the incorporation of the *PS-poll frame* $T^{msg}_{wl}$, the guard interval $T_g$, i.e., $V_u = T^{msg}_{wl} + T_g$, as well as the vacation interval $S_u$ on the combination of the residual timeslot duration and multi-point control protocol (MPCP) messaging time, i.e., $S_u = (N-1) \cdot T_{sl} + T^{msg}_{pon}$ in the uplink scheduling direction. In view of the data–reservation–vacation-based polling system, the dependence of the polling cycle time $T_c$ on the non-data H2H transmission ration $1 - \rho^{h2h}_u$ can be further expressed as follows:

$$T_c = \frac{N \cdot (M \cdot V_u + RTT)}{1 - \rho^{h2h}_u} \tag{8}$$

where $\rho^{h2h}_u$ is the uplink aggregated H2H traffic load intensity and $\rho^{h2h}_u \in [0,1)$, $RTT$ is usually represented by $2 \cdot T_{prop}$, which is the propagation delay from the OLT $l_k$ to the integrated ONU-MPP $o_i$. The mean end-to-end H2H packet delay crossing from the associated STA to the sustainable OLT in the CO is decomposed into the WMN frontend mean packet delay and TWDM-PON backhaul mean packet delay.

Here, for the sake of simplicity, it is clear that the converged ONU-MPP in the polling cycle time $T_c$ starts with the sleep phase, and then goes through the active phase and ends with the sleep phase. Each ONU-MPP designed to $M$ associated STAs is modeled as the *M/G/1* queue paradigm with a reservation interval and vacation interval, while each MP/MAP is in the WMN frontend *M/M/1* model. In the multi-hop WMN subnetwork, all the SATs firstly access the near MAP, then relay the multiple MPs, and finally arrive at the accommodated ONU-MPP in the first come first served (FCFS) manner. Similarly to the envisioned directed connectivity graph *G(N, E)* proposed in [30], from a perspective of mesh topology, the mean packet delay stemming from the source node represented by the STA $s_{i,j}$ to the destination node in terms of the ONU-MMP $o_i$ can be computed as follows:

$$\overline{D}^{j,i}_{wmn} = \sum_{h=1}^{H}\left(d^{tras}_h + d^{prop}_h + d^{syn}_h + d^{que}_h\right) \tag{9}$$

where $d_h^{tras}$, $d_h^{prop}$, $d_h^{sys}$, and $d_h^{que}$ are the transmission delay over the point-to-point (P2P) link, propagation delay over the P2P link, synchronization delay, and queuing delay over the MAP/MP, respectively. In addition, $H$ elaborates on the total number of hops in an effective critical path according to the delay-aware routing algorithm (DARA) [31]. In particular, when it comes to the fact that the transmission distance between the nodes is extremely adjacent in the WMN frontend, the propagation delay $d_h^{prop}$ from the STA $s_{i,j}$ to the ONU-MPP $o_i$ can be neglected. The effect of transmission delay $d_h^{tras}$, synchronization delay $d_h^{sys}$, and queuing delay $d_h^{que}$ on the mean packet delay is critical, and accordingly can be computed as $\frac{1}{\mu_h \cdot C_h}$, $\frac{1}{2\mu_h \cdot C_h}$, and $\frac{\rho_h}{\mu_h \cdot C_h - \lambda_h}$, where $C_h$ is link capacity, $\rho_h$ denotes link load intensity, $\lambda_h$ indicates packet arrival rate, and $\mu_h$ represents service rate. Therefore, Equation (9) can be further rewritten as Equation (10).

$$\overline{D}^{j,i}_{wmn} = \sum_{h=1}^{H}\left(\frac{1}{\mu_h \cdot C_h} + \frac{1}{2\mu_h \cdot C_h} + \frac{\rho_h}{\mu_h \cdot C_h - \lambda_h}\right) \tag{10}$$

On the other hand, both the frame queuing delay and frame processing delay over the converged ONU-MPP $o_i$ also should be taken into consideration, except for the propagation delay between the ONU-MPP $o_i$ and OLT $l_k$ in the TWDM-PON optical backhaul with a gated service. It is worth mentioning that both $T_a^i \neq 0$ and $T_s^i \neq 0$ can provide a power-efficient service for the end-users. We can also assume that the US aggregate traffic arrival rate in the OLT $l_k$ is $\lambda_u$, and therefore $\frac{\lambda_u}{N}$ is the US traffic arrival rate of the integrated ONU-MPP $o_i$ and its associated STAs. Given the paramount importance of the delay analysis, the first- and second-order moments of the US data frame service time is denoted by $\overline{X_u}$ and $\overline{X_u^2}$, while the first- and second-order moments of the US data frame reservation time are $\overline{V_u}$ and $\overline{V_u^2}$, respectively. In addition, the US data frames are assumed to arrive at the converged ONU-MPP $o_i$ buffer by right of the Poisson Process and $\rho_u^{h2h} = \lambda_u \cdot \overline{X_u}$ can be expected to be achieved. The US end-to-end delay $\overline{D}_{pon}$ in the backhaul TWDM-PON is decomposed into the average frame queue delay $\overline{Q_u}$, average frame processing service delay $\overline{X_u}$, and frame propagation delay $T_{prop}$, i.e., $\overline{D}_{pon}^{i,k} = \overline{Q_u} + \overline{X_u} + T_{prop}$. Furthermore, the queue delay $\overline{Q_u}$ in the envisioned energy-saving FiWi access network can be written as

$$\overline{Q_u} = \frac{N}{N - \rho_u^{h2h}}\left(\overline{R_u} + \overline{Y_u}\right) \tag{11}$$

where $\overline{R_u}$ and $\overline{Y_u}$ define the mean frame residual delay and the mean frame reservation and vacation delay, respectively. Reference [14] serves as a basis for the mean frame queuing delay $\overline{Q_u}$, which can be rewritten as

$$\begin{aligned}\overline{Q_u} = & \frac{\rho_u^{h2h} \cdot \overline{X_u^2}}{2(N-\rho_u^{h2h}) \cdot \overline{X_u}} + \frac{(1-\rho_u^{h2h}) V_u}{2(N-\rho_u^{h2h})} + \frac{(N-1)^2 \cdot (MV_u+RTT)}{2(N-\rho_u^{h2h}) \cdot (1-\rho_u^{h2h})} \\ & + \frac{[MV_u(N-\rho_u^{h2h}) + (N-1)RTT] \cdot (MN+M-\rho_u^{h2h})}{M(N-\rho_u^{h2h}) \cdot (1-\rho_u^{h2h})} - \frac{(M^2+M-2) V_u}{2M(N-\rho_u^{h2h})}\end{aligned} \tag{12}$$

In the envisioned joint bandwidth allocation and power-saving scheme, the mean US frame end-to-end delay of the optical backhaul subnetwork is computed by adding up the average frame queue delay, average frame processing service delay, and frame propagation delay.

$$\begin{aligned}\overline{D}_{pon}^{i,k} = & \frac{\rho_u^{h2h} \cdot \overline{X_u^2}}{2(N-\rho_u^{h2h}) \cdot \overline{X_u}} + \frac{(1-\rho_u^{h2h}) V_u}{2(N-\rho_u^{h2h})} + \frac{(N-1)^2 \cdot (MV_u+RTT)}{2(N-\rho_u^{h2h}) \cdot (1-\rho_u^{h2h})} \\ & + \frac{[MV_u(N-\rho_u^{h2h}) + (N-1)RTT] \cdot (MN+M-\rho_u^{h2h})}{M(N-\rho_u^{h2h}) \cdot (1-\rho_u^{h2h})} - \frac{(M^2+M-2) V_u}{2M(N-\rho_u^{h2h})} + \overline{X_u} + T_{prop}\end{aligned} \tag{13}$$

Finally, after combining Equations (10) and (13), the end-to-end US frame delay in the envisioned FiWi broadband access network can be derived as follows:

$$\overline{D}_u^{e2e} = \overline{D}_{wmn}^{j,i} + \overline{D}_{pon}^{i,k} \tag{14}$$

From Equation (14), the effect of the total number of STAs $M$, the total number of ONU-MPP $N$, the total number of the hop $H$, the aggregate frame US arrival rate $\lambda_u$, the reservation interval $V_u$, and propagation delay $T_{prop}$ on the end-to-end frame delay is obvious.

To the best of our knowledge, the network QoS in terms of the packet loss ratio increases along with the increased communication distance between the RRHs and end-users in the emerging CRAN, and the QoE value accordingly decreases [25]. Similarly, with the increasing multi-hops between the ONU-MPP and end-users, the mean end-to-end frame delay rises gradually, and the QoE value decreases gradually with the increase in the mean end-to-end frame delay. From the perspective of the network, according to the average data obtained in the test, a corresponding mean opinion score (MOS) value under different QoS conditions is first obtained, and then the numerical calculation software MATLAB is used to fit the results according to the specific function model, and, finally, the expression between the QoE and network delay is obtained. According to the investigation result achieved in Reference [32], the dependence of the QoE value $Q$ on the mean end-to-end frame delay in the FiWi access network $\overline{D}_u^{e2e}$ can be given by

$$Q = 2.03e^{-2.94\overline{D}_u^{e2e}} + 2.39. \tag{15}$$

## 5. Numerical Analysis and Discussion

In this section, the performance evaluation of the PoF-enabled energy harvest over the envisioned FiWi access network, which considers not only the QoE-guaranteed network operation as a function of the end-to-end traffic delay but also the minimization of energy consumption via joint bandwidth allocation and a sleep scheduling scheme, is presented and discussed based on the valid parameter settings.

### 5.1. Parameter Settings

As shown in Table 1, the parameter setting is presented before the experimental simulation. From the perspective of a QoE-guaranteed network operation alongside energy consumption minimization, providing the polling cycle time of the TWDM-PON is vitally challenging, having a relation with the total number of converged ONU-MPP in each branch, the number of STAs accommodated by each ONU-MPP, the reservation interval, the round-trip time between the OLT and ONU-MPP, and the US aggregated traffic load intensity, which are denoted by $N$, $M$, $V_u$, $RTT$, and $\rho_u^{h2h}$, respectively. For the sake of simplicity, each integrated ONU-MPP can accommodate $M$ STAs in our simulation area. On the other hand, the harvested electrical power in the ONU-MPP is designed to guarantee a minimum energy consumption in order to make the broadband access network operate efficiently without an external power supply. Therefore, the transmitted optical power of the OLT $l_k$, i.e., $P_{l_k}^{tx}$ depends on both $\rho_u^{h2h}$ and $N$. In addition, the communication distance $d_{l_k,o_i}$ and fiber attenuation factor $\alpha$ correcting to the optical fiber power loss is separately defined as 100 km and 3 dB/km due to the multimode fiber being deployed in the optical backhaul network. For a network reach of 100 km, the $RTT$ can be assumed to equal 1 ms, and the propagation delay $T_{prop}$ is equal to 0.5 ms. Then, both the photoelectric conversion efficiency $\xi$ and link resource utilization $\beta_{l_k,o_i}$ related to the conversation efficiency equaled 0.6 and 1%, respectively. Given the paramount importance of energy efficiency in both the envisioned FiWi access networks and the traditional PON access network, the power consumption of the ONU-MPP and ONU in the active and sleep state was set as 5552 mW and 5052 mW as well as 758 mW and 750 mW, respectively.

**Table 1.** Parameters settings used in the envisioned FiWi access network.

| Parameter | Description | Value |
|---|---|---|
| $N$ | Number of integrated ONU-MPPs | 16, 32, 64 |
| $M$ | Number of STAs associated with ONU-MPP | 4–20 |
| $H$ | Number of multi-hops | 4–30 |
| $\rho_u^{h2h}$ | US aggregated traffic load | (0, 1) |
| $T_{wl}^{msg}$ | WMN message time | 0.512 μs |
| $T_g$ | Guard time between the consecution STA's subslot | 1–100 μs |
| $T_c$ | Polling cycle time | ms |
| $\overline{X_u}$, $\overline{X_u^2}$ | First- and second-order moments of frame service time | 5.09 μs, 21.44 μs$^2$ |
| $T_{prop}$ | Propagation delay between OLT and ONU-MPP | 0.1–0.5 ms |
| $P^a_{onu\text{-}mpp}$, $P^a_{onu}$ | Power consumption in active state | 5552 mW, 505 mW |
| $P^s_{onu\text{-}mpp}$ $P^s_{onu}$ | Power consumption in sleep state | 758 mW, 750 mW |
| $P^{tx}_{l_k}$ | Transmitted optical power of the OLT $l_k$ | W |
| $d_{l_k,o_i}$ | Communication distance between OLT $l_k$ and ONU-MPP $o_i$ | 100 km |
| $\xi$ | Photoelectric conversion efficiency | 0.6 |
| $\alpha$ | Optical fiber attenuation factor | 3 dB/km |
| $\beta_{l_k,o_i}$ | Link resource utilization between OLT $l_k$ and ONU-MPP $o_i$ | 1% |

## 5.2. Performance Evaluation

Figure 3 illustrates the energy efficiency vs. the total number of emerging integrated ONU-MPP or single traditional ONU for different access network architectures. Clearly, from Figure 3, the energy efficiency firstly increases along with the increasing total number of integrated ONU-MPP or tradition ONU until the number reaches 30, and then rises gradually. In addition, the energy efficiency of the integrated ONU-MPP is superior to that of the traditional access network paradigm. According to a further simplified form of Equation (3), we know that the energy efficiency of the proposed power-saving method entirely depends on the total number of ONU-MPP and the power consumption in different operating states. We can also assume that the maximum energy efficiency of both access networks is 86.3% and 85.2% due to the total $N$ in Equation (3) coming close to positive infinity. More specifically, when $N$ equals 30, the energy efficiency of the two paradigms is 83.5% and 82.3%; when $N$ reaches 60, their energy efficiencies are approximately 86.3% and 85.2%, respectively.

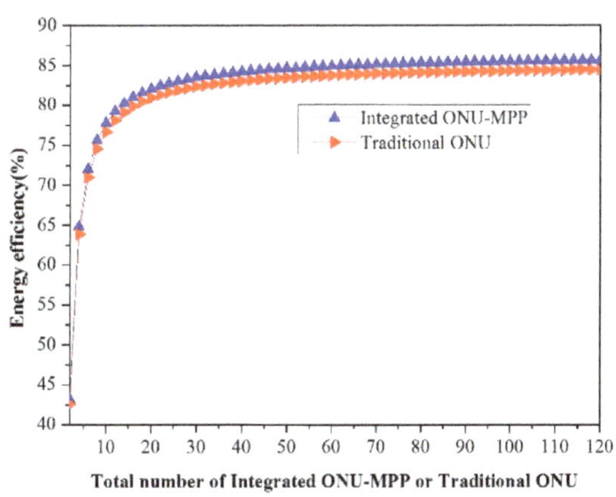


**Figure 3.** Energy efficiency $\eta$ vs. the total number of integrated ONU-MPPs or traditional ONU $N$.

The polling cycle time is calculated for dynamic resource allocation in the envisioned FiWi access networks, in which the dependence of $T_c$ on $N$, $M$, $T_{wl}^{msg}$, $T_g$, $RTT$, and $\rho_u^{h2h}$ is obvious, and Equation (8) provides the in-depth technical guidelines in our simulation. Figures 4 and 5 describe the polling cycle time $T_c$ as a function of $N$, $RTT$, $M$, and $T_g$. Given the fact that $T_{wl}^{msg}$ is constant due to the size of the PS-poll frame, the Beacon frame is set to 64 bytes. Here, for the sake of tractability, both $N$ and $RTT$ are variable in the presence of $M$ and $T_g$ is set to 8 and 0.001 ms, respectively. We can observe from Figure 4 that the polling cycle time $T_c$ rises gradually with the increasing US aggregated traffic load intensity $\rho_u^{h2h}$ for any given configuration. After exceeding the load intensity of 0.5, the polling cycle time $T_c$ sharply increases accordingly. More specifically, the polling cycle time of $RTT$ equaling 0.2 ms is significantly lower than that of the $RTT$ setting of 1 ms in the same amount of ONU-MPPs. In other words, the larger number of the ONU-MPP, or the longer optical network reach, the larger the polling cycle time. It is import to note, however, that $T_c$ could be less than 10 ms when the $N$ is 16, the $RTT$ is set to 0.2 ms, and the load intensity $\rho_u^{h2h}$ is less than 0.7. When the frame delay requested by the user equipment is tolerant, it is not necessary to set the value of $T_c$ lower. On the other hand, when $N$ and $RTT$ is regularly set to 32 and 0.2 ms, respectively, $M$ and $T_g$ is critical to $T_c$, as depicted in Figure 5. With the increase in the US aggregated traffic load intensity, the polling cycle time rises considerably; especially, $T_c$ could be very short, i.e., has less than a 10 m response time, even if the network architecture has 16 STAs accommodated in each ONU-MPP, the guard time is set to 1, and is operating at a utilization factor of 0.6. However, the $T_c$ could be adjusted by extending the reservation interval $V_u$ according to the end-user tolerant frame delay. In addition, as the $T_g$ increase varying from 1 µs to 100 µs , $T_c$ almost increases doubly. In a word, we can notice from Figures 4 and 5 that both $N$ and $RTT$ have a more significant influence on $T_c$ than both $M$ and $T_g$.

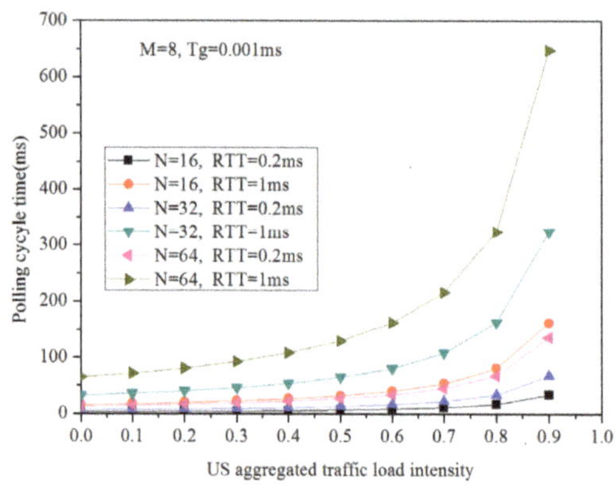


**Figure 4.** $T_c$ vs. $\rho_u^{h2h}$ for variable values of $N$ and $RTT$.

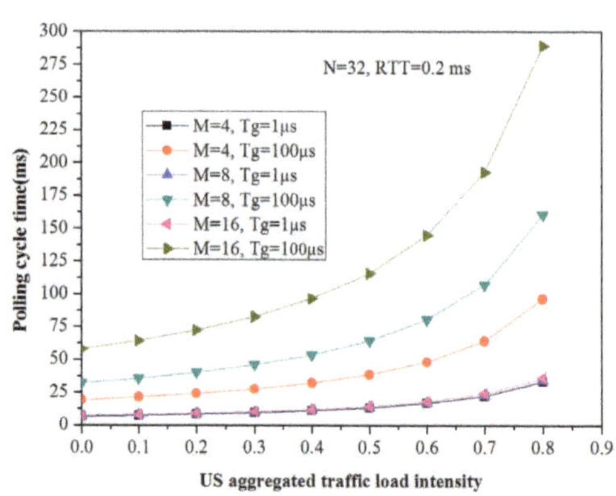


**Figure 5.** $T_c$ vs. $\rho_u^{h2h}$ for variables of $M$ and $T_g$.

In Figure 6, we clearly observe that the average transmitted optical power of the OLT vs. the total number of integrated ONU-MPPs for different QoE values is described in greater detail based on the above Equations (1) and (7). As the total number of integrated ONU-MPPs increases, the averaged transmission optical power of the OLT increases under the same QoE level. This is because the larger value of $N$ will lead to a longer interval of $T_c$, as depicted in Figure 4, which then contributes to a higher energy consumption of the ONU-MPP, and finally being conducive to a higher transmitted optical power, applying the PoF technology to supply the energy of the ONU-MPP. In view of the QoE-guaranteed network operation, the transmitted optical power is not restricted to $N$ but to the QoE level. Similarly, as can be seen from Figure 6, the transmitted optical power increases as the QoE level increases even in the same number of integrated ONU-MPPs. Notably, the larger the value of the QoE level, the shorter the end-to-end frame delay that the US scheduling scheme can achieve from Equation (7), so that the accommodated ONU-MPPs is still active for an extended period of time, and can consume more energy via increasing the power of emitting light. In turn, the transmitted optical power is significantly decreased under the lower QoE level. More specifically, when the values of $N$ exceed 20, the transmitted optical power tends to be stable for the requested QoE level; the result is the same as the description from Figure 2. Meanwhile, the explicit values of the QoE values for the 2nd, 3rd, and 4th levels are defined as 6.55 W, 6.59 W, and 6.62 W, respectively.

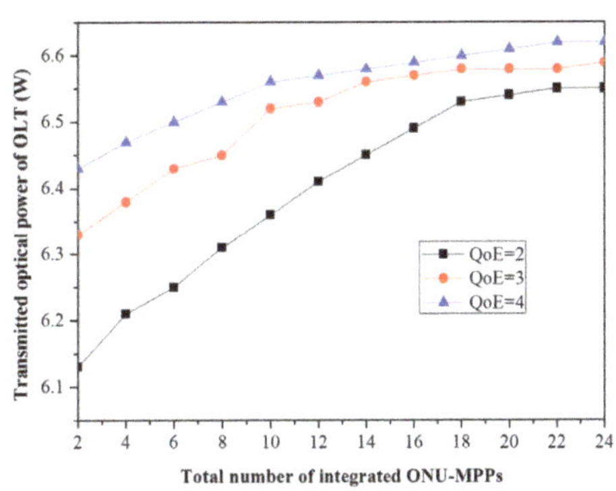


**Figure 6.** Different QoE values.

It is challenging to ensure that our proposed verification method is created with a QoE value as a function with an end-to-end US frame delay in mind. It is worth mentioning that the wireless frontend hops $H$ and uplink load intensity $\rho_u^{h2h}$ codetermines the end-to-end US frame delay. Here, we can assume that the total number of integrated ONU-MPPs, $N$, is 32; the total number of STAs provided by each ONU-MPP $M$ can set as 8; the guard interval $T_g$ is denoted as 1 μs; and the second-order moment of frame service time $\overline{X_u^2}$ equals 21.44 $\mu s^2$. As can be seen from Figure 7, when the US-aggregated traffic load intensity increases, the QoE value decreases gradually; this experimental find was calculated based on Equations (14) and (15). This is because the larger value of the US load intensity makes the end-to-end delay increase, and further decreases the QoE level. Besides, as the number of wireless hops $H$ increases in the same $\rho_u^{h2h}$ situation, the QoE value decreases sharply. For example, when the $\rho_u^{h2h}$ varies from 0.1 to 0.7 by taking into consideration the WMN frontend consisting of a total number hops of 28, the end-to-end frame delay is less than 10 ms when computed via Equation (15), and in turn a QoE value higher than the 4th level under Equation (5). Importantly, the network operators provide mobile users with an acceptable QoE level by ensuring the number of multiple hops subjected to the threshold.

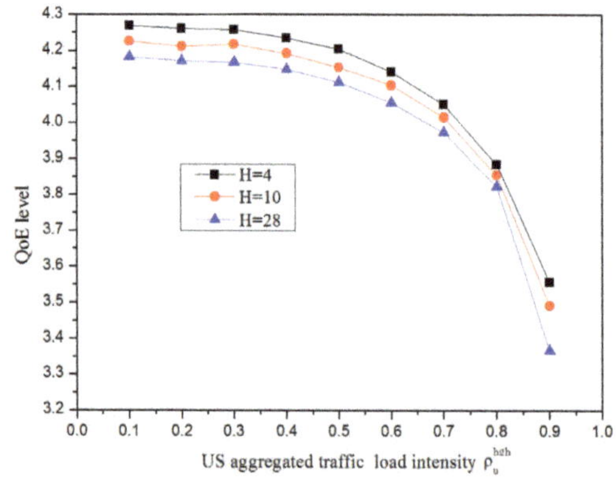


**Figure 7.** Impact of $\rho_u^{h2h}$ on the QoE for different hops.

## 6. Conclusions

In this paper, an envisioned FiWi broadband access network, integrating the WMN frontend subnetwork and TWDM-PON optical backhaul and adapting PoF technology, is proposed. To evaluate network performance, we took the joint bandwidth allocation and ONU-MPP sleep scheduling strategy, PoF-enabled electrical power harvest, and a correlation between multi-hops and the QoE value into consideration. Experimental results show that a business-driven ONU-MPP sleep paradigm outperforms the traditional power-saving scheme. The polling cycle time as a function of US-aggregated traffic load intensity was discussed in detail. In addition, the transmitted optical power can adjust dynamically according to both the QoE level and the total number of integrated ONU-MPPs. Finally, the network operators can achieve an energy-saving target by adjusting the suitable number of hops in the WMN.

**Author Contributions:** This work was a collaborative development by all authors. R.W. provided the funding acquisition, idea, and performance analysis; C.H. wrote the original manuscript and designed the original draft. All authors have read and agreed to the published version of the manuscript.

**Funding:** This work was supported by the NSFC (Grants 61771082, 61801065, and 61871062), and in part by the Science and Technology Research Program of the Chongqing Municipal Education Commission (Grants KJQN201900609, KJQN201800615).

**Acknowledgments:** The authors would like to thank all colleagues who were involved in this study. Specially, we would like to thank Zefu Tan.

**Conflicts of Interest:** The authors declare no conflict of interest.

## References

1. Vereecken, W.; Heddeghem, W.V.; Deruyck, M.; Puype, B.; Lannoo, B.; Joseph, W.; Colle, D.; Martens, L.; Demeester, P. Power Consumption in Telecommunication Networks: Overview and Reduction Strategies. *IEEE Commun. Mag.* **2011**, *49*, 62–69. [CrossRef]
2. Fu, S.; Wen, H.; Wu, J.; Wu, B. Cross-Networks Energy Efficiency Tradeoff: From Wired Networks to Wireless Networks. *IEEE Access* **2017**, *5*, 15–26. [CrossRef]
3. Marcus, M. 5G and "IMT for 2020 and beyond". *IEEE Wirel. Commun.* **2015**, *22*, 2–3. [CrossRef]
4. Maier, M.; Levesque, M.; Ivanescu, L. NG-PONs 1&2 and Beyond: The Dawn of the Uber-FiWi Network. *IEEE Netw.* **2012**, *26*, 15–21.
5. Aurzada, F.; Levesque, M.; Maier, M.; Reisslein, M. FiWi Access Networks Based on Next-Generation PON and Gigabit-Class WLAN Technologies: A Capacity and Delay Analysis. *IEEE/ACM Trans. Netw.* **2014**, *22*, 1176–1189. [CrossRef]

6. Sarkar, S.; Yen, H.H.; Dixit, S.; Mukherjee, B. Hybrid Wireless Optical Broadband Access Network (WOBAN): Network Planning using Lagrangean Relaxation. *IEEE/ACM Trans. Netw.* **2009**, *17*, 1094–1105. [CrossRef]
7. Huang, M.; Chen, Y.; Peng, P.; Wang, H.; Chang, G. A Full Field-of-View Self-Steering Beamformer for 5G mm-Wave Fiber-Wireless Mobile Fronthaul. *J. Lightwave Technol.* **2020**, *38*, 1221–1229. [CrossRef]
8. Dat, P.T.; Kanno, A.; Yamamoto, N.; Kawanishi, T. Seamless Convergence of Fiber and Wireless Systems for 5G and Beyond Networks. *J. Lightwave Technol.* **2019**, *37*, 592–605. [CrossRef]
9. Van, D.P.; Rimal, B.P.; Chen, J.; Monti, P.; Wosinska, L.; Maier, M. Power-Saving Methods for Internet of Things over Converged Fiber-Wireless Access Networks. *IEEE Commun. Mag.* **2016**, *54*, 166–175. [CrossRef]
10. Maier, M.; Levesque, M. Dependable Fiber-Wireless (FiWi) Access Networks and Their Role in a Sustainable Third Industrial Revolution Economy. *IEEE Trans. Reliab.* **2014**, *63*, 386–400. [CrossRef]
11. Sarigiannidis, A.G.; Iloridou, M.; Nicopolitidis, P.; Papadimitriou, G.; Pavlidou, F.-N.; Sarigiannidis, P.G.; Louta, M.D.; Vitsas, V.; Antonios, S.; Maria, I.; et al. Architectures and Bandwidth Allocation Schemes for Hybrid Wireless-Optical Networks. *IEEE Commun. Surv. Tut.* **2015**, *17*, 427–468. [CrossRef]
12. Hou, W.; Ning, Z.; Guo, L. Green Survivable Collaborative Edge Computing in Smart Cities. *IEEE Trans. Ind. Inform.* **2018**, *14*, 1594–1605. [CrossRef]
13. Van, D.P.; Rimal, B.P.; Andreev, S.; Tirronen, T.; Maier, M. Machine-to-Machine Communications Over FiWi Enhanced LTE Networks: A Power-Saving Framework and End-to-End Performance. *J. Lightwave Technol.* **2016**, *34*, 1062–1071. [CrossRef]
14. Raavi, S.; Andrade, M.D.; Fiandra, R.; Tornatore, M. Energy-efficient Design and Equipment Placement for Wireless Optical Broadband Access Networks. In Proceedings of the 2012 IEEE Online Conference on Green Communications (GreenCom), Piscataway, NJ, USA, 25–28 September 2012; pp. 1–6.
15. Van, D.P.; Rimal, B.P.; Maier, M.; Valcarenghi, L. ECO-FiWi: An Energy Conservation Scheme for Integrated Fiber-Wireless Access Networks. *IEEE Trans. Wirel. Commun.* **2016**, *15*, 3979–3994. [CrossRef]
16. Chowdhury, P.; Tornatore, M.; Sarkar, S.; Mukherjee, B. Building a Green Wireless Optical Broadband Access Network (WOBAN). *J. Lightwave Technol.* **2010**, *28*, 2219–2229. [CrossRef]
17. Coimbra, J.; Schtz, G.; Correia, N. Network Game based Routing for Energy Efficient Fibre-Wireless Access Networks. In Proceedings of the IEEE International Conference on Communications, Ottawa, ON, Canada, 10–15 June 2012; pp. 1958–1963.
18. Kantarci, B.; Naas, N.; Mouftah, H. Energy-efficient DBA and QoS in FiWi networks constrained to metro-access convergence. In Proceedings of the 14th International Conference on Transparent Optical Networks, Dalian, China, 24–27 August 2012; pp. 1–4.
19. Schutz, G.; Correia, N. Design of QoS-aware Energy-efficient Fiber Wireless Access Networks. *IEEE/OSA J. Opt. Commun. Netw.* **2012**, *4*, 586–594. [CrossRef]
20. Gong, X.; Hou, W.; Guo, L.; Zhang, L. Dynamic Energy-saving Algorithm in Green Hybrid Wireless Optical Broadband Access Network. *Optik* **2012**, *124*, 1874–1881. [CrossRef]
21. Liu, J.; Guo, H.; Fadlullah, Z.M.; Kato, N. Energy Consumption Minimization for FiWi Enhanced LTE-A HetNets with UE Connection Constraint. *IEEE Commun. Mag.* **2016**, *54*, 56–62. [CrossRef]
22. Guo, H.; Liu, J.; Fadlullah, Z.M.; Kato, N. On Minimizing Energy Consumption in FiWi Enhanced LTE-A HetNets. *IEEE Trans. Emerg. Top. Comput.* **2018**, *6*, 579–591. [CrossRef]
23. Ahmed, M.; Ahmad, I.; Habibi, D. Service Class Resource Management for Green Wireless-Optical Broadband Access Networks (WOBAN). *J. Lightwave Technol.* **2015**, *33*, 7–18. [CrossRef]
24. Miyanabe, K.; Rodrigues, T.G.; Lee, Y.; Nishiyama, H.; Kato, N. An Internet of Things Traffic-Based Power Saving Scheme in Cloud-Radio Access Network. *IEEE Internet Things J.* **2019**, *6*, 3087–3096. [CrossRef]
25. Suto, K.; Miyanabe, K.; Nishiyama, H.; Kato, N.; Ujikawa, H.; Suzuki, K.-I. QoE-Guaranteed and Power-Efficient Network Operation for Cloud Radio Access Network with Power Over Fiber. *IEEE Trans. Comput. Soc. Syst.* **2015**, *2*, 127–136. [CrossRef]
26. Miyanabe, K.; Suto, K.; Fadlullah, Z.M.; Nishiyama, H.; Kato, N.; Ujikawa, H.; Suzuki, K.-I. A cloud radio access network with power over fiber toward 5G networks: QoE-guaranteed design and operation. *IEEE Wirel. Commun.* **2015**, *22*, 58–64. [CrossRef]
27. Togashi, K.; Nishiyama, H.; Kato, N.; Ujikawa, H.; Suzuki, K.-I.; Yoshimoto, N. Cross Layer Analysis on ONU Energy Consumption in Smart FiWi Networks. *IEEE Wirel. Commun. Lett.* **2013**, *2*, 695–698. [CrossRef]
28. Han, P.; Guo, L.; Liu, Y.; Hou, J.; Han, X. Joint Wireless and Optical Power States Scheduling for Green Multi-Radio Fiber-Wireless Access Network. *J. Lightwave. Technol.* **2016**, *34*, 2610–2613. [CrossRef]

29. Nishiyama, H.; Togashi, K.; Kawamoto, Y.; Kato, N. A Cooperative ONU Sleep Method for Reducing Latency and Energy Consumption of STA in Smart-FiWi Networks. *IEEE Trans. Parallel Distrib. Syst.* **2015**, *26*, 2621–2629. [CrossRef]
30. Fadlullah, Z.M.; Nishiyama, H.; Kato, N.; Ujikawa, H.; Suzuki, K.-I.; Yoshimoto, N. Smart FiWi Networks: Challenges and Solutions for QoS and Green Communications. *IEEE Intell. Syst.* **2013**, *28*, 86–91. [CrossRef]
31. Sarkar, S.; Yen, H.-H.; Dixit, S.; Mukherjee, B. A Novel Delay-aware Routing Algorithm (DARA) for a Hybrid Wireless-Optical Broadband Access Network (WOBAN). *IEEE Netw.* **2008**, *22*, 20–28. [CrossRef]
32. Fiedler, M.; Hossfeld, T.; Tran-Gia, P. A Generic Quantitative Relationship between Quality of Experience and Quality of Service. *IEEE Netw.* **2010**, *24*, 36–41. [CrossRef]

sensors

MDPI

Article

# Invalid-Resource-Aware Spectrum Assignment for Advanced-Reservation Traffic in Elastic Optical Network

**Shufang He [1], Yang Qiu [1,*] and Jing Xu [2]**

[1] Key Laboratory of State Ethnic Affairs Commission for Electronic and Information Engineering, College of Electrical & Information Engineering, Southwest Minzu University, Chengdu 610041, China; shufang_hh@163.com

[2] Optical Communications Laboratory, Ocean College, Zhejiang University, Zheda Road 1, Zhoushan 316021, China; jxu-optics@zju.edu.cn

* Correspondence: yqiu@swun.edu.cn

Received: 7 July 2020; Accepted: 27 July 2020; Published: 28 July 2020

check for updates

**Abstract:** Elastic optical networks (EONs) can make service accommodation more flexible and precise by employing efficient routing and spectrum allocation (RSA) algorithms. In order to improve the efficiency of RSA algorithms, the advanced-reservation technique was introduced into designing RSA algorithms. However, few of these advanced-reservation-based RSA algorithms were focused on the unavailable spectrum resources in EONs. In this paper, we propose an Advanced-Reservation-based Invalid-Spectrum-Aware (AR-ISA) resource allocation algorithm to improve the networking performance and the resource alignment of EONs. By employing a new index, Invalid Spectrum Rate (ISR), to record the proportion of unavailable spectrum resources in EONs, the proposed AR-ISA algorithm set a network load threshold to trigger the postponement of an arriving service. Compared with the traditional slack-based AR mechanism, the proposed algorithm has more concerns about the current spectrum usage of the path designated by the service than the conflicts between AR services and other existing services. To further increase the networking performance, the proposed algorithm adopts defragmentation to increase the number of available spectrum resources when postponing a service. Theoretical analysis and simulation results show that the proposed AR-ISA algorithm has obvious effectiveness in reducing the service blocking rate and increasing the spectrum alignment rate.

**Keywords:** elastic optical network; invalid spectrum rate; advanced reservation; defragmentation; blocking probability; spectrum alignment rate

---

## 1. Introduction

Recently, with the emergence and development of a variety of novel network services, such as high definition video delivering and cloud computing, network services have become varied and require more bandwidth. All these increase people's demands for heterogeneous kinds of data transmission, especially for high-speed ones, and the network traffic grows rapidly as a result [1]. Optical network technologies have been considered as a promising solution to high-rate transmission. However, the traditional wavelength division multiplexing (WDM) optical network can hardly adapt to diverse service requests due to its intrinsic limitation in spectrum allocation. Since the WDM optical network allocates constant wavelength to different services with diverse bandwidth requirements, it lacks flexibility in resource allocation and has a limited spectrum utilization [2]. Therefore, the elastic optical network (EON) technique has been proposed to increase the flexibility and efficiency of resource allocation [3]. By employing orthogonal frequency division multiplexing technology [4], EON breaks the rigid allocation limitations of the traditional WDM network and can allocate adaptive spectrum resources to different services according to their individual bandwidth requirements. This can greatly improve the flexibility and efficiency of an optical network.

The routing and spectrum assignment (RSA) algorithm is the key technology to realize the flexibility and efficiency of EONs [5–9]. Therefore, many sophisticated RSA algorithms have been proposed for EONs [10–20]. In [10,11], researchers introduced classic routing strategies and typical resource allocation algorithms, such as the shortest-path-routing strategy and first-fitting or random fitting mechanisms into the RSA design. In [12,13], distance-adaptive modulation techniques were introduced into RSA algorithms to allow the flexible selection of a modulation format for a service request according to its transmission distance. In [14], minimum and maximum entropy-based RSA algorithms were introduced into EONs, which brought about improvement in supporting statically growing services and reduced spectrum fragments. In [15], W.B. Jia et al. applied the prediction of reverse neural network to improve the efficiency of the RSA and reduced the blocking rate of EONs. In [16], the awareness of Quality-of-Transmission was considered in the RSA algorithm to enhance the performance of EONs.

However, these typical RSA algorithms ignored some small, isolated, and unoccupied spectrum bands (a.k.a. spectrum fragments) which remained after the allocation of spectrum resources in EONs [17]. Since the spectrum fragments could hardly be utilized by the subsequent services due to the constraints in resource allocation and thus greatly affected the utilization efficiency of EONs, many spectrum defragmentation algorithms were proposed [18–30]. In [18], a hitless spectrum retuning technique was introduced into the defragmentation algorithm. In [19], a multipath technique was employed to improve the efficiency of the defragmentation algorithms. A push-pull defragmentation algorithm [20] and priority-based defragmentation algorithm [21] were proposed to increase the efficiency of the spectrum defragmentation in EONs. In [22], the researchers solved the problem of serial spectrum defragmentation by employing a dependency graph, which reduced the network spectrum fragments through the original routing and spectrum optimization scheme, and then spectrum defragmentation was performed by the dependency graph. In [23], the authors sorted the free spectrum blocks in the network and selected a spectrum block with a smaller size to provision the existing service so that the spectrum fragments in the network could be reduced. In [24,25], schemes based on an auxiliary graph were proposed to solve the problem of spectrum defragmentation, in which the problem was transformed into the problem of finding the largest independent set in the constructed auxiliary graph. Besides this, fragmentation-aware RSA algorithms [26,27] and group-based RSA algorithms [28–31] were proposed to reduce the spectrum fragments generated in EONs so as to increase their efficiency in resource utilization. Additionally, in [32,33] the survivability issue was considered in designing RSA algorithms.

Although the above RSA algorithms succeeded in increasing the allocation flexibility and utilization efficiency of EONs, most of them neglected the timing characteristics of the service requests in EONs. Thus, the advanced-reservation technique [34] was introduced into RSA algorithms to improve their efficiency in time-scheduling, so that the arriving services can be accommodated at the optimal time and the spectrum resources can be further optimized. Thus, many research works have been carried out on the advanced-reservation-based RSA algorithms [35–42]. In [35], the researchers proposed the concept of spectrum resource usage (SRU), and considered the time domain and frequency domain simultaneously to minimize SRU for the statically scheduled AR requests. Similar to [35], the researchers in [36] introduced a parameter to measure each available frequency-time block of AR requests. Besides this, dynamic scheduling methods were adopted in [37–39]. In [37], the notion of the degree of spectrum-time fragmentation (DSTF) was adopted in dynamic scheduling. In [38], the researchers solved the conflict between the AR and other traditional requests during scheduling. In [39], the request scheduling in the time domain was investigated in addition to the routing, modulation, and spectrum assignment (RMSA). In [40], the researchers proposed an advanced-reservation technique based on flexible time-windows. They explored the effects of different request time-windows and service times without the consideration of time-slots. In [41], a slack-based new mechanism was adopted to allocate resources flexibly in EONs. In such a mechanism, the client specified the slack time and the reservation of resources was carried out within the start time and the slack time. With this approach, flexible service scheduling can be realized. Besides this, in order to describe the spectrum occupation in the time axis, the authors in [42] introduced a concept of Time-Spectrum Consecutiveness (TSC). By calculating the value of TSC for all the arrival times and deadline times and selecting the maximum TSC, the provisioning time and the routing path were optimized for spectrum allocation in EONs. However, the above-mentioned AR algorithms ignored the investigation of the unavailable spectrum resources in EONs, which may have a negative effect on the decision of provisioning time for a service request and thus affect the optimization of the spectrum resources as well as the networking performance (e.g., service blocking probability).

In this paper, we propose an AR-ISA algorithm to improve the networking performance (e.g., service blocking probability) and resource alignment in EONs. By employing a new index, invalid spectrum rate (ISR), to record the proportion of unavailable spectrum resources (including both the occupied and the fragmented ones) in a path, the proposed AR-ISA algorithm set a network load threshold to trigger the postponement of an arriving service. Compared to the traditional slack-based AR mechanism, the proposed algorithm has more concerns about the current spectrum usage of the path selected by the service than the conflicts between the AR services and other existing services. To further increase the networking performance, the proposed algorithm adopts spectrum defragmentation to increase the number of available spectrum resources when postponing a service. The simulation results show that the proposed AR-ISA algorithm helps reduce the service blocking rate and increase the spectrum alignment rate.

The rest of this paper is organized as follows. In Section 2, a network model is introduced for the proposed AR-ISA algorithm. In Section 3, the details of the proposed AR-ISA algorithm are analyzed. In Section 4, the performance of the proposed AR-ISA algorithm is investigated. Finally, this paper is summarized in Section 5.

## 2. The Network Modeling for the Proposed AR-ISA Algorithm

In this section, we introduce the network model for the proposed AR-ISA algorithm. Firstly, we define the physical topology of an elastic optical network as $G(V,E,S)$, in which $V$ represent the set of network nodes, $E$ represents the set of network links, and $S$ represents the set of frequency slots (FSs) on each link. For simplicity, we represent the values of $|V|$, $|E|$, $|S|$ as $N$, $L$, $F$, respectively.

In order to investigate the status of all the FSs in the network, we define a two-dimensional array as follows:

$$U = \begin{bmatrix} u_1^1 & u_2^1 & \cdots & u_i^1 \\ u_1^2 & u_2^2 & \cdots & u_i^2 \\ \vdots & \vdots & \ddots & \vdots \\ u_1^l & u_2^l & \cdots & u_i^l \end{bmatrix}. \quad (1)$$

In (1), $u_i^l$ indicates whether the $i$th FS on the $l$th link is occupied, where I ∈F and l∈L. When $u_i^l = 1$, it indicates that the corresponding FS has been occupied and cannot be allocated to other services until it is released. In contrast, $u_i^l = 0$ indicates that the FS is unoccupied. Thus, we can calculate the quantity of the occupied FSs along a path p via the following equation:

$$U_{op} = \sum u_i^l, \quad (2)$$

where $p(s,d)$ indicates the set of the links along the path $p$ for a service, with s and d representing the source and the destination nodes of the service, respectively.

Given the spectrum continuity and contiguity constraints in the resource allocation, EONs can calculate the required amount of FSs for an arriving service request. More specifically, by setting one FS as 12.5 GHz and employing the distance-adaptive modulation technique, the amount of FSs required by a service $M$ (including one guard band) with data rate $C$ can be calculated as follows:

$$M = \left\lceil \frac{C}{m \times 12.5} \right\rceil + 1, \quad (3)$$

where $m$ represents the modulation level of the service. Noticeably, the values of $m$ can be selected from 1~4 based on the relations between the modulation formats and their corresponding transmission ranges, as in the following Table 1 [43].

**Table 1.** Modulation formats vs. their corresponding transmission range.

| Transmission Range (km) | Modulation Level(bit/Hz) | Modulation Format |
|---|---|---|
| 9600 | 1 | BPSK |
| 4800 | 2 | QPSK |
| 2400 | 3 | 8-QAM |
| 1200 | 4 | 16-QAM |

Note that, due to the spectrum continuity and contiguity constraints in resource allocation, some isolated, unoccupied spectrum bands remain. These remained spectrum bands cannot provide enough available FSs for the subsequent services and thus are known as spectrum fragments. In this paper, we define the spectrum fragments as the unoccupied spectrum bands that contain contiguous FSs less than the minimum value of M, $M_{min}$. The spectrum fragments along a path $p$, $U_{fp}$, can be defined as the sum of all spectrum bands containing continuous vacant FSs less than $M_{min}$ along the path.

Given the unavailability of both the occupied and fragmented FSs in spectrum allocation, we use the new index, ISR, to calculate the proportion of both the occupied and the fragmented FSs along path $p$ to all spectrum resources of this path as follows:

$$ISR_p = \frac{U_{op} + U_{fp}}{F \cdot L(s,d)}, \quad (4)$$

where $L(s,d)$ is the number of all links included in path $p$.

In this paper, we propose the AR-ISA algorithm to improve the networking performance and the resource alignment in EONs according to the above-introduced index, ISR. When the calculated ISR exceeds a pre-set maximum value, network postponement will be triggered by the proposed algorithm. In order to further reduce the amount of unavailable spectrum resources, when a service is about to be blocked spectrum defragmentation is triggered to reduce the fragmented FSs. Since the proposed algorithm can effectively reduce the amount of invalid FSs, more available spectrum resources can be remained for the subsequent service requests. In this way, the service blocking performance can be improved.

## 3. Details of the Proposed Advanced-Reservation-Based Invalid-Spectrum-Aware Resource Allocation Algorithm

In this section, we illustrate the details of the proposed advanced-reservation-based invalid-spectrum-aware resource allocation algorithm. For a more intuitive understanding of the formulas listed above, a simple example is adopted to illustrate the proposed AR-ISA algorithm as in Figure 1. As shown in Figure 1a, the adopted network consists of five nodes and five bidirectional links, with the distance of each link marked. The usage of all the FSs at a certain time is also shown in Figure 1a, assuming that there are totally 23 FSs on each link. In the example, we assume that a new service R1(A, C, 200) arrives, where A and C are its source and destination nodes, respectively, and 200 (in the unit of bit/s) is its required data rate. By selecting the links, L1 and L2, as its routing path, R1 needs to transmit 1100 km from A to C. Therefore, its highest modulation level can be 4 and it requires $M = \frac{200}{4\times12.5} + 1 = 5$ FSs accordingly. When its modulation level is selected as 1, 2, or 3, its required FSs are 7, 9, or 17, respectively. Thus, we regard the unoccupied FS bands that contain less than 5 FSs as fragmented FSs. The amount of the fragmented FSs can be calculated as $U_{fp} = 4 + 14 = 18$ in links L1 and L2. Besides this, the amount of the occupied FSs can be calculated as $U_{op} = 15 + 9 = 24$. Thus, the value of $ISR_p$ is $\frac{U_{op}+U_{fp}}{F \cdot L(s,d)} = \frac{24+18}{23\times2} = 0.913$. Not enough available FSs on links L1 and L2 can be found for R1 at the moment, as shown in Figure 1a. By assuming to trigger the postponement of a service request with an$ISR_p$ larger than 0.7 in this illustrative example, the proposed algorithm may postpone R1 and some occupied FSs may be released, as in Figure 1b. Then, the available FSs on links L1 and L2 can be found for R1, as shown in Figure 1c. Noticeably, spectrum defragmentation can be triggered if there are still unavailable FSs found after postponing services.

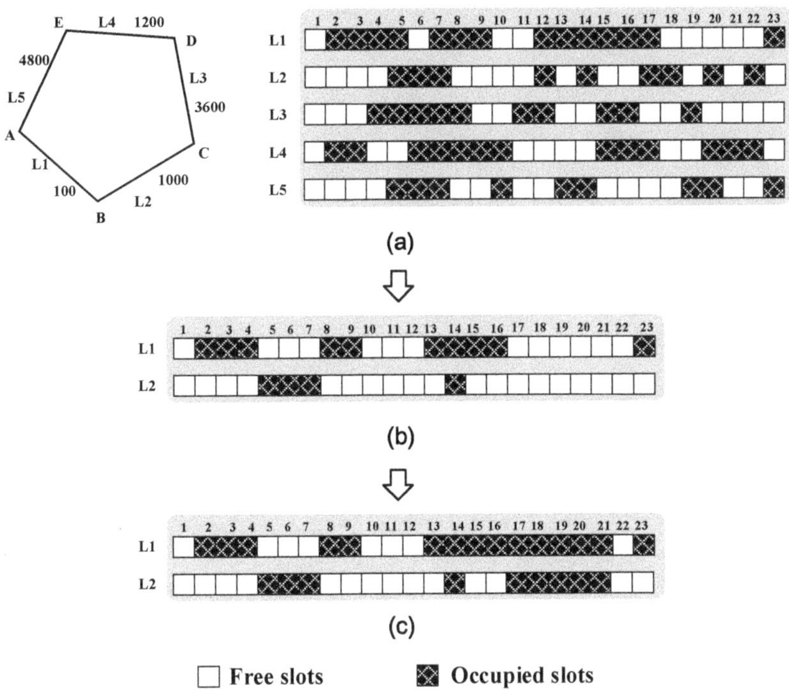


**Figure 1.** An illustrative example for the proposed Advanced-Reservation-based Invalid-Spectrum-Aware (AR-ISA) algorithm. (**a**) The network topology employed and the usage of all frequency slots (FSs) at a certain time, (**b**) after FSs are released on links L1 and L2, (**c**) available FSs on links L1 and L2 found for the service.

The details of the proposed AR-ISA algorithm can he found in the following Algorithm 1. The proposed algorithm calculates K shortest paths for each node pair in the network and sets a maximum invalid spectrum ratio (MISR) before any service arrives. For simplicity, we use R(s, d, B, $t_h$) to represent a service request, in which *s* and *d* represent the source and destination nodes of R, respectively; B is the data rate of R; and $t_h$ is the serving duration of R. When a new service R(s, d, B, $t_h$) arrives, K pre-computed paths are selected as candidate paths for R, and the modulation level as well as the amount of the needed FSs on each path is calculated according to Equation (3). For each candidate path, the proposed algorithm calculates the ISR of the path to trigger the postponement of R if the calculated ISR is larger than the preset MISR. Then, the proposed algorithm tries to seek available FSs on the path. If any available FSs are found on any candidate path, we accommodate R according to the first-found available FSs and their corresponding path. Otherwise, spectrum defragmentation is triggered for R before the spectrum resources on the K candidate paths are investigated again. If no available FSs are found for R, it will be blocked. Otherwise, it will be accommodated by the first-found available FSs. Note that the postpone time in the proposed algorithm is related to the time interval between two successive services.

**Algorithm 1.** Details of the proposed AR-ISA algorithm.

**AR-ISA Algorithm**

```
Compute K shortest-paths for each node pair in the network
Set a MISR for the algorithm
while network is running do
When a service R arrives
  Select K pre-computed paths for R as its candidate paths
  Calculate the modulate level and the amount of the needed FSs on each candidate path for R
  for one path p in the K candidate paths do
    Calculate the ISR along p
    If the calculated ISR is larger than pre-set MISR then
       Postpone R according to the time interval between two successive services
  end if
     Find available FSs on P for R
if available FSs are found on p then
Allocate the found FSs to R
     end if
  break
end for
  If spectrum allocation for R is unsuccessful then
     Do spectrum fragmentation for the existing services
  for one path pin the K candidate paths do
  Find available FSs on P for R
  if available FSs are found on p then
  Allocate the found FSs to R
end if
     break
  end for
  end if
  If spectrum allocation for R is unsuccessful then
     Block R
  end if
  Update the network
end while
```

The time complexity of the AR-ISA algorithm is mainly determined by its adopted path routing algorithm and spectrum allocation strategy. As for the path routing algorithm, the proposed algorithm employs the Dijkstra algorithm and link deleting strategy to find K shortest paths for each node pair. Thus, its complexity can be calculated as $O(K\times|E|\times|V|^2)$, with $|E|$ and $|V|$ representing the amounts of links and nodes in the network. As for the spectrum allocation procedure, its time complexity is mainly determined by the process of finding available FSs for an arrival service and the possible spectrum defragmentation for the existing services at the worst case, and thus can be calculated as $O(2\times K\times|E|^2\times|S|^2 + N\times K\times|E|^2\times|S|^2)$, where $N$ represents the amount of existing services and $|S|$represents the amount of FS on a link. In general, the time complexity of the proposed AR-ISA algorithm is a polynomial. The acronyms used in this paper are summarized in the following Table 2.

**Table 2.** A summary of acronyms.

| Full Name | Abbreviation |
|---|---|
| EONs | Elastic optical networks |
| RSA | Routing and spectrum allocation |
| AR-ISA | Advanced-reservation-based invalid-spectrum-aware |
| ISR | Invalid spectrum rate |
| WDM | Wavelength division multiplexing |
| SRU | Spectrum resource usage |
| DSTF | Degree of spectrum-time fragmentation |
| RMSA | Routing, modulation and spectrum assignment |
| TSC | Time-spectrum consecutiveness |
| FSs | Frequency slots |
| MISR | Maximum invalid spectrum ratio |
| NSFNET | National science foundation network |

## 4. Performance Evaluation

In this section, we perform simulations in DEV-C++to evaluate the effectiveness of the proposed AR-ISR algorithm in the typical national science foundation network (NSFNET), as in Figure 2. In the network, we assume that each link consists of two unidirectional fibers, with each one containing 320 FSs. We also assume that four kinds of services with different required data rates—namely, 50 Gb/s, 100 Gb/s, 150 Gb/s, and 200 Gb/s—are accommodated by the network. Their proportion is set as 1:1:1:1. In order to emulate the dynamic traffic generation, a Poisson traffic model is adopted in the simulation. In the model, a parameter λ is used to emulate the arrival rate of the services, and a negative exponential distribution with parameter μ is used to emulate the duration times of the services. We use the formula $\lambda/\mu$ to indicate the traffic load in the network, as in [44]. In the generation of each service, its source and destination, as well as its data rate, are randomly determined from their sample spaces. Noticeably, in this simulation Formula (3) can be used to calculate that the number of FSs required by various services is 2,3,4,5,7,9,13, and 17. Table 3 summarizes all the parameter settings in the simulations.

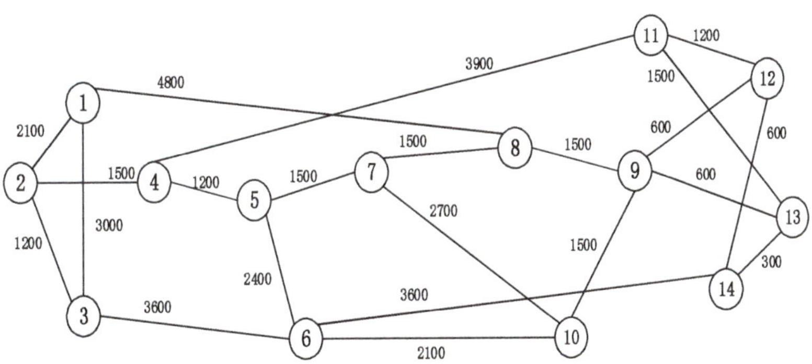


**Figure 2.** National science foundation network (NSFNET) topology.

In order to verify the effectiveness of the proposed AR-ISR algorithm in NSFNET, the proposed algorithm and an AR algorithm based on ISR without defragmentation, are investigated in the simulation. In the simulation, we set the delayed time as 0.5 times the arrival interval between two services, which is randomly generated according to the above-mentioned Poisson model. Figure 3 to Figure 4 illustrate the simulation results of the proposed algorithm on the service blocking probability and spectrum alignment rate when the MISR changes from 0.1 to 0.5. In the simulation, the service blocking probability is defined as the ratio of blocked services to all the generated services, while the spectrum alignment rate is defined as the proportion of vacant FSs with the same index on all links.

Specially, the curve MISR = ∞ represents the traditional distance-adaptive routing and spectrum allocation algorithm, as in [10].

**Table 3.** Parameter settings in simulations.

| Parameters | Settings |
| --- | --- |
| Network topology | NSF-Net |
| K, number of candidate paths | 3 |
| F, number of frequency slots | 320 |
| Bandwidth of one frequency slot | 12.5 GHz |
| Number of guard band | 1 |
| Optional modulation formats | BPSK, QPSK, 8-QAM, 16-QAM |
| Bits per symbol of different modulation formats | 1, 2, 3, 4 |
| Reachable distance of different modulation formats (km) | 9600, 4800, 2400, 1200 |
| Type of services (Gb/s) | 50, 100, 150, 200 |
| Proportion of different types of services | 1:1:1:1 |
| Arrival rate of services | Poisson distribution model |
| Duration time of each service | Negative exponential distribution |
| Number of services per simulation | $5 \times 10^6$ |

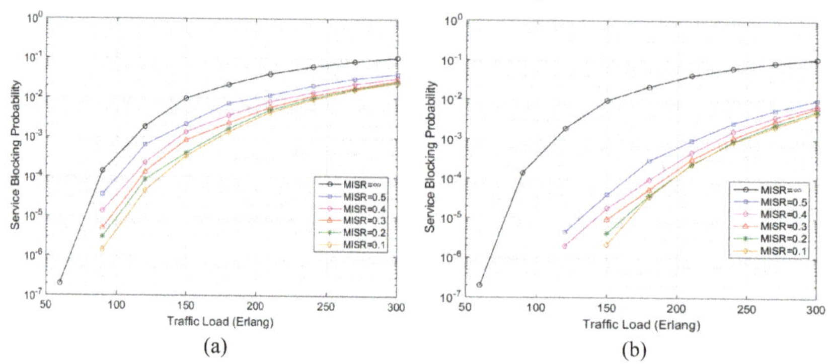


**Figure 3.** Simulation results of the service blocking probability. (**a**) Simulation results on the service blocking probability vs. traffic load with different values of maximum invalid spectrum ratio (MISR) when no spectrum defragmentation is adopted, (**b**) simulation results on the service blocking probability vs. traffic load with different values of MISR when spectrum defragmentation is adopted.

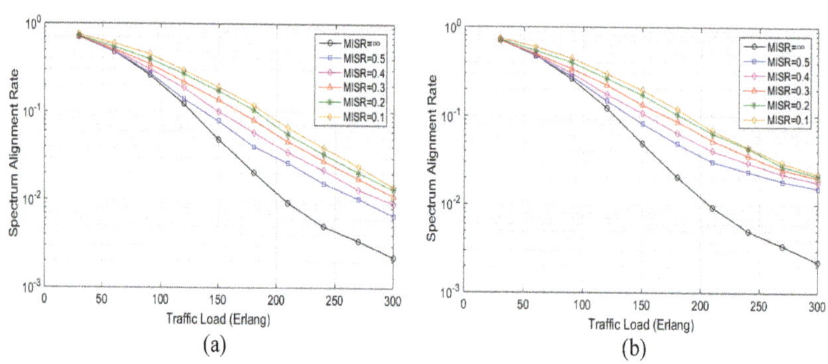


**Figure 4.** Simulation results of the spectrum alignment rate. (**a**) Simulation results of the spectrum alignment rate vs. the traffic load with different values of MISR when no spectrum defragmentation is adopted, (**b**) simulation results of the spectrum alignment rate vs. the traffic load with different values of MISR when spectrum defragmentation is adopted.

Figure 3a shows the simulation results of the service blocking probability with different values of MISR when no spectrum defragmentation is adopted in the proposed AR-ISR algorithm. By using MISR = ∞ to represent the classic distance-adaptive RSA algorithm [12], the proposed AR-ISR algorithm (with no spectrum defragmentation) gains an obvious advantage over the classic distance-adaptive algorithm in reducing the service blocking probability, as shown in Figure 3a, although such an advantage may come to saturation with the MISR decreasing from 0.5 to 0.1. For instance, the proposed AR-ISR algorithm with no spectrum defragmentation can reduce the service blocking probability by 77.55% compared to the distance-adaptive algorithm, with MISR being set as 0.5 when the traffic load is 150 Erlang. Such a reduction in the service blocking probability reaches 96.49%, with the MISR being set as 0.1 when the traffic load is 150 Erlang. As for the high traffic load (e.g., 300 Erlang), the proposed AR-ISR algorithm with no spectrum defragmentation can still reduce the service blocking probability by 61.76% compared to the distance-adaptive algorithm with the MISR being set as 0.5. Such a reduction in the service blocking probability reaches 77.03% with the MISR being set as 0.1. All the above-mentioned simulation results indicate that the proposed AR-ISR algorithm can effectively reduce the amount of blocked services by postponing the accommodation of a service according to invalid spectrum rate in the network, even without any spectrum defragmentation. It can be understood by the fact that by employing the invalid-resource-aware mechanism, the proposed AR-ISR algorithm paves a way to optimize the service scheduling and thus increases the networking performance due to the increased number of available FSs.

Figure 3b shows the simulation results of the service blocking probability with different values of MISR when spectrum defragmentation is adopted in the proposed AR-ISR algorithm. Similar to Figure 3a, the proposed AR-ISR algorithm gains an obvious advantage over the classic distance-adaptive algorithm in reducing the service blocking probability, although such an advantage may come to saturation with the MISR decreasing from 0.5 to 0.1. For instance, the proposed AR-ISR algorithm (with spectrum defragmentation) can reduce the service blocking probability by 99.55% compared to the distance-adaptive algorithm, with the MISR being set as 0.5 when the traffic load is 150 Erlang. Such a reduction in the service blocking probability reaches 99.97%, with the MISR being set as 0.1 when the traffic load is 150 Erlang. When the traffic load reaches as high as 300 Erlang, the proposed AR-ISR algorithm (with spectrum defragmentation) can still reduce the service blocking probability by 90.50% compared to the distance-adaptive algorithm, with the MISR being set as 0.5. Such a reduction in the service blocking probability reaches 95.47% with the MISR being set as 0.1. The above simulation results verify that by employing the spectrum defragmentation mechanism, the proposed AR-ISR algorithm can further reduce the amount of blocked services by reducing the unavailable spectrum resources and improve the service blocking performance of the network. All the simulation results in Figure 3a,b verify that the proposed AR-ISR algorithm can effectively reduce the service blocking probability by adopting invalid resource awareness in service scheduling and spectrum defragmentation in service allocation.

Figure 4a depicts the simulation results of the spectrum alignment rate performance with different values of MISR when no spectrum defragmentation is adopted in the proposed AR-ISR algorithm. By using MISR = ∞ to represent the classic distance-adaptive RSA algorithm [12], the proposed AR-ISR algorithm (with no spectrum defragmentation) gains an obvious advantage over the classic distance-adaptive algorithm in increasing the spectrum alignment rate, especially with a high traffic load, as shown in Figure 4a. For instance, when the traffic is as low as 90 Erlang, the proposed AR-ISR algorithm (with no spectrum defragmentation) can enhance the spectrum alignment rate by 5.34% compared to the distance-adaptive algorithm, with the MISR being set as 0.5. Such an improvement in the spectrum alignment rate reaches 72.7% with the MISR being set as 0.1. When the traffic load reaches as high as 300 Erlang, the proposed AR-ISR algorithm (with no spectrum defragmentation) can increase the spectrum alignment rate by 196.77% compared to the distance-adaptive algorithm, with the MISR being set as 0.5. Such an improvement in spectrum alignment rate reaches 528.67% with the MISR being set as 0.1. All the simulation results indicate that the proposed AR-ISR algorithm can effectively improve the spectrum alignment rate by reducing the invalid spectrum resources even without spectrum defragmentation, especially with a high traffic load. The higher improvement in spectrum alignment rate with a higher traffic load can be understood by the fact that more occupied and fragmented FSs are generated with higher traffic loads, which makes the services' postponement more likely.

Figure 4b depicts the simulation results of the spectrum alignment rate performance with different values of MISR when spectrum defragmentation is adopted in the proposed AR-ISR algorithm. Similar to Figure 4a, the proposed AR-ISA algorithm (with spectrum defragmentation) gains an obvious advantage over the classic distance-adaptive algorithm in increasing the spectrum alignment rate, especially with a high traffic load. For instance, when the traffic is as low as 90 Erlang, the proposed AR-ISA algorithm (with spectrum defragmentation) can enhance the spectrum alignment rate by 6.37% compared to the distance-adaptive algorithm, with the MISR being set as 0.5. Such an improvement in the spectrum alignment rate reaches 71.12% with the MISR being set as 0.1. When the traffic load reaches as high as 300 Erlang, the proposed AR-ISA algorithm (with spectrum defragmentation) can increase the spectrum alignment rate by 580.32% compared to the distance-adaptive algorithm, with the MISR being set as 0.5. Such an improvement in the spectrum alignment rate reaches 881.6% with the MISR being set as 0.1. All the simulation results imply that the proposed AR-ISA algorithm can improve the spectrum alignment rate more effectively by employing the spectrum defragmentation mechanism to reduce the fragmented FSs, especially with a high traffic load. All the simulation results in Figure 4a,b verify that the proposed AR-ISA algorithm can effectively improve the spectrum alignment rate by adopting invalid resource awareness in service scheduling and spectrum defragmentation in service allocation. Noticeably, fiber nonlinearities may distort the performance of the proposed AR-ISA algorithm in EONs, especially for high-rate long-range transmission [45]. Some nonlinear compensation techniques, such as full-field nonlinear compensation with probabilistically shaped constellations [46], single-channel digital back-propagation [47], optical phase conjugation [48], nonlinearity-tailored detection [49], and nonlinear Fourier transform [50], can be adopted in the EONs to guarantee the effectiveness of the proposed AR-ISA algorithm.

## 5. Conclusions

In this paper, we propose an AR-ISA algorithm to improve the networking performance and the resource alignment of EONs. By employing a new index, ISR, to record the proportion of unavailable spectrum resources (including both the occupied and the fragmented ones) along a path, the proposed AR-ISA algorithm set a network load threshold to trigger the postponement of an arriving service. Compared to the traditional AR algorithm, the proposed algorithm has more concerns about the current spectrum usage of the path the service selected. To further optimize the unavailable resources, the proposed algorithm adopts defragmentation to increase the number of available spectrum resources when postponing a service. The simulation results verify that by employing ISR to optimize the postponement of the service accommodation and the defragmentation technique to increase the available FSs, the proposed AR-ISA algorithm improves the service blocking performance and the spectrum alignment rate.

**Author Contributions:** Conceptualization, S.H. and Y.Q.; methodology, S.H.; software, S.H.; validation, S.H., Y.Q., and J.X.; formal analysis, S.H. and Y.Q.; investigation, S.H. and Y.Q.; resources, S.H., Y.Q. and J.X.; data curation, S.H. and Y.Q.; writing—original draft preparation, S.H.; writing—review and editing, S.H., Y.Q., and J.X.; visualization, S.H.; supervision, Y.Q.; project administration, Y.Q.; funding acquisition, Y.Q. All authors have read and agreed to the published version of the manuscript.

**Funding:** This work was partly supported by the National Natural Science Foundation of China (Nos. 61705190 and 61971378); the Fundamental Research Funds for the Central Universities, Southwest Minzu University (2020PTJS19004); the Fundamental Project of Science and Technology Department of Sichuan Province (No.2018JY0338); and the Graduate Innovative Research Projects(No. CX2020SZ88).

**Acknowledgments:** We appreciate the reviewers' suggestions on this paper.

**Conflicts of Interest:** The authors declare no conflict of interest.

## References

1. Sahu, H.; Hungyo, M. Introduction to SDN and NFV. In *Innovations in Software-Defined Networking and Network Functions Virtualization*; IGI Global: Hershey, PA, USA, 2018; pp. 1–25.
2. Hsu, C.F.; Chang, Y.C.; Sie, S.C. Graph-model-based dynamic routing and spectrum assignment in elastic optical networks. *IEEE/OSA J. Opt. Commun. Netw.* **2016**, *8*, 507–520. [CrossRef]
3. Gerstel, O.; Jinno, M.; Lord, A.; Yoo, S. Elastic optical networking: A new dawn for the optical layer? *IEEE Commun. Mag.* **2012**, *50*, s12–s20. [CrossRef]
4. Christodoulopoulos, K.; Tomkos, I.; Varvarigos, E. Elastic bandwidth allocation in flexible OFDM-based optical networks. *J. Lightwave Technol.* **2011**, *29*, 1354–1366. [CrossRef]
5. Liu, X.; Gong, L.; Zhu, Z. Design integrated RSA for multicast in elastic optical networks with a layered approach. In Proceedings of the IEEE Global Communications Conference (GLOBECOM), Atlanta, GA, USA, 9–13 December 2013; pp. 2346–2351.
6. Wang, Y.; Cao, X.; Pan, Y. A study of the routing and spectrum allocation in spectrum-sliced elastic optical path networks. In Proceedings of the IEEE INFOCOM, Shanghai, China, 10–15 April 2011; pp. 1503–1511.
7. Mahala, N.; Thangaraj, J. Spectrum Assignment using Prediction in Elastic Optical Networks. In Proceedings of the 2019 10th International Conference on Computing, Communication and Networking Technologies (ICCCNT), Kanpur, India, 6–8 July 2019.
8. Yaghubi-Namaad, M.; Rahbar, A.G.; Alizadeh, B. Adaptive modulation and flexible resource allocation in space-division- multiplexed elastic optical networks. *IEEE/OSA J. Opt. Commun. Netw.* **2018**, *10*, 240–251. [CrossRef]
9. Arpanaei, F.; Ardalani, N.; Beyranvand, H.; Alavian, S.A. Three-Dimensional Resource Allocation in Space Division Multiplexing Elastic Optical Networks. *IEEE/OSA J. Opt. Commun. Netw.* **2018**, *10*, 959–974. [CrossRef]
10. Zheng, W.; Jin, Y.; Sun, W.; Guo, W.; Hu, W. On the spectrum efficiency of bandwidth-variable optical OFDM transport networks. In Proceedings of the Conference on Optical Fiber Communication (OFC/NFOEC), Collocated National Fiber Optic Engineers Conference, San Diego, CA, USA, 21–25 March 2010.
11. Chatterjee, B.C.; Sarma, N.; Oki, E. Routing and Spectrum Allocation in Elastic Optical Networks: A Tutorial. *IEEE Commun. Surv. Tutor.* **2015**, *17*, 1776–1800. [CrossRef]
12. Jinno, M.; Kozicki, B.; Takara, H.; Watanabe, A.; Sone, Y.; Tanaka, T.; Hirano, A. Distance-adaptive spectrum resource allocation in spectrum-sliced elastic optical path network. *IEEE Commun. Mag.* **2010**, *48*, 138–145. [CrossRef]
13. Ding, Z.; Xu, Z.; Zeng, X.; Ma, T.; Yang, F. Hybrid routing and spectrum assignment algorithms based on distance-adaptation combined co-evolution and heuristics in elastic optical networks. *Opt. Eng.* **2014**, *53*, 046105. [CrossRef]
14. Wright, P.; Parker, M.C.; Lord, A. Minimum- and maximum-entropy routing and spectrum assignment for flexgrid elastic optical networking. *IEEE/OSA J. Opt. Commun. Netw.* **2015**, *7*, A66–A72. [CrossRef]
15. Jia, W.B.; Xu, Z.Q.; Ding, Z.; Wang, K. An efficient routing and spectrum assignment algorithm using prediction for elastic optical networks. In Proceedings of the International Conference on Information System and Artificial Intelligence (ISAI), HongKong, China, 24–26 June 2016.
16. Beyranvand, H.; Salehi, J.A. A Quality-of-Transmission aware dynamic routing and spectrum assignment scheme for future elastic optical networks. *J. Lightwave Technol.* **2013**, *31*, 3043–3054. [CrossRef]
17. Patel, A.N.; Ji, P.N.; Jue, J.P.; Wang, T. Defragmentation of transparent Flexible optical WDM (FWDM) networks. In Proceedings of the Optical Fiber Communication Conference and Exposition and the National Fiber Optic Engineers Conference, Los Angeles, CA, USA, 6–10 March 2011.
18. Zhang, M.; Yin, Y.; Proietti, R.; Zhu, Z.; Yoo, S.J.B. Spectrum defragmentation algorithms for elastic optical networks using hitless spectrum retuning techniques. In Proceedings of the Optical Fiber Communication Conference and Exposition and the National Fiber Optic Engineers Conference (OFC/NFOEC), Anaheim, CA, USA, 17–21 March 2013.
19. Chen, X.; Jukan, A.; Gumaste, A. Multipath de-fragmentation: Achieving better spectral efficiency in elastic optical path networks. In Proceedings of the IEEE INFOCOM, Turin, Italy, 14–19 April 2013; pp. 390–394.

20. Cugini, F.; Paolucci, F.; Meloni, G.; Berrettini, G.; Secondini, M.; Fresi, F.; Sambo, N.; Poti, L.; Castoldi, P. Push-pull defragmentation without traffic disruption in flexible grid optical networks. *J. Light. Technol.* **2013**, *31*, 125–133. [CrossRef]
21. Zhang, J.; Zhao, Y.; Yu, X.; Zhao, J. Priority-based defragmentation scheme in spectrum-efficient optical networks. In Proceedings of the The 10th International Conference on Optical Internet (COIN2012), Yokohama, Kanagawa, Japan, 29–31 May 2012.
22. Zhang, M.; You, C.; Jiang, H.; Zhu, Z. Dynamic and adaptive bandwidth defragmentation in spectrum-sliced elastic optical networks with time-varying traffic. *J. Light. Technol.* **2014**, *32*, 1014–1023. [CrossRef]
23. Naik, B.K.; Dutta, R.; Shashidhara, G.; Gowrishankar, R.; Sanagapati, S.S.S.; Bhyri, S.K.; Behere, P.P. Defragmentation of flexible dense wavelength division multiplexing (FDWDM) networks using wavelength tunability criterion. In Proceedings of the IEEE International Conference on Advanced Networks and Telecommunications Systems(ANTS), Bangalore, India, 6–9 November 2016; pp. 1–6.
24. Yin, Y.; Wen, K.; Geisler, D.J.; Liu, R.; Yoo, S.J.B. Dynamic on-demand defragmentation in flexible bandwidth elastic optical networks. *Opt. Express* **2012**, *20*, 1798–1804. [CrossRef] [PubMed]
25. Shakya, S.; Cao, X. Spectral Defragmentation in Elastic Optical Path Networks using Independent Sets. In Proceedings of the Optical Fiber Communication Conference and Exposition and the National Fiber Optic Engineers Conference (OFC/NFOEC), Anaheim, CA, USA, 17–21 March 2013.
26. Sone, Y.; Hirano, A.; Kadohata, A.; Jinno, M.; Ishida, O. Routing and spectrum assignment algorithm maximizes spectrum utilization in optical networks. In Proceedings of the 37th European Conference and Exhibition on Optical Communication, Geneva, Switzerland, 18–22 September 2011.
27. Wang, R.; Mukherjee, B. Spectrum management in heterogeneous bandwidth networks. *Opt. Switch. Netw.* **2014**, *11*, 83–91. [CrossRef]
28. Qiu, Y.; Xu, J. Efficient Hybrid Grouping Spectrum Assignment to Suppress Spectrum Fragments in Flexible Grid Optical Networks. *J. Light. Technol.* **2017**, *35*, 2823–2832. [CrossRef]
29. Qiu, Y. An efficient spectrum assignment algorithm based on variable-grouping mechanism for flexgrid optical networks. *Opt. Switch. Netw.* **2017**, *24*, 39–46. [CrossRef]
30. Qiu, Y. Group-based spectrum assignment in dynamic flex-grid optical networks. *Opt. Fiber Technol.* **2013**, *19*, 437–445. [CrossRef]
31. Qiu, Y. Resource-partitioned spectrum assignment to realize efficient multicasting for flexible grid optical networks. *Opt. Fiber Technol.* **2019**, *53*, 102067. [CrossRef]
32. Ruan, L.; Xiao, N. Survivable multipath routing and spectrum allocation in OFDM-based flexible optical networks. *IEEE/OSA J. Opt. Commun. Netw.* **2013**, *5*, 172–182. [CrossRef]
33. Liu, M.; Tornatore, M.; Mukherjee, B. Survivable traffic grooming in elastic optical networks—Shared protection. *J. Lightwave Technol.* **2013**, *31*, 903–909. [CrossRef]
34. Patel, A.; Jue, J. Routing and scheduling for variable bandwidth advance reservation. *IEEE/OSA J. Opt. Commun. Netw.* **2011**, *3*, 912–923. [CrossRef]
35. Chen, H.; Zhao, Y.; Zhang, J. Static provisioning for advanced reservation in elastic optical networks. In Proceedings of the 2017 16th International Conference on Optical Communications and Networks (ICOCN), Wuzhen, China, 7–10 August 2017.
36. Li, X.; Yuan, J.; Zhang, Q.; Ren, Z.; Yang, L. Farsighted Spectrum Resource Assignment Method for Advance Reservation Requests in Elastic Optical Networks. *IEEE Access* **2019**, *7*, 167836–167846. [CrossRef]
37. Li, Y.; Zhang, Q.; Xin, X.; Tian, Q.; Liu, N.; Tao, Y.; Cao, G.; Tian, F.; Wang, Y.; Wang, Z.; et al. Dynamic Service Provisioning Algorithm Based on Degree of Spectrum-Time Fragmentation of AR Requests in Elastic Optical Networks. In Proceedings of the 2019 18th International Conference on Optical Communications and Networks (ICOCN), Huangshan, China, 5–8 August 2019.
38. Sugihara, S.; Hirota, Y.; Fujii, S.; Tode, H.; Watanabe, T. Dynamic Resource Allocation for Immediate and Advance Reservation in Space-Division-Multiplexing-Based Elastic Optical Networks. *IEEE/OSA J. Opt. Commun. Netw.* **2017**, *9*, 183–197. [CrossRef]
39. Lu, W.; Zhu, Z. Dynamic Service Provisioning of Advance Reservation Requests in Elastic Optical Networks. *J. Light. Technol.* **2013**, *31*, 1621–1627. [CrossRef]
40. Kaushik, N.R.; Figueira, S.M.; Chiappari, S.A. Flexible Time-Windows for Advance Reservation Scheduling. In Proceedings of the 14th IEEE International Symposium on Modeling, Analysis, and Simulation, Monterey, CA, USA, 11–14 September 2006; Volume 1, pp. 218–225.

41. Hu, C.; Huai, J.; Wo, T. Flexible Resource Reservation Using Slack Time for Service Grid. In Proceedings of the 12th International Conference on Parallel and Distributed Systems—(ICPADS'06), Minneapolis, MN, USA, 12–15 July 2006; Volume 1, pp. 327–334.
42. Chen, H.; Zhao, Y.; Zhang, J.; He, R.; Wang, W.; Wu, J.; Wang, Y.; Ji, Y.; Zheng, H.; Lin, Y.; et al. Time-Spectrum Consecutiveness Based Scheduling with Advance Reservation in Elastic Optical Networks. *IEEE Commun. Lett.* **2015**, *19*, 70–73. [CrossRef]
43. Bocoi, A.; Schuster, M.; Rambach, F.; Kiese, M.; Bunge, C.A.; Spinnler, B. Reach-dependent capacity in optical networks enabled by OFDM. In Proceedings of the Conference on Optical Fiber Communication—Incudes Post Deadline Papers, San Diego, CA, USA, 22–26 March 2009.
44. Fan, Z.; Qiu, Y.; Chan, C.K. Dynamic Multipath Routing with Traffic Grooming in OFDM Based Elastic Optical Path Networks. *J. Lightwave Technol.* **2015**, *33*, 275–281. [CrossRef]
45. Zhao, J.; Wymeersch, H.; Agrell, E. Nonlinear impairment-aware static resource allocation in elastic optical networks. *J. Lightwave Technol.* **2015**, *33*, 4554–4564. [CrossRef]
46. Semrau, D.; Xu, T.; Shevchenko, N.A.; Paskov, M.; Alvarado, A.; Killey, R.I.; Bayvel, P. Achievable information rates estimates in optically amplified transmission systems using nonlinearity compensation and probabilistic shaping. *Opt. Lett.* **2017**, *42*, 121–124. [CrossRef]
47. Karanov, B.; Xu, T.; Shevchenko, N.A.; Lavery, D.; Killey, R.I.; Bayvel, P. Span length and information rate optimisation in optical transmission systems using single-channel digital backpropagation. *Opt. Express* **2017**, *25*, 25353–25362. [CrossRef]
48. Fabbri, S.; Stephens, M.F.; McCarthy, M.E.; Perentos, A.; Phillips, I.D.; Lavery, D.; Liga, G.; Maher, R.; Harper, P.; Doran, N.J.; et al. 4 Tbit/s transmission reach enhancement using 10 × 400 Gbit/s super-channels and polarization insensitive dual band optical phase conjugation. *J. Lightwave Technol.* **2016**, *34*, 1–7.
49. Liga, G.; Alvarado, A.; Agrell, E.; Secondini, M.; Killey, R.I.; Bayvel, P. Optimum detection in presence of nonlinear distortions with memory. In Proceedings of the 2015 European Conference on Optical Communication (ECOC), Valencia, Spain, 27 September–1 October 2015.
50. Prilepsky, J.E.; Derevyanko, S.A.; Blow, K.J.; Gabitov, I.; Turitsyn, S.K. Nonlinear Inverse Synthesis and Eigenvalue Division Multiplexing in Optical Fiber Channels. *Phys. Rev. Lett.* **2014**, *113*, 013901. [CrossRef] [PubMed]

sensors

MDPI

Article

# Hybrid Routing, Modulation, Spectrum and Core Allocation Based on Mapping Scheme

**Edson Rodrigues** *,†,‡, **Eduardo Cerqueira** ‡, **Denis Rosário** ‡ and **Helder Oliveira** ‡

Federal University of Pará, Belém 66075-110, Brazil; cerqueira@ufpa.br (E.C.); denis@ufpa.br (D.R.); heldermay@ufpa.br (H.O.)

* Correspondence: edson.rodrigues@itec.ufpa.br

† Current address: Rua Augusto Corrêa 01, Belém 66075-110, Brazil.

‡ These authors contributed equally to this work.

Received: 30 August 2020; Accepted: 5 October 2020; Published: 9 November 2020

**Abstract:** With the persistently growing popularity of internet traffic, telecom operators are forced to provide high-capacity, cost-efficient, and performance-adaptive connectivity solutions to fulfill the requirements and increase their returns. However, optical networks that make up the core of the Internet gradually reached physical transmission limits. In an attempt to provide new solutions emerged, the Space-Division Multiplexing Elastic Optical Network emerged as one of the best ways to deal with the network depletion. However, it is necessary to establish lightpaths using routing, modulation, spectrum, and core allocation (RMSCA) algorithms to establish connections in these networks. This article proposes a crosstalk-aware RMSCA algorithm that uses a multi-path and mapping scheme for improving resource allocation. The results show that the proposed algorithm decreases the blocking ratio by up to four orders of magnitude compared with other RMSCA algorithms in the literature.

**Keywords:** Elastic Optical Network; space division multiplexing; routing

## 1. Introduction

The constant growth in the number of Internet users and greater applications with high bandwidth demands have made the traffic passing through the backbone of the Internet increase exponentially in recent years [1]. Furthermore, it is expected that the traffic will continue to grow, pushing the current network to its physical limit [2]. Despite Wavelength-Division Multiplexing (WDM) technology meeting the demand for a period, it does not allow the spectrum of the network to be used in the best possible way due to its rigid coarse granularity, leading to a waste in spectrum caused by the size of requests that do not fit precisely to frequency slots [3]. To overcome the limitations of the transport data network that employs WDM, the Elastic Optical Network (EON) has been proposed, which has increased flexibility for the allocation of a bandwidth [4,5]. In EONs, the fixed spectrum is allocated in small units called slots, allowing flexible allocation to accommodate the connections' bandwidth demands, grouping several of these units. In this way, the residual band is minimized, leading to more efficient use of the spectrum, and separating each lightpath established by a band called Filter Guard Band (FGB) [6].

To substantially expand the network's physical capacity and increase the possibilities of data transport, the spacial strategy has been explored through the use of multi-core fibers, implementing Space-Division Multiplexing (SDM) technology to the EONs [7,8]. The increase in the flexibility and availability of resources, with the advent of SDM-EON technology, gives rise to new problems, which need to be solved for the network's best use. One of these problems is related to the routing and allocation of resources. The addition of more cores on fiber increases the complexity in allocating resources as it adds a dimension to be considered when choosing paths for network requests. In SDM-EON, two constraints must be considered to establish a connection from a source to the destination: the continuity of frequency slots, which checks whether the requested slots are continuous in all links in the path, avoiding core switching, and contiguity, which ensures that all slots are allocated in contiguous bands of the spectrum same core [4].

When allocating resources in SDM-EON, in addition to the continuity and contiguity constraints, the transmission rate can also be considered, i.e., increasing the bit rate encoded by each optical symbol. The factor that limits the use of modulation formats is their sensitivity to losses, thus the greater the number of bits per symbol, the greater the probability of an error in its decoding [9]. This effect makes transmissions over a longer distance more complex due to losses caused by signal attenuation and interference. From this concept, the use of adaptive modulation in optical networks arose, able to regulate the number of bits per symbol to be used according to the distance to be traveled [9]. The greater the number of bits transmitted per symbol, the smaller the signal's reach. The multi-core utilization and adaptive modulation bring a new problem to resource allocation for requests, the Routing, Modulation, Spectrum, and Core Allocation (RMSCA) problem [4].

Allocating requests that preserve continuity and contiguity conditions in a single path can be difficult when the traffic load is high. In allocating and releasing spaces, requests create small slices of bandwidth in the spectrum that are not enough for some requests, rejecting those that cannot be accommodated on the network. In an attempt to increase the number of connections established and improve the spectrum's use, providing more than one path for a request presents good results compared to using only one path [10]. The use of multi-paths presents challenges known as the time of arrival of each path. To deal with this problem, the differential delay is added as one of the path selection conditions, where the difference in arrival time between the first and the last path must respect the acceptable limit [11].

This paper's contribution is the introduction of an RMSCA algorithm called Hybrid Routing, Modulation, Spectrum and Core Allocation based on the Mapping Scheme in SDM-EONs (*PANORAMIC*). The proposed algorithm aims to reduce the number of blocked requests using hybrid routing and adaptive modulation. We call hybrid routing a policy, which first tries to allocate a single path for a request and then allocates multiple paths only if the allocation of a single path is impossible. The proposed algorithm uses the spectrum mapping scheme to find paths that can accommodate requests that use adaptive modulation. Finally, we simulated an online and dynamic scenario, which showed the benefits of the *PANORAMIC* in terms of blocking bandwidth and Energy Efficiency while keeping the crosstalk per slot and fragmentation at an acceptable level compared to state-of-the-art RMSCA algorithms. The results obtained here extend those found in preliminary investigation [12]. While previous work [12] proposes an RMSCA algorithm using single routing, we propose an RMSCA algorithm using hybrid routing in this paper. The results present that the proposed hybrid lightpath provisioning mechanism outperforms the traditional single-path routing by reducing the blocking ratio while reducing traffic that may be affected by single link failures.

This paper is organized as follows. In Section 2, we give an overview of the related work. Section 3 presents the networks used and its physical features and describes how the algorithm works. Section 4 is the performance evaluation for the algorithm. Section 5 presents the conclusions and future works.

## 2. Related Work

The space-division multiplexing elastic optical networks have driven several studies to increase the allocation of resources. However, to the best of our knowledge, this is the first work that considered multi-path routing and delay differentials to solve the routing, modulation, spectrum, and core allocation problem in SDM-EON.

Zhu et al. [13] propose a dynamic Routing, Modulation Level, and Spectrum Assignment (RMLSA) based on both online and offline path computation with several path selection policies. The provisioning resources use a hybrid single-/multi-path routing scheme. However, they do not consider delay differential, which may lead to an unreal case where the network can support any multi-path. Moreover, the article is not concerned with the increasing current traffic and consequently does not consider fibers with multiple cores.

Monghaddam et al. [14] present a Mixed Integer Linear Programming (MILP) and a heuristic algorithm for dealing with the crosstalk-aware RMCSA and scheduling problem. In their scenario, they consider static traffic, and paths are calculated in advance using the K-shortest path algorithm. The MILP makes sure that only one single path is assigned for each call, then adaptive modulation is applied, and the strict crosstalk for each slot in the path is calculated. Then, the maximum modulation is selected if the crosstalk threshold allows the connection without any signal degradation. The objective is to reduce the resources used by reducing the number of slots allocated. The heuristic equilibrates the load on links as the use of frequency slices index. Despite using the spatial dimension, they do not consider the multi-path as a strategy to increase the number of acceptances and reduce the waste in the spectrum and use a spectrum scan that does not favor the best use of the spectrum.

Yin et al. [15] introduce a survivable crosstalk-aware multi-path strategy in SDM-EONs considering multi-path protection. They used the super-channels approach, utilizing cost-saving by reducing the number of required slots lasers when the transponders architecture of spatial super-channel. They used Bhandari's disjoint path algorithm to calculate multiple routes in advance. Then, they verified that a connection can be established, in case the crosstalk for the path is acceptable. However, the authors disregard differential delay and multiple modulation formats to transmission, causing a spectrum wastage.

Oliveira and da Fonseca [16] introduce an algorithm that provides paths exclusively when a protection path can also be found. For such a case, they adopt the single -ath strategy at first, then, in the case where a single path cannot be allocated, the adoption of multi-paths happens. In both cases, Shared Backup Path Protection (SBPP) is utilized to guarantee the network's survivability. They use Dijkstra's algorithm to find the shortest path for both single- and multi-path cases, but they do not consider adaptive modulation, utilizing only one modulation format, which leads to spectrum underutilization, contributing to increasing the blocking probability.

Yousefi et al. [17] introduce two state-of-the-art novel metrics: holding time and coefficient of the variant metric. Based on these metrics, they propose three algorithms to solve the fragmentation problem and improve blocking probability in the Routing, Spectrum, and Core Allocation (RSCA) problem. The proposed algorithms utilize the spectrum waste base algorithm to find a suitable rectangle for each connection to fit the K-Shortest Path, considering the fragmentation in each algorithm in order to reduce the fragmentation ratio. Adaptive modulation and multi-path are not used in algorithms proposed by the authors.

Yousefi et al. [18] introduce six algorithms aiming to solve the fragmentation problem and improve blocking probability, dividing the connection into multiple paths. The algorithms consider External Fragmentation Metric and splitting of lightpaths to reduce blocking probability. When fragmentation is not an obstacle, a connection may be established if the K-shortest path algorithm finds an available path. They use the FindFirstFreeGap algorithm to allocate frequency slots to a request. Their algorithms work for Routing and Spectrum Assignment (RSA) for dynamic scenarios, but the RMSCA problem cannot be supported because they do not use any core or adaptive modulation.

Yousefi and Rahbar [19] introduce three novel algorithms to solve fragmentation problems and improve the blocking probability for SDM-EON. The algorithms can reduce blocking probability by controlling fragmentation in cores, since they try different ways of assigning a spectrum for a connection. However, they only consider the RSCA problem without considering different modulations, and only one algorithm utilizes a multi-path mechanism, but differential delay constraint is neglected.

Yousefi and Rahbar [20] develop three algorithms—MINCROSS, MINFRAG, and MODFRAGCROSS—to solve crosstalk and maintain the physical layer security level and reduce blocking probability. The authors consider the impairment and physical layer security leak attack caused by inter-core crosstalk. They try to control blocking probability by preventing fragmentation and crosstalk and used an algorithm called ExactFit to find a free gap that can allocate the request utilizing the exact number of frequency slots. In MINCROSS, they prioritize the crosstalk as the main condition when selecting the path, while in MINFRAG they consider fragmentation as first conditional. In MODFRAGCROSS, they keep both crosstalk and fragmentation at acceptable levels. Although the work considers crosstalk per-slot and fragmentation metrics, they do not apply modulation and multi-path, so the spectrum is not used in the best way.

Zhu et al. [21] propose a multi-path, fragmentation-aware routing modulation and spectrum assignment for advance and immediate reservations. The authors proposed splitting advance and immediate reservation requests into smaller parts, transferring each of these parts along with single- or multi-paths, taking into account the delay differential addressing with additional electronic buffering in the higher layer of the destination node. They propose a time and spectrum fragmentation occurrence measurement. They applied adaptive modulation but did not use a multi-core fiber.

Moghaddam et al. [22] formulate a mixed-integer, linear programming MILP in networks with dedicated and shared path protection schemes. A heuristic algorithm to solve a large scale problem is also proposed. The proposed algorithms support the network with both dedicated and shared path protection and consider the crosstalk strictly. The K-shortest path calculates the path between two points with adaptive modulation, but the multi-path mechanism is not utilized in this approach.

Moura and Da Fonseca [23] propose four RMSCA algorithms based on image processing algorithms. They utilize two image processing algorithms—Connected Component Labeling (CCL) and Inscribed Rectangles Algorithm (IRA)—to reduce computational complexity. For ImageRCMLSA, CCL was used combined with the best fit policy for reducing spectrum waste. Modulations are calculated by the second of the transmission distance and the estimated crosstalk, and then a fitting policy is applied. However, requests that cannot find a single lightpath are not allowed to split into paths to find available paths.

Dharmaweera et al. [24] present an analytical optimization formulation for small networks and a heuristic for large networks to increase spectral efficiency. Both use the multi-path technique to allocate requests that could not find a single path from point to another, and also apply traffic grooming to avoid the waste caused by the number of guard bands utilized in optical connections. The authors use the Yen Shortest Path algorithm to find the path to calls. However, they did not consider adaptive modulations during the calculation between source to destination.

Malekzadeh and Shahkooh [25] introduce an ILP and a heuristic to routing, modulation, and spectrum allocation in EONs. In ILP, they aim to minimize the total spectrum utilized in the network, utilizing the K-shortest path algorithm to find the shortest path. The proposed heuristic uses the single- and multi-path routing. A threshold to calls with multi-path is set and, if this limit is passed, then the request is blocked. However, they did not consider the space dimension in their work.

Table 1 compares the algorithm proposed in this paper to various others described in the literature concerning the use of space-division multiplexing, the use of adaptive modulation, the employment of the multi-path, and consideration of the delay differential. Based on our analysis of the state-of-the-art, we conclude that few studies consider the study of delay differential when multi-path is used. To the best of our knowledge, this is the first article to solve the routing, modulation, spectrum, and core allocation problem through multi-path considering the differential delay.

**Table 1.** Related works comparison with the *PANORAMIC* algorithm.

| Approach | SDM | Modulation | Multi-path | Delay Differential |
|---|---|---|---|---|
| Zhu et al. [13] | | ✓ | ✓ | |
| Moghaddam et al. [14] | ✓ | ✓ | | |
| Yin et al. [15] | ✓ | | ✓ | |
| Oliveira and da Fonseca [16] | ✓ | | ✓ | |
| Yousefi et al. [17] | ✓ | | | |
| Yousefi et al. [18] | | | ✓ | |
| Yousefi and Rahbar [19] | ✓ | | ✓ | |
| Yousefi and Rahbar [20] | ✓ | | | |
| Zhu et al [21] | | ✓ | ✓ | ✓ |
| Moghaddam et al. [22] | ✓ | ✓ | | |
| Moura and Da Fonseca [23] | ✓ | ✓ | | |
| Dharmaweera et al. [24] | ✓ | | ✓ | ✓ |
| Malekzadeh and Shahkooh [25] | | ✓ | ✓ | |
| *PANORAMIC* | ✓ | ✓ | ✓ | ✓ |

## 3. Algorithm

This section introduces Hybrid Routing, Modulation, Spectrum, and Core Allocation based on Mapping Scheme (*PANORAMIC*). The algorithm aims to deal with the RMLSCA problem, utilizing a mapping scheme that verifies the topology with all frequency slots in each core, trying to find a free gap capable of fitting a request. When the requirements are met, the connection is established. In the routing process, modulation formats are chosen according to the transmission distance, allowing signal decoding.

### *3.1. Network and System Model*

We adopted an optical network operating with a spatially flexible reconfigurable optical add/drop multiplexers capable of a wavelength-selective switch, and space-wavelength granularity, with Multiple-Input Multiple-Output (MIMO) transceivers. Furthermore, we consider Multi-Flow Transponders (MFTs) to split into multiple sub transponders to facilitate allocation.

The network topology is composed of bidirectional fibers with seven cores arranged in a hexagonal array. Each of these cores has 320 slices of frequency slots with a transmission capacity of 12.5 Gbps. We represent the network model considering actual distances from one point to another. We kept the continuity and contiguity restrictions, which means that the equipment does not allow core- and slots-switching to occur during the transmission. The calculation of the number of slots required for transmission is done based on the number of bits per symbol that the modulation applied to the transmission can transmit. Thus, the greater the modulation, the greater the amount of data transmitted in a slot, reducing the total number of slots used for transmission. In addition, a slot is always allocated for the FGB, which separates the transmissions.

Decoding the signal is essential for the transmission to be carried out successfully. Thus, several modulations were used in order to improve the use of the network spectrum. However, the modulation choice must be made, so that the maximum distance that the signal can arrive with quality that can be read. The most efficient modulation level for the transmission must be selected according to its characteristics [26]. Table 2 shows the modulation characteristics considered in our network scenario.

**Table 2.** Modulation Characteristics.

| Modulation Level | # Bits Per Symbol | Slot Capacity (Gb/s) | Maximum Distance (km) |
|---|---|---|---|
| 64QAM | 6 | 75 | 125 |
| 32QAM | 5 | 62.5 | 250 |
| 16QAM | 4 | 50 | 500 |
| 8QAM | 3 | 37.5 | 1000 |
| QPSK | 2 | 25 | 2000 |
| BPSK | 1 | 12.5 | 4000 |

### 3.2. Algorithm PANORAMIC

*PANORAMIC* is an RMLSCA algorithm that performs in SDM-EONs, with different loads, scenarios, and topologies. The algorithm's objective is to increase the number of connections established in the network and improve the use of spectrum, avoiding the waste of frequency slots that can be idle after several connections and disconnections. The algorithm does not switch cores during the transmission, to prevent signal conversion from optical to electrical, keeping the continuity and contiguity restrictions. The Mapping Scheme used in our proposal offers a map of the available resources in the network, leading to an increase in the number of paths. The algorithm receives the network's traffic as the input and, after running the routing, returns the possible lightpath with sufficient resources to accommodate requests. Table 3 shows the notation that will be used to describe the algorithm.

**Table 3.** Notation.

| Notation | Definition |
|---|---|
| $s$ | Source node |
| $d$ | Destination node |
| $b$ | Demand in slots |
| $e \in E$ | Link from the network |
| $v \in V$ | Node from the network |
| $c \in C$ | Core from the network |
| $s \in S$ | Slot from the network |
| $r(s,d,b)$ | Request from $s$ to $d$ with bandwidth demand of $b$ |
| $m \in M$ | The set of modulations $M = \{1,2,3,4,5,6\}$ according to the Table 2 |
| $g(v,e,c,s)$ | A virtual graph that maps the slot $s$ across the network |
| $G = \{g(v,e,c,s)\}$ | The set of virtual graphs |
| $p \in P$ | Path for each request, whether single or multi-path |
| $J_{max}$ | Delay Differential Threshold |
| $Map_{cs}$ | Map Matrix for each fiber link |
| $b_m$ | Number of slots required for transmission according to the modulation applied. |

Algorithm 1 receives a request and, with information such as source node, destination node, and bandwidth, then starts the routing process. Although it is a hybrid routing and multiple paths can be used, the PANORAMIC first looks for a single path in the network that can support the transmission. In Line 1, the algorithm is tested for each modulation level, then the number of sufficient frequency slots is calculated based on the number of bits that the modulation can transmit (Line 2). Then, in Line 3, the mapping algorithm is called for, and the number of frequency slots required is used as a parameter for the mapping, which return a set of graphs generated, which are essential to the process of finding the shortest available path for transmission. If the path is found, the connection is accepted and, soon after, established. Otherwise, a similar process is taken to find multi-paths for the connection. For the multi-path, modulations are tested in the previous process for a single path, from the highest to the lowest. The data are then divided into two, and the mapping takes place for each of the blocks to be transmitted. The mapping algorithm checks the spectrum of the network and returns the available spaces capable of supporting the transmission. Then, the difference in time of arrival of the paths is calculated and, according to what is established in [27], the difference between the paths cannot be greater than 15 ms, or 3000 Km, thus, if the difference between the paths is greater than the upper-bound, the connection is not accepted.

**Algorithm 1:** PANORAMIC

```
Input  : r(s,d,b), V,E,C,S,M
Output: Lightpath
for all m ∈ M do
    b_m: Estimate demand in slots for modulation;
    G = Mapping(V,E,C,S,b_m);
    for all g ∈ G do
        P: Find the shortest available path on mapped network;
        if P ≠ ø then
            Accept request;
            return established path;
        end
    end
end
if path = ø then
    for all m ∈ M do
        b_m: Estimate demand in slots for modulation;
        Split demand in two paths;
        G = Mapping(V,E,C,S,b_m);
        for all g ∈ G do
            P = Find the two disjoint shortest available paths on mapped network;
            if | p_1 − p_2 ∈ P| < J_max then
                Accept request;
                return established path;
            end
        end
    end
end
Block request;
return;
```

In the Mapping Algorithm (Algorithm 2), the network spectrum is mapped in a binary matrix format, where contiguous slots are scanned and, if they can accommodate the request, the initial index is mapped as available. Figure 1 represents three states of the mapping for a specific link (Line 1), with core-to-core sequence mapping (Line 2). First, represented in red, the block of slots required for the requisition is checked from left to right (Line 3) and, as all contiguous slots are available, that is, these slots can accommodate the requisition, it is represented in the mapping as available in Line 6. In the second case, represented in blue, only two of the three required slots are available in the main matrix, so this set is represented as unavailable (Line 9) in the mapped matrix, signaling that, although some slots are still available, they do not fully support the request. The third case, represented in green, shows where all the slots are occupied, so this set is mapped in the matrix as unavailable.

**Algorithm 2:** MAPPING

```
   Input  : V, E, C, S, d_m
   Output : Mapped network
1  for all e ∈ E do
2      for all c ∈ C do
3          for all s ∈ S do
4              Map_cs: Binary Matrix Mapped;
5              if all s + b_m slots are available then
6                  Map_cs = True;
7              end
8              if s + b_m aren't available then
9                  Map_cs = False;
10             end
11             G = g(v, e, c, s);
12         end
13     end
14 end
15 return G;
```

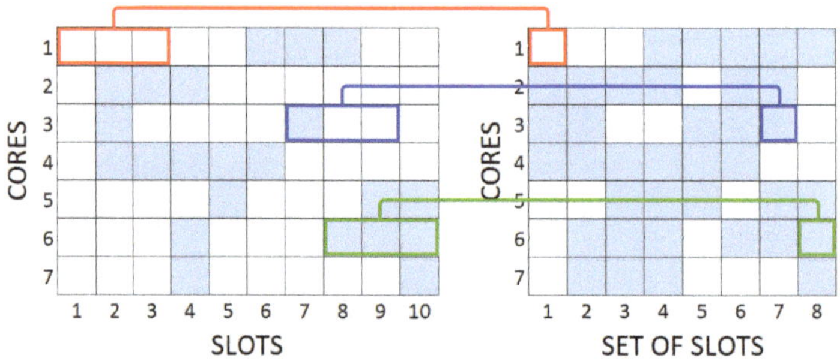


**Figure 1.** Spectrum Matrix and Mapped Matrix.

Then in Line 11, for each of the mapped slots, a different graph is created to indicate the links where the core and slots continuity can be obeyed, this way, when a mapped matrix slot is unavailable, the generated graph does not recognize that mapped link as part of the topology, represented in Figure 2 by the red link. The set of graphs generated return to the main algorithm in Line 15, where the choice of the shortest path for transmission will be made, causing only the available paths to be found by the algorithm.

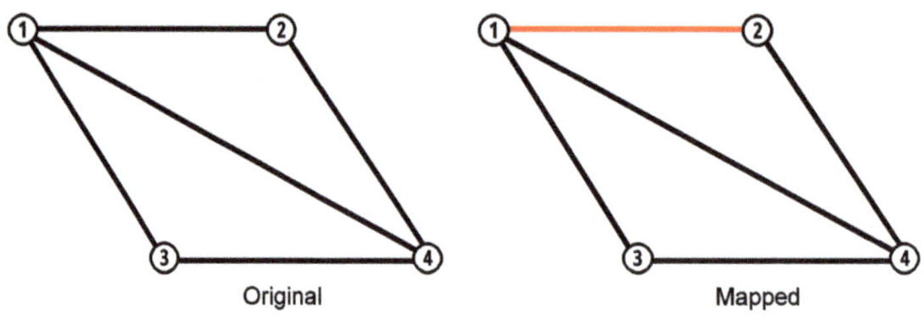


**Figure 2.** Mapped Graph.

The complexity of the *PANORAMIC* algorithm is analyzed as follows: The complexity of mapping the network is $O(E+V)$. For a single path, in the worst case, the chosen modulation will be the one with the lowest spectral efficiency, with it being necessary to transform the multigraph into graphs $M$ times. In the worst case, to find the path, the Dijkstra algorithm runs $M$ times in $C \times (N - b_m)$ graphs, where Dijkstra's amortized complexity is $O(E + VlogV)$, obtaining a complexity of $O(E + VlogV)$. The algorithm will not find a single path in the worst case, making it necessary to find a multipath. For this, the Suurballe algorithm is run $M$ times in $C \times (N - b_m)$ graphs, where Suurballe complexity is $O(E + VlogV)$. As $M$, $N$, $C$ and $b_m$ can be expressed as constants where the complexity of the *PANORAMIC* algorithm is $O(E + VlogV)$.

## 4. Evaluation Performance

In this section, we present the evaluation for the proposed algorithm *PANORAMIC*, compared to other works in the literature with simulation parameters, with metrics involved at the evaluation for different scenarios.

### *4.1. Simulation Description and Metrics*

To assess the performance of the RMSCA algorithms, simulation experiments were conducted employing the FlexGridSim [28]. Requests followed a Poisson process and were uniformly distributed between all pairs of nodes. At least ten replications were generated for each scenario. We varied the network load from 50 to 1000 erlangs, and we used 100,000 connection requests in each simulation. Confidence intervals were derived using the independent replication method, and a 95% confidence level was adopted. Simulations for the different algorithms used the same set of seeds. Seven types of requests were employed by 25/50/125/200/500/750/1000 Gbps. The used fiber is composed of seven cores, and each fiber core was divided into 320 slots. The topologies used in the simulations were the USA (Figure 3), and NSF (Figure 4). The USA topology has 24 nodes, and 43 links and the NSF topology has 14 nodes and 19 links. The numbers on the links represent the length of the link in kilometers.

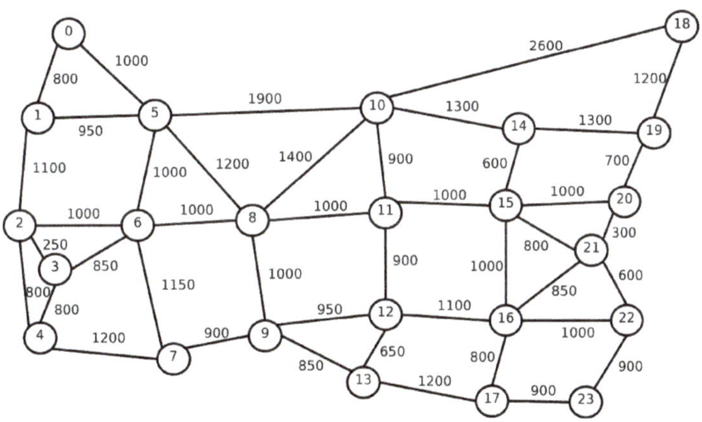

Figure 3. USA Topology.

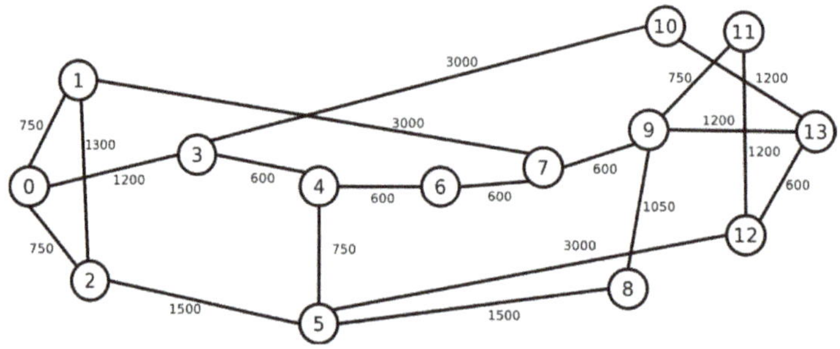


**Figure 4.** NSF Topology.

The metrics considered for evaluation of the algorithms are the Bandwidth Blocking Ratio (BBR), the Crosstalk per Slot (CpS), Energy Efficiency (EE), Fragmentation Ratio, and Modulation Format Percentage (MFP). The BBR value is calculated by the sum of all blocked bandwidth ($\alpha$) during the simulation period, divided by the sum of all requested bandwidth ($\beta$) over the same period. The *BBR* is denoted by Equation (1).

$$BBR = \frac{\sum \alpha}{\sum \beta} \tag{1}$$

The *CpS* is the result of the division between the number of allocated slots that have the same frequency ($\gamma$), both in the core being considered and in the adjacent cores of a link, and the total number of slots used in the same link ($\varphi$), calculated by periodic divisions ($T$) over periods of the same interval. *CpS* is denoted by Equation (2).

$$CpS = \frac{\sum \frac{\sum \gamma}{\sum \varphi}}{T} \tag{2}$$

The *EE* calculation is made by dividing the sum of all bandwidth values (in Mbps) of the requests accepted in the network ($\gamma$), by the sum of the energy consumption values (in Joules) of the transponders, switches and optical amplifiers present in the routes of each accepted request ($\epsilon$). The *EE* is denoted by Equation (3).

$$EE = \frac{\sum \gamma}{\sum \epsilon} \tag{3}$$

The *FR* calculation is made by the maximum number of contiguous slots available ($\mu$) to the number of available slots in the link ($\delta$). The *FR* is denoted by Equation

$$FR = \frac{Max(\sum \mu)}{\sum \delta} \tag{4}$$

The *MFP* is calculated by incrementing a counter for each time the modulation was used in a transmission ($T_m$); in the case of using multi-path, it was added for each of the disjoint paths. In the end, the number of times each modulation was used was divided by the total number of transmissions (*Total*) in single or multi-path. The *MFP* is denoted by Equation

$$MFP = \frac{T_m}{Total} \tag{5}$$

### 4.2. Simulation Results

In this subsection's figures, the curves labeled *PANORAMIC* show the results generated by the *PANORAMIC* algorithm proposed in this article. The curves with the label MINFRAG present the results obtained using the algorithm proposed in [20]. The curves labeled MINCROSS illustrate the results regarding the use of the algorithm proposed in [20]. The curves labeled ImageRCMLSA present the results regarding the use of the algorithm proposed in [23].

Figures 5 and 6 present the BBR for USA and NSF topologies. For USA topology (Figure 3), while the *PANORAMIC* algorithm starts blocking requests only under loads of 600 erlangs, the ImageRCMLSA, MINCROSS and MINFRAG algorithms start blocking requests of about 150, 50 and 50 erlangs, respectively. For NSF topology (Figure 4), *PANORAMIC* algorithm starts blocking requests under loads of 525 erlangs, performing better than ImageRCMLSA, MINFRAG, and MINCROSS algorithms. While ImageRCMLSA starts blocking requests under loads of 200 erlangs, MINFRAG and MINCROSS algorithms start blocking requests under 50 erlangs. The results shown emphasize the efficiency of the mapping scheme used in the *PANORAMIC* algorithm. The mapping increases the number of free paths found in the network compared to the method used in ImageRCMLSA, MINCROSS, and MINFRAG, which offers a simpler finding path model. The K-Shortest Path (KSP) algorithm used in these algorithms considers the whole network, not only the free paths, taking the situation where the path found is already occupied. This means that, despite the network congestion, our algorithm still accommodates requests by allowing the demand to be divided into multiple-paths, where before it would not be possible to transmit using a single path, increasing the number of accepted requests as well as the quantity transmitted data, in addition to enhancing the efficiency of spectral use. However, due to the high number of connections established on the network, the interference caused by crosstalk tends to increase, but proposed algorithm has satisfactory results, as can be seen later. Topology USA has high connectivity between nodes, which favors the use of the disjoint paths strategy, while NSF Topology does not have this characteristic. However, the results showed similar curves for both topologies.

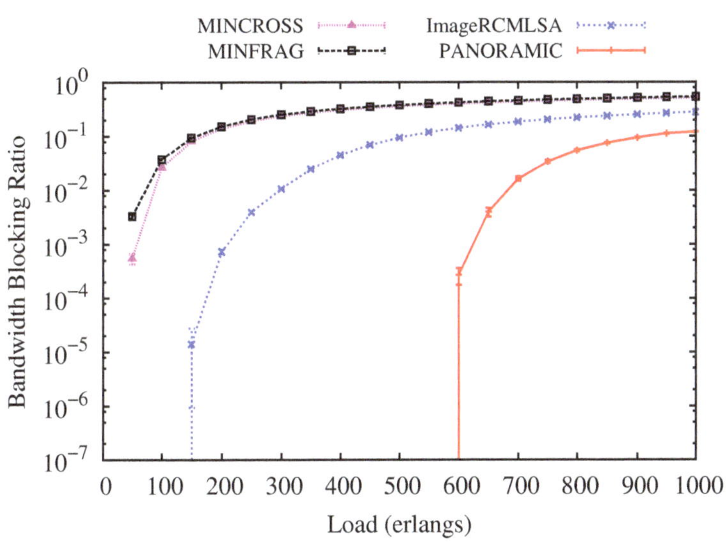


**Figure 5.** Bandwidth Blocking for USA topology.

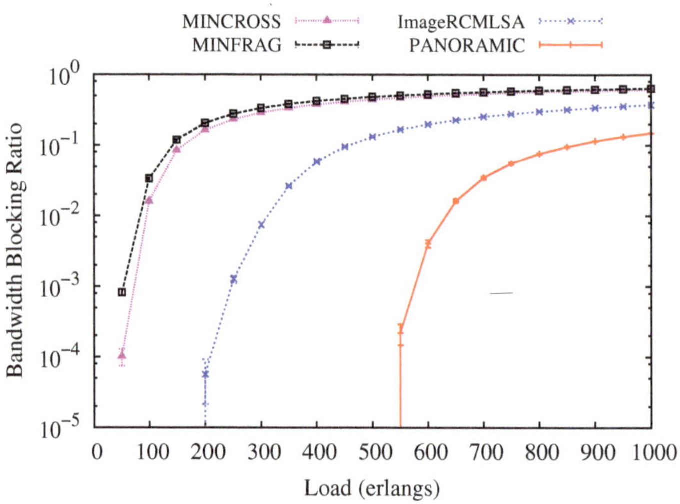


**Figure 6.** Bandwidth Blocking for NSF topology.

Figures 7 and 8 show an evaluation of the Energy Efficiency performed by *PANORAMIC* , ImageRCMLSA, MINCROSS, and MINFRAG for NSF and USA topologies. Comparatively, the USA topology has a greater number of nodes than the NSF topology, thus the energy consumption is different, interfering in the energy efficiency for transmissions. For NSF topology (Figure 8), the *PANORAMIC* algorithm produces more energy efficiency than the compared algorithm under all loads simulated, demonstrating that the effort to make the connection makes the total data transmitted worthwhile. For USA topology (Figure 7), the *PANORAMIC* algorithm also has a better performance than ImageRCMLSA, MINCROSS, and MINFRAG. The energy efficiency is higher under all loads where algorithms were performed. However, when it comes to higher loads, the energy efficiency of all algorithms shows closer results.

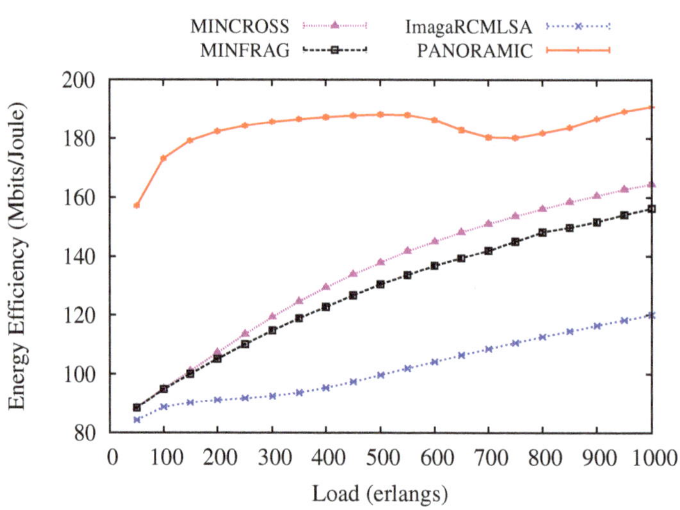


**Figure 7.** Energy Efficiency for USA topology.

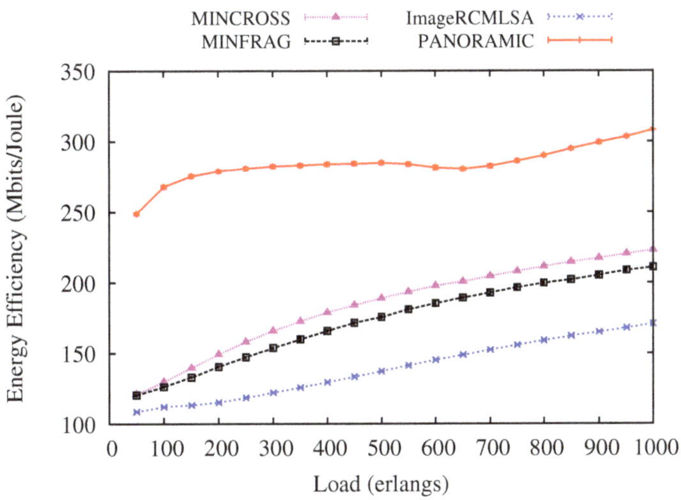


**Figure 8.** Energy Efficiency for NSF topology.

Figures 9 and 10 show results for Fragmentation to all simulated algorithms for NSF and USA topologies. For NSF topology, the *PANORAMIC* algorithm presents lower fragmentation than the compared algorithm up to 400 erlangs. Then, the results are very similar for all algorithms. However, ImageRCMLSA keeps fragmentation, with intermediate values between the algorithms used, while MINCROSS and MINFRAG present the lowest fragmentation, respectively. For USA topology, fragmentation for *PANORAMIC* has better results up to 450 erlangs, then values keep close, and under higher loads, the *PANORAMIC* algorithm presents the highest fragmentation among all algorithms. ImageRCMLSA presents acceptable results for fragmentation since it is not the main objective of this algorithm. The MINCROSS and MINFRAG algorithms present the best results for fragmentation, reaching both algorithms' proposal, which deals with fragmentation control at some level of routing. When comparing the different topologies, the curve is very similar, concluding that there was no such big difference due to the physical characteristics of the topologies.

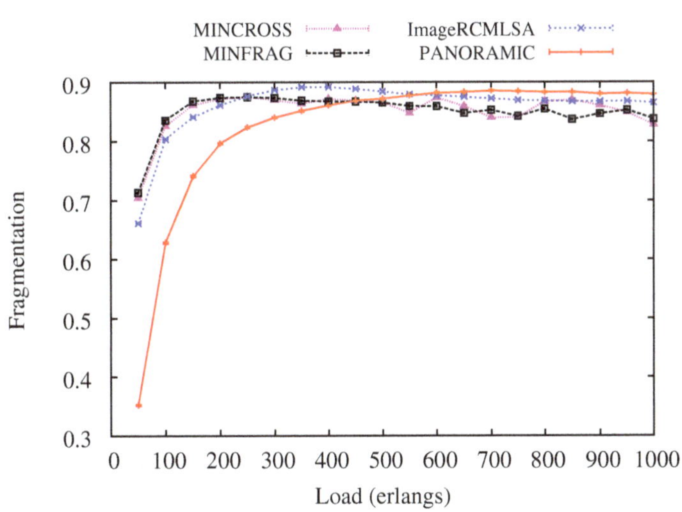


**Figure 9.** Fragmentation Ratio for NSF topology.

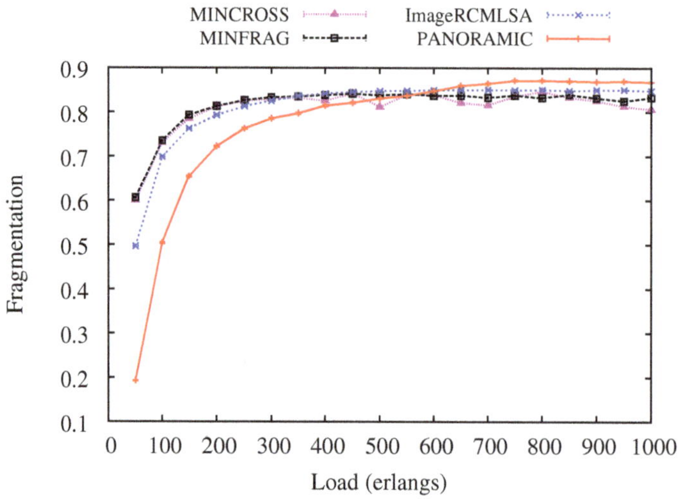


**Figure 10.** Fragmentation Ratio for USA topology.

Figures 11 and 12 illustrate the Crosstalk per Slot (CpS) values analyzed for NSF and USA topologies. For NSF topology (Figure 11), the *PANORAMIC* algorithm has constant values for all loads simulated, keeping results between the values for the ImageRCMLSA, which is the algorithm that has higher values for CpS, and two of the algorithms compared MINCROSS and MINFRAG. As proposed, the MINCROSS algorithm archives the lowest CpS between all algorithms, and the MINFRAG presents higher CpS than MINCROSS and lower than *PANORAMIC* .

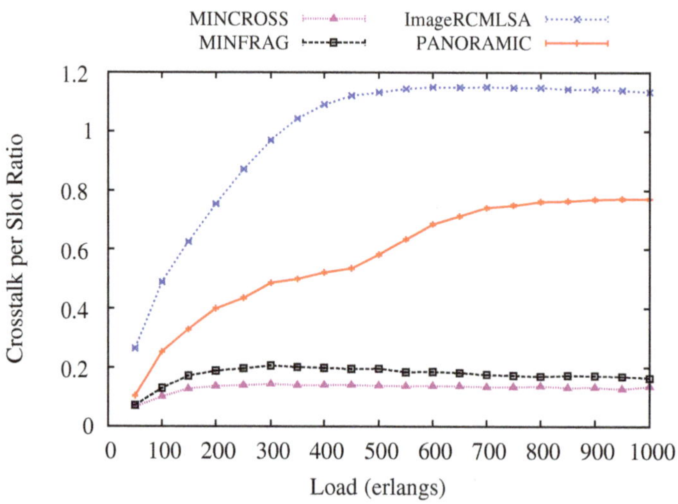


**Figure 11.** Crosstalk per Slot for NSF topology.

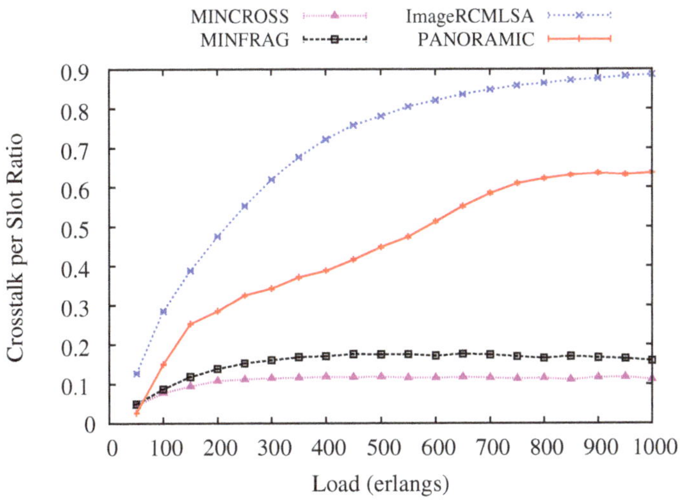


**Figure 12.** Crosstalk per Slot for USA topology.

For USA topology (Figure 12), the results are very similar to obtained for NSF topology. ImageRCMLSA presents the highest CpS during all simulated loads, followed by the *PANORAMIC* algorithm, with medium values between the compared algorithms, and the MINFRAG and MINCROSS algorithm, respectively. These results show that MINCROSS is the most effective to reduce CpS between all algorithms. However, two factors may be caused by this: the goal of the algorithm is to reduce CpS and this algorithm blocks many requests, which implies that there are not as many established connections in the network as in the other algorithms, which means that there is little interference between cores. The ImageRCMLSA algorithm presents the highest CpS because of the way the gap of frequency slots for each request is selected, allowing the connection to allocate spaces in adjacent cores, which leads to more interference. Our algorithm's goal is to accept a satisfactory number of requests and still keep an acceptable value of CpS.

Figures 13 and 14 present the arrangement of modulations used for each of the executed algorithms. For the Topology USA, the BPSK modulation is more used than the other modulations, representing more than 30% of the requests accepted in all loads of all algorithms. The USA topology has several connections, which implies a greater number of more distant transmissions. As the applied modulation is chosen, mainly based on the transmission distance, the BPSK modulation is predominant. In contrast, the 16 and 32QAM modulations reach intermediate distances, and therefore contemplate a small portion of the total transmissions with less than 2% in all USA topology situations. The modulations are slightly more distributed for the NSF topology, with QPSK and 64QAM modulations representing a large portion of the modulations. The NSF topology does not have as many links as the USA topology, and the distance between the nodes is average. Due to these network characteristics, modulations 16QAM and 32QAM were little used, representing less than 1% of accepted requests.

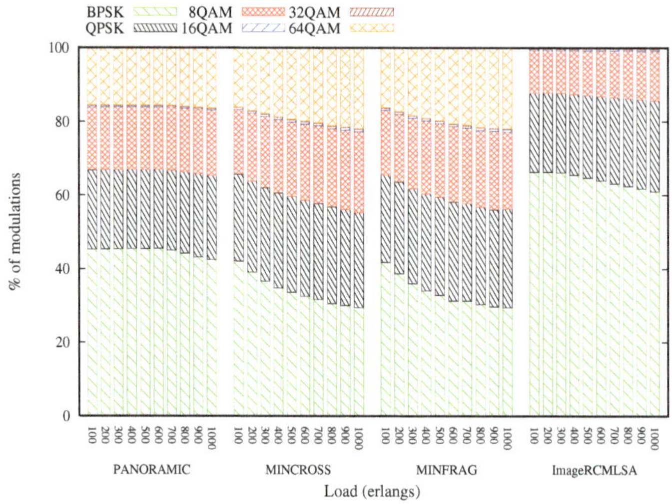


**Figure 13.** Modulation Format Percentage for USA topology.

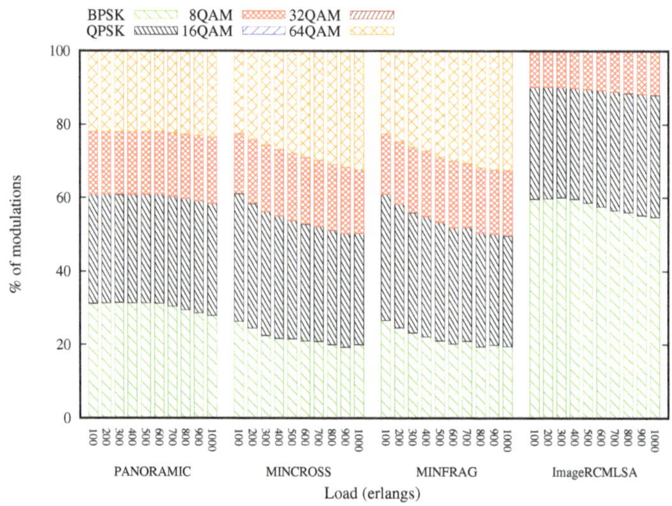


**Figure 14.** Modulation Format Percentage for NSF topology.

## 5. Conclusions and Future Work

This article presented *PANORAMIC* , an RMSCA algorithm that uses adaptive modulation with a spectrum mapping scheme capable of recognizing gaps of free frequency slots and using the hybrid single/multi-path technique for greater resource allocation. The *PANORAMIC* algorithm aims to reduce the bandwidth blocking rate by increasing the number of requests accepted by the network, in addition to improving the use of the network spectrum by reducing the number of idle frequency slots. For this, the *PANORAMIC* algorithm implements the multi-path technique for rejected requests at the first moment, therefore, for those requests that could not be allocated in a single path, the multi-path is sought in order to provide resources for the call, always considering the limitations generated by the differential delay caused by the possible difference in distance between the allocated paths. The *PANORAMIC* algorithm proved to be efficient in reducing the bandwidth blocking rate by accepting more requests, as well as promoting an improvement in energy efficiency for both network topologies used through the use of adaptive modulation, which increases the amount of data transmitted by used slots. In addition, the proposed algorithm also shows good results for the crosstalk rate per slot when comparing the number of connections established in the network with the other algorithms. As a future work, we aim to develop a machine learning model that considers the crosstalk and the spectrum fragmentation to improve resource allocation.

**Author Contributions:** Conceptualization, H.O. and E.R.; methodology, E.R.; formal analysis, E.R.; writing—original draft preparation, E.R., H.O., E.C., and D.R.; writing—review and editing, H.O., E.C., and D.R. All authors have read and agreed to the published version of the manuscript.

**Funding:** This work is founded by PAPQ-PROPESP-UFPA.

**Conflicts of Interest:** "The authors declare no conflict of interest."

## Abbreviations

The following abbreviations are used in this manuscript:

| | |
|---|---|
| BBR | Bandwidth Blocking Ratio |
| CCL | Connected Component Labeling |
| CpS | Crosstalk per Slot |
| EE | Energy Efficiency |
| EON | Elastic Optical Network |
| FGB | FilterGuard Band |
| FR | Fragmentation Ratio |
| IRA | Inscribed Rectangles Algorithm |
| MCF | Multi-Core Fiber |
| MFP | Modulation Format Percentage |
| MILP | Mixed Integer Linear Programming |
| MIMO | Multiple-Input Multiple-Output |
| PANORAMIC | Hybrid Routing, Modulation, Spectrum ANd CORe Allocation BAsed on Mapping SCheme |
| RSA | Routing and Spectrum Assignment |
| RMLSA | Routing, Modulation Level and Spectrum Assigment |
| RMSCA | Routing, Modulation, Spectrum and Core Allocation |
| RSCA | Routing, Spectrum and Core Allocation |
| RWA | Routing and Wavelength Allocation |
| SBPP | Shared Backup Path Protection |
| SDM | Space-Division Multiplexing |
| SDM-EON | Space-Division Multiplexing Elastic Optical Network |
| WDM | Wavelength Division Multiplexing |

## References


1. Cisco, U. Cisco Annual Internet Report (2018–2023) White Paper. 2020. Available online: https://www.cisco.com/c/en/us/solutions/collateral/executive-perspectives/annual-internet-report/white-paper-c11-741490.html (accessed on 2 April 2020).
2. Banaszek, K.; Kunz, L.; Jachura, M.; Jarzyna, M. Quantum limits in optical communications. *J. Light. Technol.* **2020**, *38*, 2741–2754. [CrossRef]
3. Oliveira, H.M.; da Fonseca, N.L. Protection in elastic optical networks using Failure-Independent Path Protecting p-cycles. *Opt. Switch. Netw.* **2020**, *35*, 100535, [CrossRef]
4. Brasileiro, Í.; Costa, L.; Drummond, A. A survey on Crosstalk and Routing, Modulation Selection, Core and Spectrum Allocation in Elastic Optical Networks. *arXiv* **2019**, arXiv:1907.08538.
5. Gerstel, O.; Jinno, M.; Lord, A.; Yoo, S.J.B. Elastic optical networking: A new dawn for the optical layer? *IEEE Commun. Mag.* **2012**, *50*, s12–s20. [CrossRef]
6. Chatterjee, B.C.; Ba, S.; Oki, E. Fragmentation problems and management approaches in elastic optical networks: a survey. *IEEE Commun. Surv. Tutor.* **2017**, *20*, 183–210. [CrossRef]

7. Tode, H.; Hirota, Y. Routing, Spectrum and Core Assignment on SDM Optical Networks (Invited). In Proceedings of the Optical Fiber Communication Conference, Anaheim, CA, USA, 20–24 March 2016.
8. Oliveira, H.M.; da Fonseca, N.L. Routing, spectrum and core assignment algorithms for protection of space division multiplexing elastic optical networks. *J. Netw. Comput. Appl.* **2019**, *128*, 78–89. [CrossRef]
9. Yaghubi-Namaad, M.; Rahbar, A.G.; Alizadeh, B. Adaptive modulation and flexible resource allocation in space-division- multiplexed elastic optical networks. *J. Opt. Commun. Netw.* **2018**, *10*, 240–251. [CrossRef]
10. Goścień, R.; Walkowiak, K.; Tornatore, M. Survivable multipath routing of anycast and unicast traffic in elastic optical networks. *J. Opt. Commun. Netw.* **2016**, *8*, 343–355. [CrossRef]
11. Alvizu, R.; Maier, G.; Tornatore, M.; Pióro, M. Differential delay constrained multipath routing for SDN and optical networks. *Electron. Notes Discret. Math.* **2016**, *52*, 277–284. [CrossRef]
12. Rodrigues, E.; Rosário, D.; Cerqueira, E.; Oliveira, H. Routing, Modulation, Spectrum and Core Allocation Based on Mapping Scheme. In Proceedings of the 2020 IEEE Symposium on Computers and Communications (ISCC), Rennes, France, 8–10 July 2020; pp. 1–6.
13. Zhu, Z.; Lu, W.; Zhang, L.; Ansari, N. Dynamic Service Provisioning in Elastic Optical Networks With Hybrid Single-/Multi-Path Routing. *J. Light. Technol.* **2013**, *31*, 15–22. [CrossRef]
14. Moghaddam, E.E.; Beyranvand, H.; Salehi, J.A. Crosstalk-aware Routing, Modulation Level, Core and Spectrum Assignment, and Scheduling in SDM-based Elastic Optical Networks. In Proceedings of the 2018 9th International Symposium on Telecommunications (IST), Tehran, Iran, 17–19 December 2018; pp. 160–165.
15. Yin, S.; Zhang, Z.; Chen, Y.; Ma, R.; Huang, S. A Survivable XT-Aware Multipath Strategy for SDM-EONs. In Proceedings of the 2019 Asia Communications and Photonics Conference (ACP), Chengdu, China, 2–5 November 2019; pp. 1–3.
16. Oliveira, H.M.N.S.; da Fonseca, N.L.S. Multipath Routing, Spectrum and Core Allocation in Protected SDM Elastic Optical Networks. In Proceedings of the 2019 IEEE Global Communications Conference (GLOBECOM), Waikoloa, HI, USA, 9–13 December 2019; pp. 1–6.
17. Yousefi, F.; Ghaffarpour Rahbar, A.; Ghadesi, A. Fragmentation and time aware algorithms in spectrum and spatial assignment for space division multiplexed elastic optical networks (SDM-EON). *Comput. Netw.* **2020**, *174*, 107232. [CrossRef]
18. Yousefi, F.; Ghaffarpour Rahbar, A.; Yaghubi-Namaad, M. Fragmentation-aware algorithms for multipath routing and spectrum assignment in elastic optical networks. *Opt. Fiber Technol.* **2019**, *53*, 102019. [CrossRef]
19. Yousefi, F.; Rahbar, A.G. Novel fragmentation-aware algorithms for multipath routing and spectrum assignment in elastic optical networks-space division multiplexing (EON-SDM). *Opt. Fiber Technol.* **2018**, *46*, 287–296. [CrossRef]
20. Yousefi, F.; Rahbar, A.G. Novel crosstalk, fragmentation-aware algorithms in space division multiplexed-Elastic Optical Networks (SDM-EON) with considering physical layer security. *Opt. Switch. Netw.* **2020**, *37*, 100566. [CrossRef]
21. Zhu, R.; Jue, J.P.; Yousefpour, A.; Zhao, Y.; Yang, H.; Zhang, J.; Yu, X.; Wang, N. Multi-Path Fragmentation-Aware Advance Reservation Provisioning in Elastic Optical Networks. In Proceedings of the 2016 IEEE Global Communications Conference (GLOBECOM), Washington, DC, USA, 4–8 December 2016; pp. 1–6.
22. Moghaddam, E.E.; Beyranvand, H.; Salehi, J.A. Crosstalk-Aware Resource Allocation in Survivable Space-Division-Multiplexed Elastic Optical Networks Supporting Hybrid Dedicated and Shared Path Protection. *J. Light. Technol.* **2020**, *38*, 1095–1102. [CrossRef]
23. Moura, P.M.; Da Fonseca, N.L.S. Routing, core, modulation level, and spectrum assignment based on image processing algorithms. *IEEE/OSA J. Opt. Commun. Netw.* **2018**, *10*, 947–958. [CrossRef]
24. Dharmaweera, M.N.; Zhao, J.; Yan, L.; Karlsson, M.; Agrell, E. Traffic-grooming and multipath-routing-enabled impairment-aware elastic optical networks. *IEEE/OSA J. Opt. Commun. Netw.* **2016**, *8*, 58–70. [CrossRef]
25. Malekzadeh, K.; Shahkooh, S.A. Hybrid single and multi-path routing and distance adaptive modulation level spectrum allocation in OFDM-based elastic optical networks. In *Electromagnetics Research Symposium*; Springer: New York, NY, USA, 2017; pp. 2443–2449.
26. Oliveira, H.M.N.S.; da Fonseca, N.L.S. Routing, Spectrum, Core and Modulation Level Assignment Algorithm for Protected SDM Optical Networks. In Proceedings of the 2017 IEEE Global Communications Conference, Singapore, 4–8 December 2017; pp. 1–6.

27. Lu, W.; Zhou, X.; Gong, L.; Zhang, M.; Zhu, Z. Dynamic Multi-Path Service Provisioning under Differential Delay Constraint in Elastic Optical Networks. *IEEE Commun. Lett.* **2013**, *17*, 158–161. [CrossRef]
28. Moura, P.M.; Drummond, A.C. FlexGridSim: Flexible Grid Optical Network Simulator. Available online: http://www.lrc.ic.unicamp.br/FlexGridSim/ (accessed on 12 August 2020).

**Sample Availability:** Samples of the compounds are available from the authors.

**Publisher's Note:** MDPI stays neutral with regard to jurisdictional claims in published maps and institutional affiliations.

sensors

MDPI

*Review*

# Virtual Network Embedding for Multi-Domain Heterogeneous Converged Optical Networks: Issues and Challenges

**Yue Zong [1,*], Chuan Feng [1], Yingying Guan [1], Yejun Liu [2] and Lei Guo [2]**

1 School of Computer Science and Engineering, Northeastern University, Shenyang 110819, China; fcyaokuaile@163.com (C.F.); guanyingying@stumail.neu.edu.cn (Y.G.)

2 School of Communication and Information Engineering, Chongqing University of Posts and Telecommunications, Chongqing 400065, China; yjliu@cqupt.edu.cn (Y.L.); guolei@cqupt.edu.cn (L.G.)

* Correspondence: zongyue2012@gmail.com

Received: 13 April 2020; Accepted: 3 May 2020; Published: 6 May 2020

**Abstract:** The emerging 5G applications and the connectivity of billions of devices have driven the investigation of multi-domain heterogeneous converged optical networks. To support emerging applications with their diverse quality of service requirements, network slicing has been proposed as a promising technology. Network virtualization is an enabler for network slicing, where the physical network can be partitioned into different configurable slices in the multi-domain heterogeneous converged optical networks. An efficient resource allocation mechanism for multiple virtual networks in network virtualization is one of the main challenges referred as virtual network embedding (VNE). This paper is a survey on the state-of-the-art works for the VNE problem towards multi-domain heterogeneous converged optical networks, providing the discussion on future research issues and challenges. In this paper, we describe VNE in multi-domain heterogeneous converged optical networks with enabling network orchestration technologies and analyze the literature about VNE algorithms with various network considerations for each network domain. The basic VNE problem with various motivations and performance metrics for general scenarios is discussed. A VNE algorithm taxonomy is presented and discussed by classifying the major VNE algorithms into three categories according to existing literature. We analyze and compare the attributes of algorithms such as node and link embedding methods, objectives, and network architecture, which can give a selection or baseline for future work of VNE. Finally, we explore some broader perspectives in future research issues and challenges on 5G scenario, field trail deployment, and machine learning-based algorithms.

**Keywords:** virtual network embedding; converged optical networks; network slicing; machine learning; software-defined network

## 1. Introduction

The exponential growth of the emerging of dynamic applications and the billions of devices in the Internet of Things (IoT) with sensing, computing, and communication capabilities have driven the investigation of network architecture. Current network architectures fail to address the diverse performance requirements in terms of latency, scalability, availability, and reliability [1]. To overcome the issues and support more heterogeneous applications, network slicing is considered as a promising technology formed by partitioning or combining a set of network resources, and abstracting it to users [2,3]. Network virtualization and orchestration are key processes for network slicing, where software-defined network (SDN) and network function virtualization (NFV) are the key enabling technologies for network orchestration [4,5]. In addition, SDN can manage and deploy the service requirements automatically by decoupling the transmitting layer and control layer [6,7]. It is specified to leverage the benefits of network virtualization to allow high flexibility among various mobile and IoT services in multi-domain heterogeneous converged optical networks.

Infrastructure provider (InP) and service provider (SP) have been decoupled in the network virtualization environment to enable multiple virtual networks (VNs) coexisting and sharing substrate resources (e.g., node computing resources and link resources) [8,9]. Efficient resource allocation for both virtual nodes and links is one of the major challenges which refers as virtual network embedding (VNE) [10]. Many existing works have solved the VNE resource allocation problem by integer linear program (ILP) [11–13]. Efficient VNE algorithms have been proposed in many works to improve the performance [14,15], which include two-stage algorithms by efficient node ranking and link assignment method, coordinated VNE approaches, and machine learning (ML)-based algorithms. Many existing works have focused on specific domain network architecture of heterogeneous converged optical networks for the VNE approaches, such as wireless network, fiber-wireless (FiWi) access network, and optical data center network (ODCN).

Existing surveys and literature have considered various details and network features for VNE under different network scenarios (e.g., elastic optical network) [16–18]. In [16], the authors have focused on algorithmic aspects for VNE for cloud networks. However, these surveys have not focused on VNE for multi-domain heterogeneous converged optical networks and the future research issues on implementation and intelligent algorithms. In this paper, the representative references in the latest popular top journals and conferences about VNE and network slicing are discussed. Furthermore, multi-domain heterogeneous converged optical network architecture has been described and the differences among them are discussed, e.g., radio resource for wireless channel, spectrum characteristics for optical network, and various substrate nodes. Thereby, the characteristics of specific single domain network for VNE have been discussed. We provide a brief survey of the basic VNE problem formulas and a taxonomy of VNE approaches on existing works. Issues and challenges have been discussed for the road on VNE in the future.

The organization of the paper is as follows. VNE in multi-domain heterogeneous converged optical networks and key enabling technologies are discussed in Section 2. In Section 3, the basic VNE problem and major metrics are presented. In Section 4, we give a VNE algorithm taxonomy for existing works. The future issues and challenges on the road of VNE are discussed in Section 5. Finally, we conclude the paper in Section 6.

## 2. Virtual Network Embedding in Multi-Domain Heterogeneous Converged Optical Networks

For emerging enhanced mobile broadband (eMBB), massive machine communications, and ultra-reliable and ultra-low latency communications 5G scenarios, network slicing as a promising technology can guarantee the requirements and efficient resource utilization. The diagram of multi-domain heterogeneous converged optical network architecture in Figure 1 is composed of wireless access network domain, metro network domain, core network domain, and edge computing/data center domain. Wireless access networks have been architectured to support a number of diverse vertical applications of end users and converged the requirements into metro and core networks. Edge computing and data centers provide computing capacity to guarantee the implementation of VNE. Many 5G research works and demonstration projects (e.g., 5GNORMA, 5GEx, 5GinFIRE, and 5G!Pagoda) have addressed the realization of 5G slicing mainly on wireless domain through the combination of key enabling technologies of SDN and NFV [19]. Thereby, SDN and NFV are described in the following as key enabling technologies to provide and guarantee the deployment and implementation of VNE. Although there exist few researches about VNE for 5G multi-domain networks, many of the existing works have focused on VNE for single domain. We review the VNE in wireless network, FiWi access network, and optical network single domain in the following.

### 2.1. Key Enabling Technologies

VNE implementation, resource allocation, and scheduling are key points in multi-domain heterogeneous networks. SDN and NFV have been proposed as key enabling technologies to achieve network slicing orchestration for many 5G researches and demonstration projects. SDN can decouple the infrastructure layer and the control layer, which is considered a key enabling technology to implement network virtualization (e.g., deployment VNE algorithms) and manage services for network operators [20–22]. The SDN controller enables configuring network remotely to enhance the network flexibility and service provisioning, which can provide a closer tie between application requirements and the combination of resources (e.g., optical transport network, IP layer, computing, and storage) [23]. SDN is an ideal platform for implementation of network virtualization since it can flexibly offer end-to-end network slices according to the requirements of different applications through hypervisors such as Flowvisor, OpenVirteX, FlowN, and AutoVFlow [24].

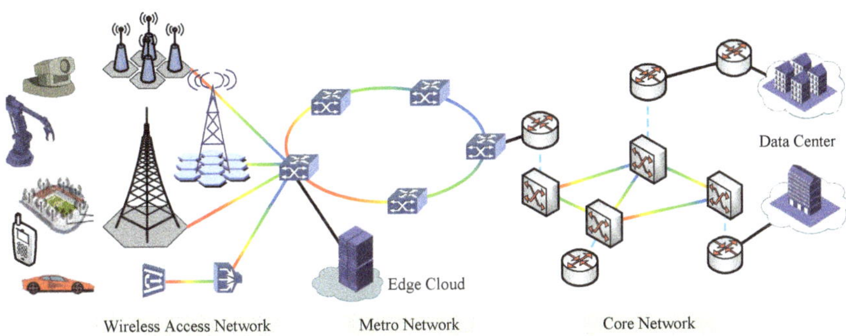


**Figure 1.** Architecture of multi-domain heterogeneous converged optical networks.

NFV plays an important role in the realization of virtual network functions and network services by decoupling network functions from the dedicated physical devices, implementing them as software on industrial standard high volume servers. Service chaining composed of virtualized network functions can make both data and control plane functions flexible so that the traffic of certain users or applications only traverses a particular set of functions [25]. Virtual network functions placement can be regarded as a special VNE issue. Thereby, the complexity of VNE problem for network management and deployment in 5G multi-domain network architecture has increased, especially in the context of large number of VN requests. Management and orchestration (MANO) framework is leveraged as a critical automatically solution to manage and orchestrate network virtualization [26,27].

### 2.2. Wireless Network

Due to the development of emerging 5G applications (e.g., IoT and Internet of Vehicles (IoV)), the VNE problem in wireless network domain of 5G has gained more attention due to the growing popularity of 5G applications. The implementation and efficient slicing of wireless domain are essential to provide services for users as it is closed to the massive users side in 5G networks. The concept of the cloud-radio access network (C-RAN) has been proposed to decouple digital units (DUs) and radio units (RUs) of base stations (BSs) and centralize DUs into central offices [28]. C-RAN is supposed to increase the capacity by 1000x to handle the growing number of connected devices and increasing data rates, which can ease the implementation of advanced radio coordination techniques, e.g., coordinated multi-point (CoMP) Transmission/Reception. In addition, the revolution of IoT is reshaping the modern industrial systems, where industrial wireless networks (IWNs) refer to the pervasive deployment of devices with sensing, processing, and connecting capabilities [24]. The example of VNE in IWNs is shown in Figure 2, where massive devices deployed in the access layer perform monitoring and controlling. Those OpenFlow-enabled access points (APs) with mesh topology compose the data plane for packet accessing and transferring to edge and cloud computing. The substrate network is controlled by the controller of the control layer to satisfy the industrial virtual network in the application layer.

According to the European Telecommunications Standards Institute (ETSI) [29], mobile edge computing (MEC) is defined as "Mobile edge computing provides an IT service environment and cloud computing capabilities at the edge of the mobile network, within the radio access network (RAN) and in close proximity to mobile subscribers." To extend cloud computing services to the edge of networks leveraging mobile base stations, MEC is an emergent architecture which can be applied to mobile, wireless, and wireline scenarios, using software and hardware platforms, located at the network edge in the vicinity of end users [30].

Wireless sensor networks (WSN) are regarded as the basic constituents of IoT that can facilitate the interaction of users (humans or machines) with their environment and react to real-world events. To create large-scale sensor platforms, WSN virtualization is envisioned as an important technology to satisfy efficient usage of network resources. The authors of [31] have analyzed the importance and approaches for sensor node-level virtualization and network-level virtualization of WSN. To facilitate the QoS provisioning for different applications with strict demands on latency and reliability, an application-driven virtual network embedding scheme has been proposed for flexible network resource allocation of industrial WSNs [32].

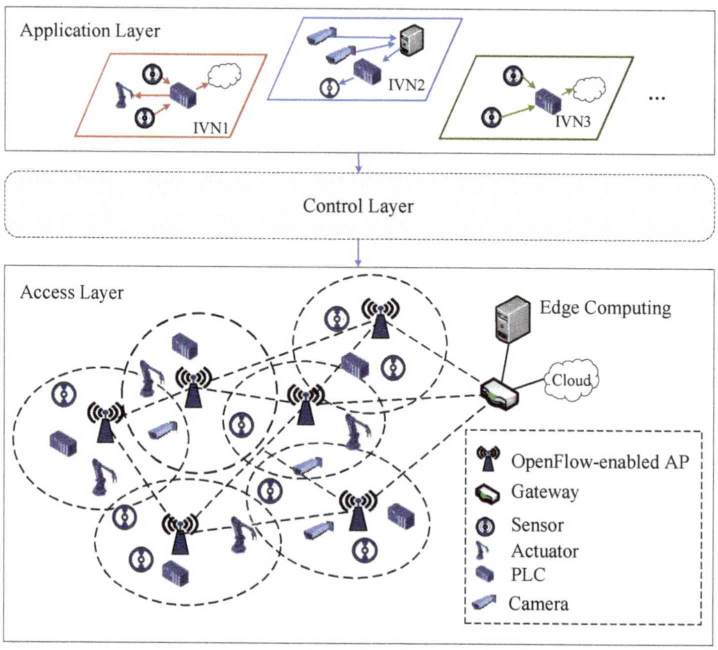


**Figure 2.** Illustration of slice-based virtualization for industrial wireless networks (IWNs).

In addition, many unique characteristics of wireless networks need to be considered for VNE [32–35]. Node mapping is only partially deterministic, as the AP is simply selected according to the location of users, but gateway (GW) has to be determined based on throughput optimization [34]. Mobility for node (re-)mapping has been considered in [35]. Due to the broadcast nature of wireless channels, link mapping needs to consider the specific multiple access mechanisms. The authors of [32] propose an approach based on anypath routing technique to reduce the resources consumed by re-transmission.

### 2.3. Fiber-Wireless Access Network

The FiWi access network is a network converged by wireless and optical components, and is an essential part of 5G networks as it guarantees the 5G service requirements and converges them into the core network. It gains more popularity for its advantages of high capacity, long distance, and flexible access ability [36–38]. Network virtualization for FiWi is required to overcome the bottleneck of joint wireless and optical resource allocation.

Additional network characteristics of FiWi need to be considered, where the substrate nodes include optical line terminal (OLT), optical network unit (ONU), wireless router, and wireless gateway. The substrate links are composed of fiber link, cable, and wireless link. Additional link features are supposed to be considered for link embedding, e.g., the channels of wireless radio interfaces. The illustration of VNE in FiWi access network [39] is shown in Figure 3. Efficient resource management for both optical and wireless resources of the SDN/NFV-based converged network has been discussed in [40] to guarantee the specific delay and bandwidth requirements of the multiple services of network slices. Furthermore, to centralize control and allocate network and computing resources of converging edge computing over FiWi network, the authors of [41] propose two VNE algorithms to obtain higher revenue and profit ratio.

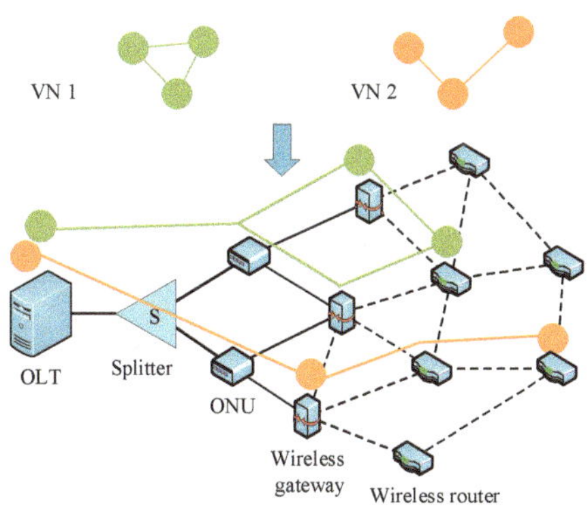


**Figure 3.** Illustration of virtual network embedding (VNE) in fiber-wireless (FiWi) access network.

### 2.4. Optical Network

Optical network is the fundamental part in 5G multi-domain heterogeneous network to ensure the high bandwidth and low latency transmission. Many existing works have proposed efficient VNE allocation schemes from various aspects to guarantee the converged massive 5G services performance requirements for network virtualization [42–45]. A novel dynamic VNE approach based on an auxiliary graph is proposed to improve network utilization and performance by adjusting the weights of the edges of the auxiliary graph on fixed-grid DWDM network [43]. The proposed VNE algorithms for migration in [44] have improved network utilization and energy consumption efficiently.

To facilitate the flexible allocation of the fiber spectrum, elastic optical network (EON) is an emerging technology by leveraging finer-grained channel spacing, tunable modulation formats and forward error correction overheads, and baud-rate assignment [46]. In [11,14,45], the authors have proposed efficient approaches to solve VNE in EON for network slicing to guarantee the service requirements. Spatial resources of optical network are also considered in some works [47–49], which refer to fiber cores or modes in multi-core fibers or multi-mode fibers, or even single-mode fiber bundles. The proposed genetic algorithm in [48] has obtained the optimal VNE schemes with core allocation to efficiently by designing tailor-made encoding scheme, crossover, and mutation operators. Some additional constraints for VNE should be considered for optical network domain such as spectrum continuity, spectrum contiguity, and physical layer impairments [50–52].

Data centers (DCs) have become an efficient and promising infrastructure to provide data storage and computing capacity. Geographical distribution data center networks connected with optical network guarantee the requirements of 5G network services and applications (e.g., video streaming) [53–55]. Furthermore, network virtualization in ODCNs can be classified into intra- [17,18] and inter-ODCNs [16,56–58]. Three provisioning schemes have been proposed in [59] by constructing a virtual auxiliary graph that decomposes the physical infrastructure into several layered graphs, according to the spectrum slot requirements of a virtual optical network request. Network services deployment and orchestration for network slice in inter-ODCNs have been developed in [60]. OpenStack-based orchestrator deploys the VMs for IT requirements by the path computation engine and contacts with the OpenDaylight SDN controller to guarantee the network configuration. The illustration of VNE in inter-ODCN and the network architecture are shown in Figure 4.

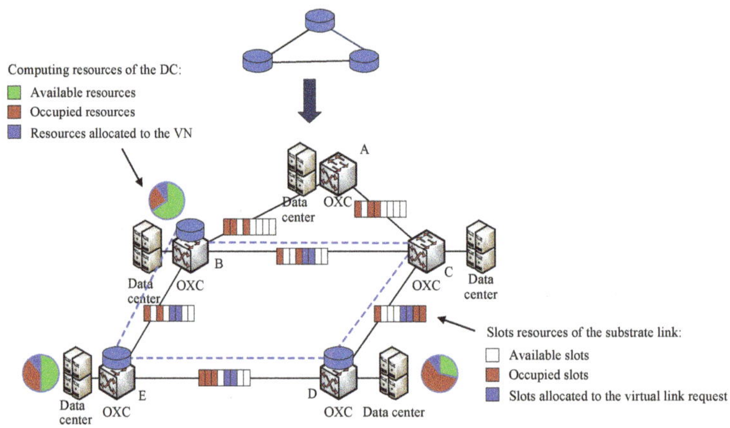


**Figure 4.** Illustration of VNE in inter-optical data center networks (ODCNs).

## 3. Virtual Network Embedding Problem

This section introduces the substrate network, virtual network, and general VNE problem description and formulation with restrictions for resource allocation. Furthermore, the major objectives and efficiency metrics are formulated to evaluate the performance of VNE problem for resource allocation.

### 3.1. Substrate Network

Similar to the work in [13], the substrate network is described by an undirected graph $G^S = (N^S, E^S, A_N^S, A_E^S)$, where $N^S$ is the set of substrate nodes and $E^S$ refers to the set of substrate links. Substrate nodes and links are associated with their attributes, denoted by $A_N^S$ and $A_E^S$, respectively. For each substrate node $n \in N^S$, the node attributes usually consider CPU capacity $C_n^S$ and location $Loc_n^S$. For substrate link $e^S(m,n)$ between the substrate node $m$ and $n$, the typical attribute is bandwidth capacity $b_e^S$ [61] and wireless channel [37]. Additional link constraints (i.e., wavelength, spectrum continuity, and spectrum contiguity) need to be considered when the substrate link is optical fiber [51,52]. An important issue has to be taken into account while the actual effect of users mobility for wireless networks [62].

### 3.2. Virtual Network

The undirected graph, $G^V = (N^V, E^V, A_N^V, A_E^V)$, describes the set of virtual network requests [63], where $N^V$ and $E^V$ refer to the sets of virtual nodes and links, respectively. The network topology, $(N^V, E^V)$, is a logical network that should be configured as a sub-network of the substrate network.

Typically, the substrate network should satisfy the attributes associated with virtual nodes and links described as $A_N^V$ and $A_E^V$, respectively. In addition, for the virtual node $v \in N_r^V$ of the $r^{th}$ VN, the requested node attributes for embedding are CPU capacity request $c_r^v$, the location $Loc_r^v$, and maximum location distance $\rho_r^v$. For each virtual link $e_r(v,u) \in E_r^V$, bandwidth requirement, $B_{e_r}^V$, is the major considered virtual link attribute.

### 3.3. Virtual Network Embedding

The problem of mapping virtual network to substrate network can be defined as a mapping $M$: $G^V(N^V, E^V) \rightarrow G^S(N^S, E^S)$, from $G^V$ to a subset of $G^S$ [11,14,15]. Figure 5d shows the example of VN embedding solution for virtual network requests in Figure 5a,b embedded in initial substrate network shown in Figure 5c. The numbers over the links represent the link capacity and the numbers in rectangles represent CPU resource of virtual network and substrate network. VNE can be decomposed into two steps:

1. Virtual Node Embedding (VNoE): $f: N_r^V \rightarrow N^S$.

   Virtual nodes need to be embedded to different substrate nodes that satisfy the node resource and location constraints, which are described by Equations (1)–(4), where $\delta_n^{v,r} \in \{0,1\}$. If virtual node $v$ of $r^{th}$ VN is embedded into substrate node $n$, $\delta_n^{v,r} = 1$. Equation (1) guarantees that all virtual nodes that are accommodated by the substrate node $n$ cannot exceed the total substrate computing resource. Each virtual node $v$ can only play host once to a unique substrate node shown in Equation (2). Each substrate node $n$ can only host one virtual node of the same VN request described by Equation (3). The distance constraint for each virtual node is described by Equation (4), where dis($\cdot$) refers the distance between the locations of substrate node $n$ and virtual node $v$.

$$\sum_{r \in R} \sum_{v \in N_r^V} c_r^v \cdot \delta_n^{v,r} \leq C_n^S \qquad \forall n \in N^S \tag{1}$$

$$\sum_{v \in N_r^V} \delta_n^{v,r} \leq 1 \qquad \forall n \in N^S,\ \forall r \in R \tag{2}$$

$$\sum_{n \in N^S} \delta_n^{v,r} = 1 \qquad \forall v \in N_r^V,\ \forall r \in R \tag{3}$$

$$dis(Loc_r^v, Loc_n^S) \leq \rho_r^v \quad \forall n \in N^S, \forall r \in R,\ v \in N_r^V \tag{4}$$

2. Virtual Link Embedding (VLiE): $f: E_r^V \rightarrow E^S$.

   Virtual links embedded to loop-free paths on the substrate network that satisfy the link bandwidth resource requirements and the total virtual link requirements cannot exceed the bandwidth resource of substrate link $e^S(m,n)$, as shown in Equation (5). Binary variable $f(e_r(v,u), e^S(m,n))$ equals 1, if substrate link $e^S$ is embedded by virtual link $e_r$. Flow conservation constraint is shown in Equation (6). According to features of substrate links, additional link constraints should be considered, i.e., optical wavelength, spectrum continuity in EON [50–52], and wireless channel, expected anypath transmission time of anypath [24].

$$\sum_{r \in R} \sum_{e_r \in E_r^V} f(e_r(v,u), e^S(m,n)) \cdot B_{e_r}^V \leq b_e^S, \quad \forall e^S \in E^S \tag{5}$$

$$\sum_{n \in N^S} f(e_r(v,u), e^S(m,n)) - \sum_{n \in N^S} f(e_r(v,u), e^S(n,m)) = \delta_m^{v,r} - \delta_m^{u,r}, \ \forall e_r(v,u) \in E_r^V, \forall r \in R \tag{6}$$

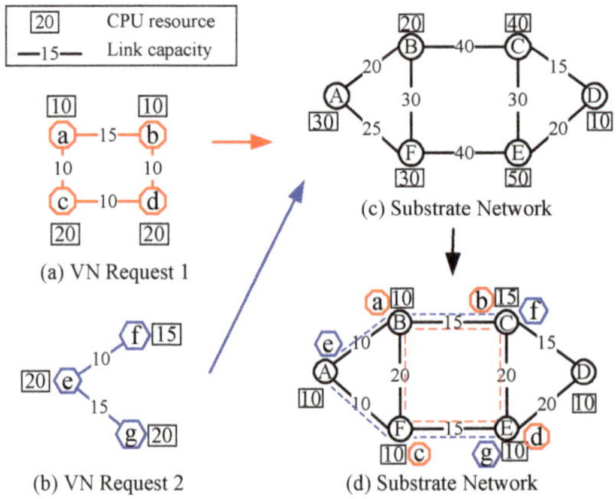


**Figure 5.** Example of VN embedding solution.

### 3.4. *Main Objectives and Metrics*

The VNE problem needs to evaluate the performance for VNE approaches and satisfy the diverse 5G service requirements. Thereby, one or some metrics are considered as objectives according to the motivation of literatures. In this section, the major objectives and metrics are described and analyzed.

#### 3.4.1. Profit

Similar to work in [64,65], the revenue of serving a VN request is defined by summing up the required CPU and bandwidth resource for VN requests as shown in Equation (7), where $\alpha$ and $\beta$ are the weights. Although revenue is the metric that InP will gain by accepting VN requests, it is not very considerable without knowing the cost of InP. The cost of VNE is defined by summing up all CPU and bandwidth resources of the substrate network resources allocated for VN requests. The embedding cost for VN request $r$ is described by Equation (8), where $\alpha_c$ and $\beta_c$ are the weights for CPU and bandwidth costs, respectively. In addition, the whole profit for InP for serving all the VN requests is defined by Equation (9). The revenue/cost ratio described by Equation (10) indicates the percentage between revenue and cost, where it also can reflect the profit for InP.

$$Revenue(G_r^V) = \alpha \cdot \sum_{v \in N_r^V} c_r^v + \beta \cdot \sum_{e_r \in E_r^V} B_{e_r}^V \tag{7}$$

$$Cost(G_r^V) = \alpha_c \cdot \sum_{v \in N_r^V} c_r^v + \beta_c \cdot \sum_{e_r \in E_r^V} \sum_{e^S \in E^S} f(e_r(v,u), e^S(m,n)) \cdot B_{e_r}^V \tag{8}$$

$$Prf(G^V) = \sum_{r \in R} Revenue(G_r^V) - \sum_{r \in R} Cost(G_r^V) \tag{9}$$

$$Revenue/Cost\ Ratio = \frac{\sum_{r \in R} Revenue(G_r^V)}{\sum_{r \in R} Cost(G_r^V)} \tag{10}$$

#### 3.4.2. Acceptance Ratio

The acceptance ratio is also related with profit, as the revenue is calculated if the VN request is accepted [64]. As described in Equation (11), it measures the number of VN requests which are completely embedded, where $Acc_r^V \in \{0,1\}$ refers VN request $r$ is accepted or not. For some online problem, usually blocking ratio is considered, where $Blocking\ Ratio = 1 - Acceptance\ Ratio$.

$$Acceptance\ \ Ratio = \frac{\sum_{G_r^V \in G^V} Acc_r^V}{|G^V|} \tag{11}$$

### 3.4.3. Resource Utilization

Resource utilization is defined by summing the occupied substrate resources (node and link) for the embedded VN requests divided by the total amount of resources. This metric can take into account the resource usage. Node resource utilization $RU_{Com}$ and link resource utilization $RU_{Link}$ are shown in Equations (12) and (13) by summing the requirements of the embedded node or link for VN requests divided by the total node or link capacity of the substrate network [66]. For some online scenarios, the lifetime of VN requests is considered.

$$RU_{Com} = \frac{\sum_{n \in N^S} \sum_{r \in R} \sum_{v \in N_r^V} c_r^v \cdot \delta_n^{v,r}}{\sum_{n \in N^S} C_n^S} \tag{12}$$

$$RU_{Link} = \frac{\sum_{e^S \in E^S} \sum_{r \in R} \sum_{e_r \in E_r^V} f(e_r(v,u), e^S(m,n)) \cdot B_{e_r}^V}{\sum_{e^S \in E^S} b_e^S} \tag{13}$$

### 3.4.4. Latency

As the emerging 5G services (e.g., IoV) have strict latency requirements, the VNE problem must guarantee these targets [67]. In the existing literature, latency requirements have been modeled as constraints applied to the virtual links of VN (i.e., each virtual link is mapped to satisfy a given latency target) [45]. The authors of [68–70] focus on latency-aware algorithms by considering the time (e.g., propagation time or delay) from one embedded substrate node to another to satisfy the requirements of virtual links. For simplicity, path length is used to represent the latency metric in [69], which sums up the length of substrate links where a virtual link request is embedded. In connection with the longer length of corresponding path, more resources are consumed in substrate network and users will suffer longer latency.

Especially for 5G end users that require ultra-low latency services (e.g., video broadcast service, gaming service), link latency should obtain more consideration. The authors of [71] have formulated the end-to-end delay in a fronthaul network as $D_{e2e} = D_{proc} + D_{prop} + D_{link} + D_{queue}$. The total processing delay $D_{proc}$ is a fixed value required to forward a packet. $D_{prop}$ is the propagation delay, which is determined by the fiber length. The serialization delay $D_{link}$ is proportional to the frame size and inversely proportional to the link bandwidth capacity. Queuing delay $D_{queue}$ is caused by the competition among fronthaul packets. The authors of [45] focus on latency model in EON, where the latency of lightpath is shown as $L_p = L_n + len(p)L_{prop} + n_{amp}L_{amp} + (|p| + 1)L_{roadm}$. $L_n$ means the latency at terminal node considering FEC modules and transponders. Propagation delay $len(p)L_{prop}$ is the major latency contribution for a lightpath, which $L_{prop}$ amounts to $\approx$ 4.9 µs per kilometer of fiber and $len(p)$ is the physical length of lightpath. The latency of amplifiers $n_{amp}L_{amp}$ is considered, where $n_{amp}$ is the number of amplifiers on a lightpath $p$. The latency of component reconfigurable optical add-drop multiplexer is shown by $L_{roadm}$ and $|p|$ is the number of substrate optical links on the lightpath. The differences between the two equations mentioned above are based on the network components of substrate links.

### 3.4.5. Energy Efficiency

Energy consumption of network infrastructures in network virtualization has been focused due to the rising energy costs and ecological awareness. Without compromising the network performance (e.g., InPs), switching off or sleeping power-consuming elements by consolidating requests is considered as the primary approaches to minimize the energy consumption in [66,69,72]. Various network components in network architecture (e.g., servers, routers, and transponders) are considered in [73–75]. In general, the ratio between running nodes and the total number of substrate nodes is taken into account as a performance metric. The authors of [76] have considered migration to re-optimize the energy consumption with consideration of interruption time and bandwidth waste of migration.

### 3.4.6. Survivability

Failures can affect a large number of VN requests, which can be divided into two categories: node failure [77,78] and link failure [79]. VN survivability is the ability that a VN continuously provides services in compliance with the given requirements in present failures and other events [79,80]. Link survivability can be classified into two categories: protection [81,82] and restoration [83]. The authors of [84,85] have focused on the link recovery in C-RAN.

To solve the issue, the proposed survivable algorithms should consider some performance metrics as follows.

- Number of backups: The metric counts the number of backup resources that is reserved for a VN. Additional substrate resources have to be reserved to serve the VN request when failures happen. Path Redundancy measures the ratio between the number of backup paths to the number of direct paths. Some redundancy algorithms set up backup paths that can be used in case some parts of the network break down [86]. Therefore, the metric refers to the amount of additional resources that are used to backup the embedded network.
- Migration frequency: For node failure, migration frequency shows the performance required to achieve higher acceptance ratio and lower embedding cost of node migration [77]. The affected task node will be migrated to one backup host after node failure to reduce the cost of node migration and re-embedding the path. Link failure or path length constraint also can trigger migrations. Therefore, migration frequency should be considered as a metric to show the migration performance.

### 3.4.7. Traffic Prediction

Traffic prediction is not an objective or a metric; however, it is an important procedure to improve objective or metric performance. In many 5G scenarios and applications, network traffic prediction is playing an important role [87] for resource allocation and load balancing in management and provisioning (e.g., network management, traffic (re)-routing). Autoregressive (AR), autoregressive moving average (ARMA), autoregressive integrated moving average (ARIMA), and support vector machine (SVM) are the most common models for network prediction problems [88]. Due to the ability of processing the high-dimensional data, deep/convolutional/recurrent neural network models are widely used in the network traffic prediction to improve the accuracy of prediction [88–92]. The accuracy of the predicted results based on the algorithms needs to guarantee the change of the bandwidth explosion and service diversity. In addition, the accuracy is used to measure how efficient VNE method impacts the number of VN requests or the profit ratio for InPs.

To evaluate the prediction accuracy, mean absolute error (MAE) [93], measure square error (MSE) [94], and root mean square errors (RMSE) [88] are used to quantify the difference between the forecasted values and the actual values. MAE is an average sum of the absolute errors described in Equation (14), where $y_i$ and $\hat{y}_i$ are the observed value and the predicted value, and $N$ represents the total number of predictions. MAE is a widely used prediction accuracy measurement and a small value of it means that the predictor has high performance. MSE is a scale dependent metric by computing the average sum of squared errors as shown in Equation (15). In addition, RMSE is the square root of MSE as shown in Equation (16).

$$MAE = \frac{1}{N}\sum_{i=1}^{N}|y_i - \hat{y}_i| \tag{14}$$

$$MSE = \frac{1}{N}\sum_{i=1}^{N}(y_i - \hat{y}_i)^2 \tag{15}$$

$$RMSE = \sqrt{\frac{1}{N}\sum_{i=1}^{N}(y_i - \hat{y}_i)^2} \tag{16}$$

## 4. VNE Algorithms Taxonomy

According to the VNE problem description and formulation in Section 3, ILP with various objectives and constraints of network architecture features is an optimal VNE solution for many existing research works [11,14,69]. However, the VNE problem is known as an NP-hard problem, which cannot be solved in polynomial time when the substrate network topology or virtual network requests are scaled through ILP. To overcome the issue, many existing research works have proposed novel algorithms. The VNE problem is divided into two sub-problems as described in Section 3, we review the methods for VNoE and VLiE. In this section, we classify the existing VNE algorithms into three categories: two-stage VNE algorithms, coordinated VNE algorithms, and machine learning-based VNE algorithms.

### 4.1. Two-Stage VNE Algorithms

Two-stage VNE algorithms execute VNoE according to node ranking strategy for virtual nodes. The link assignment strategy of VLiE is executed for resource allocation after all the virtual nodes embedded. This section reviews the representative node ranking strategies for VNoE and virtual link assigment approaches for VLiE.

#### 4.1.1. Virtual Node Embedding

For virtual node embedding, greedy strategy is the most common method for virtual node resource allocation. The choice of method used in ranking the virtual node and substrate node efficiently is essential for embedding. The most considered metric for node ranking is the CPU capacity in decreasing order according to the requirement of virtual nodes and residual CPU capacity of substrate nodes [76]. Available resource node ranking method is formulated as product of node residual CPU capacity and sum of unoccupied bandwidth capacity of the neighbor links [12,95,96]. Instead of CPU capacity, the available resource method ensures that enough CPU capacity available and also considers bandwidth capacity to prepare for the subsequent link mapping stage. The authors of [97] propose a candidate-assisted algorithm by constructing candidate substrate nodes and candidate substrate paths for a virtual network request to reduce the mapping execution time.

The Markov random walk (RW) topology-aware node ranking method is proposed inspired by the PageRank algorithm [63], which is computed using a classic iterative scheme for the product of the vector of all substrate nodes resource and a one-step stochastic transition matrix of the Markov chain. To quantify the embedding potential of each node in the substrate network, the proposed global resource capacity (GRC) takes the topological attributes resources of the entire network into consideration [15]. The GRC value for each node is the sum of weighted normalized residual CPU capacity and bandwidth resource of links connected with the node. GRC node rankings are computed using an iterative scheme of the GRC vector which is composed of the calculated GRC for the nodes. An additional attribute is considered in [69,98] for the formulated modified GRC method.

#### 4.1.2. Virtual Link Embedding

After all nodes embedded in VNoE, virtual link embedding strategy is executed as fixed source–destination link requirement assignment. The most common used strategies for wired network are Dijkstra shortest path (SP) [76,95,98], K-shortest path (KSP) [12,14,36,63], and multi-commodity flow (MCF) [99]. The authors of [36] have considered link availability for survivability. Additional link restrictions of network (e.g., spectrum continuity, spectrum contiguity, and modulations) need to be considered. The authors of [12] have considered path splitting to reduce link congestion for better link resource allocation. The authors of [96,97] have proposed to construct satisfied the virtual link requirements candidate path set with reduction of mapping execution time, meanwhile without loss of the main performance indices.

### 4.2. Coordinated VNE Algorithms

Coordinated VNE approaches have been proposed to improve network resource utilization and solve the drawback and limitation of two-stage VNE approaches that it may cause embedding failure due to inefficient link capacity after all vitrtual node embedded. In [15,99], the proposed coordinated VNE algorithms consider node and link stage jointly to improve the performance metrics. The proposed coordinated VNE algorithms execute virtual link embedding according to descending order of the degree of virtual node for the virtual link [57]. Additional link restrictions (e.g., spectrum continuity, spectrum contiguity, and modulations) have been considered in EON with multi-core fiber. Due to the unique features of wireless networks, SP or MCF is not appropriate for mapping. Taking advantage of the broadcast nature of wireless channels, authors have proposed an anypath link mapping scheme to fulfill the diverse QoS requirements of VNs and reduce the resources consumed by retransmissions [32].

In order to further improve the performance, metaheuristic-based coordinated VNE algorithms are proposed. The authors of [48] propose an effective genetic algorithm for virtual optical network mapping, core allocation, and spectrum assignment in EONs using multi-core fibers. The objective is minimizing the maximum index of used frequency slots, which is regard as spectrum usage in this section. Virtual nodes mapping population, routing population, and core allocation population are used in proposed genetic algorithm. A periodical planning of embedding process is proposed in [100], where profitable VN requests are selected through an auction mechanism to maximize the revenue. The authors of [56,69] have proposed ant colony optimization (ACO)-based VNE algorithm in inter-ODCN. The proposed multi-objectives VNE algorithm based on particle swarm optimization (PSO) in [101] can improve the energy and revenue performance through the particle iteration.

### 4.3. Machine Learning Based VNE Algorithms

Two-stage and coordinated VNE algorithms will lead to a sub-optimal solution because of artificial rules. Due to the development of ML [102,103] technologies, efficient VNE algorithms based on ML have received greater concern in satisfying the increasing diversity of applications and demands and to reduce search space. The authors of [104] formalize the virtual node mapping problem by using the Markov decision process (MDP) framework and devise node mappings for the proposed MDP using the Monte Carlo tree search algorithm.

Several works have appeared on the design of VNE solution using reinforcement learning (RL), which focuses on how to interact with the environment to achieve maximum cumulative return. The authors of [105] proposed a RL-based dynamic attribute matrix representation algorithm for VNE. Substrate network node information and link information are represented by an attribute matrix and an adjacency matrix. Furthermore, a novel approach, NeuroViNE, to speed up and improve the existing VNE algorithms has been proposed in [106], where NeuroViNE relies on Hopfield network to reduce search space and preprocess the problem. The Hopfield network is a form of recurrent neural network, which can extract whole valuable subgraphs and compute a probability for each node. The authors of [107] have developed a DRL-based VNE solution called DeepViNE. The key idea is to encode substrate and virtual networks as two-dimensional images.

In IWNs, the dynamic link quality and time-varying workload of the forwarding nodes make it intractable for the optimal anypath forwarding actions computation. To better learn the environment, deep Q-learning by combining deep neural network with Q-learning is used to solve VNE in IWNs [24]. Based on the above literature review, we summarize the representative references and list typical VNE algorithms in Table 1, arranged in different categories as given in section.

**Table 1.** VNE algorithm taxonomy.

| Algorithm | Network | Request Types | Objectives | Network Control | ILP | Node Ranking | Link Assignment | Reference |
|---|---|---|---|---|---|---|---|---|
| Two-stage | General | Dynamic | Revenue | N | Y | Available resource | KSP + splitting | [12] Yu et al. (2008) |
| | | | | | N | RW | KSP | [63] Zhang et al. (2012) |
| | | | Cost | N | Y | Candidate node set | Candidate path set | [97] Cao et al. (2018) |
| | | | Energy efficiency | N | Y | Residual CPU | SP | [76] Zhang et al. (2016) |
| | | | | | | Modified GRC | SP | [98] Cao et al. (2018) |
| | FiWi | Static | Survivablility | N | Y | Residual CPU | KSP | [36] Liu et al. (2019) |
| | Inter—ODCN | Dynamic | Cost | N | Y | Available resource | SP | [95] Jiang et al. (2015) |
| | | | Acceptance | Y | Y | Available resource | Candidate path set | [96]Pagès et al. (2019) |
| | EON | Static | Spectrum usage | N | Y | Random | KSP + splitting | [14] Shahriar et al. (2019) |
| Coordinated | General | Dynamic | Revenue | N | N | GRC | SP | [15] Gong et al. (2014) |
| | | | Cost | N | Y | Available resource | MCF + splitting | [99] Chowdhury et al. (2012) |
| | | | Energy efficiency+ Revenue | N | N | Candidate node set | SP | [101] Shahin et al. (2015) |
| | | | Revenue | N | Y | N/A | N/A | [100] Jarray et al. (2015) |
| | WSN | Dynamic | Revenue | N | N | N/A | anypath | [32] Li et al. (2017) |
| | EON | Static | Cost | N | Y | N/A | SP | [57] Lin et al. (2018) |
| | | | Spectrum usage | N | N | Random | KSP | [48] Xuan et al. (2017) |
| | Inter—ODCN | Static | Energy efficiency | Y | Y | Modified GRC | SP | [69] Zong et al. (2018) |
| | | Dynamic | Acceptance | Y | Y | Residual CPU | SP | [56] Fajjari et al. (2014) |
| ML | IWN | Static | Latency | Y | N | N/A | Anypath | [24] Li et al. (2019) |
| | General | Dynamic | Revenue + cost | N | N | Residual CPU | N/A | [106] Blenk et al. (2018) |
| | | | | | | N/A | N/A | [107] Dolati et al. (2019) |
| | | | Profit | N | N | MCTS | MCF | [104] Haeri et al. (2018) |

## 5. Issues and Challenges

Although VNE algorithms are currently undergoing a comprehensive research phase, there are still numerous challenges and research problems to be addressed, especially for ML-based algorithms. The existing works mainly focus on single network domain; furthermore, the VNE problem for multi-domain networks still challenge. Similar to other emerging technologies, network slicing brings forward a significant potential toward 5G, but introduces several technical and business challenges by regarding as architecture and deployment. In this section, we discuss the future challenges and experiences learned on the road of VNE approaches. Three main fields, but not limited to that, may be focused in the near future are identified: 5G architecture network slicing, field trial deployment, and ML-based approaches.

### 5.1. 5G Architecture Network Slicing

In comparison with the wireless network and optical network architectures, there are many challenges in the 5G multi-domain heterogeneous network (e.g., signal propagation, interference, user mobility, radio access technology, and optical signal). Network slicing towards 5G is envisioned to support multi-domain heterogeneous network with a widely range diverse set of performance and requirement services. Multi-domain 5G network orchestration has obtained more consideration in network slicing. Slicing the physical network into multiple isolated logical networks to support various VN requirements has emerged as a key solution to management the network resources. The authors of [35] have considered the user mobility in 5G network scenario. Some survey works for wireless network virtualization and 5G have been proposed [108,109], which have analyzed the state-of-the-art and challenges.

To implement and manage network slices, spectrum slicing problem and efficient bandwidth resources sharing among different slices should be solved according to the requirements. To solve this issue, the authors of [110] have presented a prototype in the C-RAN using Open Air Interface platform and SDN controller to validate the feasibility of configuring multiple slices on demands. According to the presented documentation by organizations such as ETSI, the network slice manager needs to follow the following features; services management, QoS, service composition, and service sharing. The network slice manager has been developed and validated in the multimedia real-time communications over optical network considering two network slices with different QoS [111].

In addition, due to the connectivity guarantee of the heterogeneous characteristics of the IoT ubiquitous network, resource allocation and energy efficiency improvement are challenging for the 5G scenario. Intelligent VNE for IWNs pervasive devices with sensing, processing, and connecting capabilities has been described in [24]. Furthermore, many 5G scenarios by regarding as IoV need to be addressed and discussed in the future.

### 5.2. Field Trial Deployment

As many researchers have focused on the network slicing and VNE approaches by simulation [4,24,112], how to evaluate the network performance using tools and experiment is one of the challenges. Net2Plan (http://www.net2plan.com/) is an open source Java-based network planning optimization software tool, which is designed with the aim to overcome the barriers imposed by existing network planning tools to integrate customized algorithm of users. Net2Plan can define a network representation, based on abstract concepts such as nodes, links, traffic demands, routes, protection segments, shared-risk groups, and network layers [113,114]. The authors of [115] have demonstrated an open source Net2Plan extension interfacing multiple OpenStack instances for enabling multi-datacenter IT resource management, with multi-tenant slicing in an ETSI-OSM orchestrated and ONOS-controlled IP over WDM transport network. However, intelligent functions and modules are still need to be addressed in the future works for the deployment tools.

As the growing of some enabling techniques (e.g., SDN), the deployment of network virtualization has obtained more considerations [26,27], where OpenDaylight and OpenStack are the most common tools to establish the platform for resource provisioning and demonstration [116–119]. To offer the implementation of network virtualization according to the diverse requirements of applications, SDN is an ideal platform through hypervisors such as Flowvisor, OpenVirteX, FlowN, and AutoVFlow [24]. The authors of [120] have focused on implementation of automatic network slicing for microservices, where open source software Node-RED is modified and extended to design IoT services for implementation. Open Air Interface platform and FlexRan controller are used for network slicing implementation of C-RAN for eMBB and IoT slices [110,121]. For vertical services slicing and orchestration solutions in 5G infrastructures, eMBB network slices instantiated interconnecting physical and virtual functions, provisioned and configured on-demand have been proposed in [122]. The authors of [123] have described a programmable optical software-defined network testbed, which has been upgraded to offer backhaul and fronthaul transport capabilities in support of C-RAN functionalities with increased reliability. For the inter elastic ODCNs domain, proposed feasible virtualized bandwidth variable transceiver (V-BVT) architecture for network slicing implementation has been demonstrated by an experimental platform with SDN controller to maintain the coexisting and isolation features in the physical layer in [124,125].

To implement and valid slicing in 5G networks, researchers should keep their eyes on the implementation technologies and devices. Field trial deployment for network slicing on multi-domain heterogeneous 5G architectures to support 5G services still have many challenges and should obtain more concerns.

### 5.3. Machine Learning Based Management Algorithm

Due to high-bandwidth and low-latency applications increasing the burden of network, network management and resource allocation need more dynamic and self-adaptive approaches to address the problem. VNE has obtained more concerns due to its importance for 5G network slicing. Some existing works [106,107] have addressed the issues by considering ML technique to learn how to allocate resource and manage the service request itself automatically. However, more ML-based algorithms for VNE need to be proposed in the future to obtain better performance metrics (e.g., profit, latency, energy efficiency, and survivability) for network management dynamically [126].

To provide automatic embedding solutions, the authors of [127] have proposed a novel algorithm combining reinforcement learning with a novel neural network structure for general network. In [128], the authors have proposed an efficient VNE algorithm adopting parallel reinforcement learning framework with graph convolutional network. Asynchronous advantage actor–critic-based policy gradient method is selected to train the network parameters. Simulation results of the proposed algorithm outperforms the typical VNE algorithms. However, majority latest ML-based VNE algorithms are based on the general network architecture without consideration additional characteristics such as optical network nodes and links. The generality of ML-based VNE algorithms should obtain more consideration for multi-domain heterogeneous network architecture.

In addition, the analysis of traffic demands can reveal valuable information for the management to gather information interacting with SDN to centralize control network. Traffic prediction strategy is essential to predict future traffic matrix via its prior measurements, where services can be provisioned taking into account future resource needs [129]. Some existing works have focused on ML-based traffic prediction strategies [130,131]. Recurrent neural networks have been designed for sequence prediction problem to optimize resource allocation of optical backbone network, where gate recurrent units (GRU) in RNN have been considered to achieve great accuracy [93]. Convolutional neural network (CNN) and long short-term memory (LSTM) are integrated for modeling and estimating the future network traffic [131]. Furthermore, ML-based traffic prediction mechanisms for VNE need to be proposed in the future.

## 6. Conclusions

This paper has presented a survey of existing works on the VNE problem towards multi-domain heterogeneous converged optical network, which have focused on the resource allocation optimization of multiple virtual networks coexisting and sharing resource in substrate networks. We have pointed out the features of the multi-domain heterogeneous 5G network architectures, where special constraints have to be considered for VNE according to the features of various network architecture (e.g., wireless network). The basic VNE problem with motivation and performance metrics has been described in details for general network scenario. A VNE algorithm taxonomy has been proposed for analyzing the existing VNE algorithms according to two-stage, coordinated, and machine learning-based algorithms. We have analyzed the issues and challenges of VNE towards multi-domain heterogeneous network, and pointed out some promising research directions: 5G architecture network slicing and field trail deployment for VNE- and ML-based management algorithms for resource allocation.

**Author Contributions:** Conceptualization, Y.Z. and Y.L.; methodology, Y.Z. and Y.G.; Project administration, Y.L. and L.G.; Formal analysis, Y.Z. and C.F.; Writing—original draft, Y.Z.; Writing—review and editing, Y.L. and L.G. All authors have read and agreed to the published version of the manuscript

**Funding:** This work is partly supported by National Key R&D Program of China (2018YFE0206800), National Natural Science Foundation of China (61775033, 61771120).

**Conflicts of Interest:** The authors declare no conflicts of interest.

## References

1. Ordonez-Lucena, J.; Ameigeiras, P.; Lopez, D.; Ramos-Munoz, J.J.; Lorca, J.; Folgueira, J. Network Slicing for 5G with SDN/NFV: Concepts, Architectures, and Challenges. *IEEE Commun. Mag.* **2017**, *55*, 80–87. [CrossRef]
2. Foukas, X.; Patounas, G.; Elmokashfi, A.; Marina, M.K. Network Slicing in 5G: Survey and Challenges. *IEEE Commun. Mag.* **2017**, *55*, 94–100. [CrossRef]
3. Zhang, S. An Overview of Network Slicing for 5G. *IEEE Wirel. Commun.* **2019**, *26*, 111–117. [CrossRef]
4. Addad, R.; Bagaa, M.; Taleb, T.; Cadette Dutra, D.L.; Flinck, H. Optimization Model for Cross-Domain Network Slices in 5G Networks. *IEEE Trans. Mob. Comput.* **2019**, *19*, 1156–1169. [CrossRef]
5. Afolabi, I.; Taleb, T.; Samdanis, K.; Ksentini, A.; Flinck, H. Network Slicing and Softwarization: A Survey on Principles, Enabling Technologies, and Solutions. *IEEE Commun. Surv. Tutorials* **2018**, *20*, 2429–2453. [CrossRef]
6. Bizanis, N.; Kuipers, F.A. SDN and Virtualization Solutions for the Internet of Things: A Survey. *IEEE Access* **2016**, *4*, 5591–5606. [CrossRef]
7. BinSahaq, A.; Sheltami, T.; Salah, K. A Survey on Autonomic Provisioning and Management of QoS in SDN Networks. *IEEE Access* **2019**, *7*, 73384–73435. [CrossRef]
8. Wang, A.; Iyer, M.; Dutta, R.; Rouskas, G.N.; Baldine, I. Network Virtualization: Technologies, Perspectives, and Frontiers. *J. Lightwave Technol.* **2013**, *31*, 523–537. [CrossRef]
9. Chowdhury, N.M.M.K.; Boutaba, R. Network Virtualization: State of the Art and Research Challenges. *IEEE Commun. Mag.* **2009**, *47*, 20–26. [CrossRef]
10. Chowdhury, N.M.M.K.; Boutaba, R. A Survey of Network Virtualization. *Comput. Netw.* **2010**, *54*, 862–876. [CrossRef]
11. Wang, Y.; McNulty, Z.; Nguyen, H. Network Virtualization in Spectrum Sliced Elastic Optical Path Networks. *J. Lightwave Technol.* **2017**, *35*, 1962–1970. [CrossRef]
12. Yu, M.; Yi, Y.; Rexford, J.; Chiang, M. Rethinking Virtual Network Embedding: Substrate Support for Path Splitting and Migration. *ACM SIGCOMM Comput. Commun. Rev.* **2008**, *38*, 19–29. [CrossRef]
13. Cheng, X.; Su, S.; Zhang, Z.; Shuang, K.; Yang, F.; Luo, Y.; Wang, J. Virtual Network Embedding through Topology Awareness and Optimization. *Comput. Netw.* **2012**, *56*, 1797–1813. [CrossRef]

14. Shahriar, N.; Taeb, S.; Chowdhury, S.R.; Tornatore, M.; Boutaba, R.; Mitra, J.; Hemmati, M. Achieving a Fully-Flexible Virtual Network Embedding in Elastic Optical Networks. In Proceedings of the IEEE INFOCOM 2019—IEEE Conference on Computer Communications, Paris, France, 29 April–2 May 2019; pp. 1756–1764.
15. Gong, L.; Wen, Y.; Zhu, Z.; Lee, T. Toward Profit-seeking Virtual Network Embedding Algorithm via Global Resource Capacity. In Proceedings of the IEEE INFOCOM 2014 - IEEE Conference on Computer Communications, Toronto, ON, Canada, 27 April–2 May 2014; pp. 1–9.
16. Davalos, E.J.; Baran, B. A Survey on Algorithmic Aspects of Virtual Optical Network Embedding for Cloud Networks. *IEEE Access* **2018**, *6*, 20896–20906. [CrossRef]
17. Bari, M.F.; Boutaba, R.; Esteves, R.; Granville, L.Z.; Podlesny, M.; Rabbani, M.G.; Zhang, Q.; Zhani, M.F. Data Center Network Virtualization: A Survey. *IEEE Commun. Surv. Tutor.* **2013**, *15*, 909–928. [CrossRef]
18. Singh, S.; Jeong, Y.S.; Park, J.H. A Survey on Cloud Computing Security: Issues, Threats, and Solutions. *J. Netw. Comput. Appl.* **2016**, *75*, 200–222. [CrossRef]
19. Taleb, T.; Afolabi, I.; Bagaa, M. Orchestrating 5G Network Slices to Support Industrial Internet and to Shape Next-Generation Smart Factories. *IEEE Netw.* **2019**, *33*, 146–154. [CrossRef]
20. Zhou, Y.; Yin, S.; Guo, B.; Huang, H.; Li, W.; Zhang, M.; Huang, S. Experimental Demonstration of Software-Defined Optical Network for Heterogeneous Packet and Optical Networks. *Photonic Netw. Commun.* **2016**, *32*, 329–335. [CrossRef]
21. Zhou, Y.; Ramamurthy, B.; Guo, B.; Huang, S. Supporting Dynamic Bandwidth Adjustment Based on Virtual Transport Link in Software-Defined IP Over Optical Networks. *J. Opt. Commun. Netw.* **2018**, *10*, 125–137. [CrossRef]
22. Yang, H.; Zhang, J.; Zhao, Y.; Li, H.; Huang, S.; Ji, Y.; Han, J.; Lin, Y.; Lee, Y. Cross Stratum Resilience for OpenFlow-enabled Data Center Interconnection with Flexi-Grid Optical Networks. *Opt. Switch. Netw.* **2014**, *11*, 72–82. [CrossRef]
23. Han, Y.; Hyun, J.; Hong, J.W.K. Graph Abstraction based Virtual Network Management Framework for SDN. In Proceedings of the 2016 IEEE Conference on Computer Communications Workshops (INFOCOM WKSHPS), San Francisco, CA, USA, 10–14 April 2016; pp. 884–885.
24. Li, M.; Chen, C.; Hua, C.; Guan, X. Intelligent Latency-Aware Virtual Network Embedding for Industrial Wireless Networks. *IEEE IoT J.* **2019**, *6*, 1–6. [CrossRef]
25. Medhat, A.M.; Taleb, T.; Elmangoush, A.; Carella, G. A.; Covaci, S.; Magedanz, T. Service Function Chaining in Next Generation Networks: State of the Art and Research Challenges. *IEEE Commun. Mag.* **2017**, *55*, 216–223. [CrossRef]
26. Hammad, A.; Aguado, A.; Peng, S.; Vilalta, R.; Mayoral, A.; Casellas, R.; Martínez, R.; Muñoz, R.; Nejabati, R.; Simeonidou, D. On-demand Virtual Infrastructure Composition over Multi- domain and Multi-technology Networks. In Proceedings of the IEEE Optical Fiber Communications Conference and Exhibition (OFC), Anaheim, CA, USA, 20–24 March 2016; pp. 4–6.
27. Martínez, R.; Mayoral, A.; Vilalta, R.; Casellas, R.; Muñoz, R.; Pachnicke, S.; Szyrkowiec, T.; Autenrieth, A. Integrated SDN/NFV Orchestration for the Dynamic Deployment of Mobile Virtual Backhaul Networks Over a Multilayer (Packet/Optical) Aggregation Infrastructure. *J. Opt. Commun. Netw.* **2017**, *9*, A135–A142. [CrossRef]
28. Wang, X.; Cavdar, C.; Wang, L.; Tornatore, M.; Zhao, Y.; Chung, H.; Lee, H.H.; Park, S.; Mukherjee, B. Joint Allocation of Radio and Optical Resources in Virtualized Cloud RAN with CoMP. In Proceedings of the 2016 IEEE Global Communications Conference (GLOBECOM), Washington, DC, USA, 4–8 December 2016; pp. 1–5.
29. Hu, Y.C.; Patel, M.; Sabella, D.; Sprecher, N.; Young, V. Mobile Edge Computing—A key technology towards 5G. *ETSI White Pap.* **2015**, *11*, 1–16.
30. Abbas, N.; Zhang, Y.; Taherkordi, A.; Skeie, T. Mobile Edge Computing: A Survey. *IEEE IoT J.* **2018**, *5*, 450–465. [CrossRef]
31. Khan, I.; Belqasmi, F.; Glitho, R.; Crespi, N.; Morrow, M.; Polakos, P. Wireless Sensor Network Virtualization: A Survey. *IEEE Commun. Surv. Tutor.* **2016**, *18*, 553–576. [CrossRef]
32. Li, M.; Hua, C.; Chen, C.; Guan, X. Application-driven Virtual Network Embedding for Industrial Wireless Sensor Networks. In Proceedings of the 2017 IEEE International Conference on Communications (ICC), Paris, France, 21–25 May 2017; pp. 1–6.

33. Yun, D.; Ok, J.; Shin, B.; Park, S.; Yi, Y. Embedding of Virtual Network Requests over Static Wireless Multihop Networks. *Comput. Netw.* **2013**, *57*, 1139–1152. [CrossRef]
34. Lv, P.; Wang, X.; Xu, M. Virtual Access Network Embedding in Wireless Mesh Networks. *Ad Hoc Netw.* **2012**, *10*, 1362–1378. [CrossRef]
35. Guan, Y.; Zong, Y.; Liu, Y.; Guo, L.; Ning, Z.; Rodrigues, J.J.P.C. Virtual Network Embedding Supporting User Mobility in 5G Metro/Access Networks. In Proceedings of the ICC 2019—2019 IEEE International Conference on Communications (ICC), Shanghai, China, 20–24 May 2019, pp. 1–7.
36. Liu, Y.; Han, P.; Hou, J.; Zheng, J. Resource-Efficiently Survivable IoT Services Provisioning via Virtual Network Embedding in Fiber-Wireless Access Network. *IEEE Access* **2019**, *7*, 65007–65018. [CrossRef]
37. Han, P.; Liu, Y.; Guo, L. QoS Satisfaction Aware and Network Reconfiguration Enabled Resource Allocation for Virtual Network Embedding in Fiber-Wireless Access Network. *Comput. Netw.* **2018**, *143*, 30–48. [CrossRef]
38. Wang, W.; Guo, W.; Hu, W. Network Service Slicing Supporting Ubiquitous Access in Passive Optical Networks. In Proceedings of the 2018 20th International Conference on Transparent Optical Networks (ICTON), Bucharest, Romania, 1–5 July 2018; pp. 1–3.
39. Han, P.; Guo, L.; Liu, Y. Virtual Network Embedding in SDN/NFV based Fiber-Wireless Access Network. In Proceedings of the International Conference on Software Networking, Jeju, South Korea, 23–26 May 2016; pp. 1–5.
40. Mosahebfard, M.; Vardakas, J.; Ramantas, K.; Verikoukis, C. SDN/NFV-based Network Resource Management for Converged Optical-wireless Network Architectures. In Proceedings of the 2019 21st International Conference on Transparent Optical Networks (ICTON), Angers, France, 9–13 July 2019; pp. 1–4.
41. Wang, Q.; Shou, G.; Liu, J.; Liu, Y.; Hu, Y.; Guo, Z. Resource Allocation for Edge Computing over Fibre-wireless Access Networks. *IET Commun.* **2019**, *13*, 2848–2856. [CrossRef]
42. Rahman, S.; Gupta, A.; Tomatore, M.; Mukherjee, B. Dynamic Workload Migration over Optical Backbone Network to Minimize Data Center Electricity Cost. *IEEE Trans. Green Commun. Netw.* **2017**, *2*, 570–579. [CrossRef]
43. Zhang, J.; Ji, Y.; Song, M.; Li, H.; Gu, R.; Zhao, Y.; Zhang, J. Dynamic Virtual Network Embedding over Multilayer Optical Networks. *J. Opt. Commun. Netw.* **2015**, *7*, 918–927. [CrossRef]
44. Rodriguez, E.; Alkmim, G.P.; Da Fonseca, N.L.; Batista, D.M. Energy-Aware Mapping and Live Migration of Virtual Networks. *IEEE Syst. J.* **2017**, *11*, 637–648. [CrossRef]
45. Taeb, S.; Shahriar, N.; Chowdhury, S.R.; Tornatore, M.; Boutaba, R. Virtual Network Embedding with Path-based Latency Guarantees in Elastic Optical Networks. In Proceedings of the 2019 IEEE 27th International Conference on Network Protocols (ICNP), Chicago, IL, USA, 8–10 October 2019.
46. Huang, S.; Zhou, Y.; Yin, S.; Kong, Q.; Zhang, M.; Zhao, Y.; Zhang, J.; Gu, W. Fragmentation Assessment based On-line Routing and Spectrum Allocation for Intra-data-center Networks with Centralized Control. *Opt. Switch. Netw.* **2014**, *14*, 274–281. [CrossRef]
47. Klonidis, D.; Cugini, F.; Gerstel, O.; Jinno, M.; Lopez, V.; Palkopoulou, E.; Sekiya, M.; Siracusa, D.; Thouénon, G.; Betoule, C. Spectrally and Spatially Flexible Optical Network Planning and Operations. *IEEE Commun. Mag.* **2015**, *53*, 69–78. [CrossRef]
48. Xuan, H.; Wang, Y.; Xu, Z.; Hao, S.; Wang, X. Virtual Optical Network Mapping and Core Allocation in Elastic Optical Networks using Multi-Core Fibers. *Opt. Commun.* **2017**, *402*, 26–35. [CrossRef]
49. Huang, H.; Huang, S.; Yin, S.; Zhang, M.; Zhang, J.; Gu, W. Virtual Network Provisioning Over Space Division Multiplexed Optical Networks Using Few-Mode Fibers. *J. Opt. Commun. Netw.* **2016**, *8*, 726–733. [CrossRef]
50. Ou, Y.; Hammad, A.; Peng, S.; Nejabati, R.; Simeonidou, D. Online and Offline Virtualization of Optical Transceiver. *J. Opt. Commun. Netw.* **2015**, *7*, 748–760. [CrossRef]
51. Li, X.; Zhang, L.; Tang, Y.; Gao, T.; Zhang, Y.; Huang, S. On-demand Routing, Modulation Level and Spectrum Allocation (OD-RMSA) for Multicast Service Aggregation in Elastic Optical Networks. *Opt. Express* **2018**, *26*, 24506. [CrossRef]
52. Gao, T.; Li, X.; Guo, B.; Yin, S.; Li, W.; Huang, S. Spectrum-efficient Multipath Provisioning with Content Connectivity for the Survivability of Elastic Optical Datacenter Networks. *Opt. Fiber Technol.* **2017**, *36*, 353–365. [CrossRef]

53. Guo, B.; Shang, Y.; Zhang, Y.; Li, W.; Yin, S.; Zhang, Y.; Huang, S. Timeslot Switching-Based Optical Bypass in Data Center for Intrarack Elephant Flow with an Ultrafast DPDK-Enabled Timeslot Allocator. *J. Lightwave Technol.* **2019**, *37*, 2253–2260. [CrossRef]
54. Li, X.; Zhang, L.; Tang, Y.; Guo, J.; Huang, S. Distributed Sub-Tree-Based Optical Multicasting Scheme in Elastic Optical Data Center Networks. *IEEE Access* **2018**, *6*, 6464–6477. [CrossRef]
55. Li, X.; Yin, S.; Wang, X.; Zhou, Y.; Zhao, Y.; Huang, S.; Zhang, J. Content Placement With Maximum Number of End-to-Content Paths in k-Node (Edge) Content Connected Optical Datacenter Networks. *J. Opt. Commun. Netw.* **2017**, *9*, 53–66. [CrossRef]
56. Fajjari, I.; Aitsaadi, N.; Pióro, M.; Pujolle, G. A New Virtual Network Static Embedding Strategy within the Cloud's Private Backbone Network. *Comput. Netw.* **2014**, *62*, 69–88. [CrossRef]
57. Lin, R.; Luo, S.; Zhou, J.; Wang, S.; Cai, A.; Zhong, W.; Zukerman, M. Virtual Network Embedding with Adaptive Modulation in Flexi-Grid Networks. *J. Lightwave Technol.* **2018**, *36*, 3551–3563. [CrossRef]
58. Huang, H.; Guo, B.; Li, X.; Yin, S.; Zhou, Y.; Huang, S. Crosstalk-aware Virtual Network Embedding over Inter-datacenter Optical Networks with Few-mode Fibers. *Opt. Fiber Technol.* **2017**, *39*, 70–77. [CrossRef]
59. Zhu, M.; Zhang, S.; Sun, Q.; Li, G.; Chen, B.; Gu, J. Fragmentation-aware VONE in elastic optical networks. *J. Opt. Commun. Netw.* **2018**, *10*, 809–822. [CrossRef]
60. Montero, R.; Agraz, F.; Pages, A.; Spadaro, S. End-to-End 5G Service Deployment and Orchestration in Optical Networks with QoE Guarantees. In Proceedings of the 2018 20th International Conference on Transparent Optical Networks (ICTON), Bucharest, Romania, 1–5 July 2018.
61. Cao, H.; Yang, L.; Zhu, H. Novel Node-Ranking Approach and Multiple Topology Attributes-Based Embedding Algorithm for Single-Domain Virtual Network Embedding. *IEEE IoT J.* **2018**, *5*, 108–120. [CrossRef]
62. Chochlidakis, G.; Friderikos, V. Mobility Aware Virtual Network Embedding. *IEEE Trans. Mob. Comput.* **2017**, *16*, 1343–1356. [CrossRef]
63. Zhang, S.; Qian, Z.; Wu, J.; Lu, S. An Opportunistic Resource Sharing and Topology-aware Mapping Framework for Virtual Networks. In Proceedings of the 2012 IEEE INFOCOM, Orlando, FL, USA, 25–30 March 2012; pp. 2408–2416.
64. Chen, T.; Liu, J.; Tang, Q.; Huang, T.; Huo, R. Virtual Network Embedding Algorithm for Location-Based Identifier Allocation. *IEEE Access* **2019**, *7*, 31159–31169. [CrossRef]
65. Cheng, X.; Su, S.; Zhang, Z.; Wang, H.; Yang, F.; Luo, Y.; Wang, J. Virtual Network Embedding through Topology-aware Node Ranking. *ACM SIGCOMM Comput. Commun. Rev.* **2011**, *41*, 38–47. [CrossRef]
66. Zhu, M.; Sun, Q.; Zhang, S.; Gao, P.; Chen, B.; Gu, J. Energy-Aware Virtual Optical Network Embedding in Sliceable- Transponder- Enabled Elastic Optical Networks. *IEEE Access* **2019**, *7*, 41897–41912. [CrossRef]
67. Ning, Z.; Huang, J.; Wang, X.; Rodrigues, J.J.P.C.; Guo, L. Mobile Edge Computing-Enabled Internet of Vehicles: Toward Energy-Efficient Scheduling. *IEEE Netw.* **2019**, *33*, 1–8. [CrossRef]
68. Chochlidakis, G.; Friderikos, V. Low Latency Virtual Network Embedding for Mobile Networks. In Proceedings of the 2016 IEEE International Conference on Communications (ICC), Kuala Lumpur, Malaysia, 22–27 May 2016; pp. 1–6.
69. Zong, Y.; Ou, Y.; Hammad, A.; Kondepu, K.; Nejabati, R.; Simeonidou, D.; Liu, Y.; Guo, L. Location-Aware Energy Efficient Virtual Network Embedding in Software-Defined Optical Data Center Networks. *J. Opt. Commun. Netw.* **2018**, *10*, 58–70. [CrossRef]
70. Hejja, K.; Hesselbach, X. Online Power Aware Coordinated Virtual Network Embedding with 5G Delay Constraint. *J. Netw. Comput. Appl.* **2018**, *124*, 121–136. [CrossRef]
71. Song, C.; Zhang, M.; Zhan, Y.; Wang, D.; Guan, L.; Liu, W.; Zhang, L.; Xu, S. Hierarchical Edge Cloud Enabling Network Slicing for 5G Optical Fronthaul. *J. Opt. Commun. Netw.* **2019**, *11*, B60–B70. [CrossRef]
72. Nonde, L.; Elgorashi, T.E.; Elmirgahni, J.M. Virtual Network Embedding Employing Renewable Energy Sources. In Proceedings of the 2016 IEEE Global Communications Conference (GLOBECOM), Washington, DC, USA, 4–8 December 2016; pp. 1–6.
73. Zhu, M.; Gao, P.; Zhang, J.; Zeng, X.; Zhang, S. Energy Efficient Dynamic Virtual Optical Network Embedding in Sliceable-Transponder-Equipped EONs. In Proceedings of the GLOBECOM 2017—2017 IEEE Global Communications Conference, Singapore, 4–8 December 2017; pp. 1–6.
74. Nonde, L.; El-gorashi, T.E.H.; Elmirghani, J.M.H. Energy Efficient Virtual Network Embedding for Cloud Networks. *J. Lightwave Technol.* **2015**, *33*, 1828–1849. [CrossRef]

75. Xiong, Y.; Shi, J.; Yang, Y.; Lv, Y.; Rouskas, G.N. Lightpath Management in SDN-Based Elastic Optical Networks with Power Consumption Considerations. *J. Lightwave Technol.* **2018**, *36*, 1650–1660. [CrossRef]
76. Zhang, Z.; Su, S.; Shuang, K.; Li, W.; Zia, M.A. Energy Aware Virtual Network Migration. In Proceedings of the GLOBECOM 2016—2016 IEEE Global Communications Conference, Washington, DC, USA, 4–8 December 2016; pp. 1–6.
77. Guo, B.; Qiao, C.; Wang, J.; Yu, H.; Zuo, Y.; Li, J.; Chen, Z.; He, Y. Survivable Virtual Network Design and Embedding to Survive a Facility Node Failure. *J. Lightwave Technol.* **2014**, *32*, 483–493. [CrossRef]
78. Li, X.; Gao, T.; Zhang, L.; Tang, Y.; Zhang, Y.; Huang, S. Survivable K-Node (Edge) Content Connected Virtual Optical Network (KC-VON) Embedding Over Elastic Optical Data Center Networks. *IEEE Access* **2018**, *6*, 38780–38793. [CrossRef]
79. Su, Y.; Meng, X.; Kang, Q.; Han, X. Survivable Virtual Network Link Protection Method Based on Network Coding and Protection Circuit. *IEEE Access* **2018**, *6*, 67477–67493. [CrossRef]
80. Khan, A.; An, X.; Iwashina, S. Virtual Network Embedding for telco-grade Network Protection and Service Availability. *Comput. Commun.* **2016**, *84*, 25–38. [CrossRef]
81. Chowdhury, S.R.; Ahmed, R.; Khan, M.M.A.; Shahriar, N.; Boutaba, R.; Mitra, J.; Zeng, F. Dedicated Protection for Survivable Virtual Network Embedding. *IEEE Trans. Netw. Serv. Manag.* **2016**, *13*, 913–926. [CrossRef]
82. Ayoubi, S.; Chen, Y.; Assi, C. Towards Promoting Backup-Sharing in Survivable Virtual Network Design. *IEEE/ACM Trans. Netw.* **2016**, *24*, 3218–3231. [CrossRef]
83. Aguado, A.; Davis, M.; Peng, S.; Álvarez, M.V.; López, V.; Szyrkowiec, T.; Autenrieth, A.; Vilalta, R.; Mayoral, A.; Muñoz, R.; et al. Dynamic Virtual Network Reconfiguration over SDN Orchestrated Multitechnology Optical Transport Domains. *J. Lightwave Technol.* **2016**, *34*, 1933–1938. [CrossRef]
84. Kondepu, K.; Sgambelluri, A.; Sambo, N.; Giannone, F.; Castoldi, P.; Valcarenghi, L. Orchestrating Lightpath Recovery and Flexible Functional Split to Preserve Virtualized RAN Connectivity. *J. Opt. Commun. Netw.* **2018**, *10*, 843–851. [CrossRef]
85. Ramanathan, S.; Kondepu, K.; Mirkhanzadeh, B.; Razo, M.; Tacca, M.; Valcarenghi, L.; Fumagalli, A. Performance Evaluation of Two Service Recovery Strategies in Cloud-Native Radio Access Networks. In Proceedings of the 2019 21st International Conference on Transparent Optical Networks (ICTON), Angers, France, 9–13 July 2019; pp. 1–5.
86. Gao, T.; Zou, W.; Li, X.; Guo, B.; Huang, S.; Mukherjee, B. Distributed Sub-light-tree based Multicast Provisioning with Shared Protection in Elastic Optical Datacenter Networks. *Opt. Switch. Netw.* **2019**, *31*, 39–51. [CrossRef]
87. Andreoletti, D.; Troia, S.; Musumeci, F.; Giordano, S.; Maier, G.; Tornatore, M. Network Traffic Prediction based on Diffusion Convolutional Recurrent Neural Networks. In Proceedings of the IEEE INFOCOM 2019—IEEE Conference on Computer Communications Workshops (INFOCOM WKSHPS), Paris, France, 29 April–2 May 2019, pp. 1–6.
88. Cao, X.; Zhong, Y.; Zhou, Y.; Wang, J.; Zhu, C.; Zhang, W. Interactive Temporal Recurrent Convolution Network for Traffic Prediction in Data Centers. *IEEE Access* **2017**, *6*, 5276–5289. [CrossRef]
89. Singh, S.K.; Jukan, A. Machine-Learning-Based Prediction for Resource (Re)allocation in Optical Data Center Networks. *J. Opt. Commun. Netw.* **2018**, *10*, D12–D28. [CrossRef]
90. Mata, J.; de Miguel, I.; Durán, R.J.; Merayo, N.; Singh, S.K.; Jukan, A.; Chamania, M. Artificial Intelligence (AI) Methods in Optical Networks: A Comprehensive Survey. *Opt. Switch. Netw.* **2018**, *28*, 43–57. [CrossRef]
91. Li, Y.; Liu, H.; Yang, W.; Hu, D.; Wang, X.; Xu, W. Predicting Inter-Data-Center Network Traffic Using Elephant Flow and Sublink Information. *IEEE Trans. Netw. Serv. Manag.* **2016**, *13*, 782–792. [CrossRef]
92. Fadlullah, Z.M.; Tang, F.; Mao, B.; oKato, N.; Akashi, O.; Inoue, T.; Mizutani, K. State-of-the-Art Deep Learning: Evolving Machine Intelligence Toward Tomorrow's Intelligent Network Traffic Control Systems. *IEEE Commun. Surv. Tutor.* **2017**, *19*, 2432–2455. [CrossRef]
93. Troia, S.; Alvizu, R.; Zhou, Y.; Maier, G.; Pattavina, A. Deep Learning-Based Traffic Prediction for Network Optimization. In Proceedings of the International Conference on Transparent Optical Networks (ICTON), Bucharest, Romania, 1–5 July 2018; pp. 1–4.
94. Azzouni, A.; Pujolle, G. NeuTM: A Neural Network-based Framework for Traffic Matrix Prediction in SDN. In Proceedings of the NOMS 2018—2018 IEEE/IFIP Network Operations and Management Symposium, Taipei, Taiwan, 23–27 April 2018; pp. 1–5.

95. Jiang, H.; Wang, Y.; Gong, L.; Zhu, Z. Availability-Aware Survivable Virtual Network Embedding in Optical Datacenter Networks. *J. Opt. Commun. Netw.* **2015**, *7*, 1160–1171. [CrossRef]
96. Pagès, A.; Agraz, F.; Montero, R.; Landi, G.; Capitani, M.; Gallico, D.; Biancani, M.; Nejabati, R.; Simeonidou, D.; Spadaro, S. Orchestrating Virtual Slices in Data Centre Infrastructures with Optical DCN. *Opt. Fiber Technol.* **2019**, *50*, 36–49. [CrossRef]
97. Cao, H.; Zhu, Y.; Zheng, G.; Yang, L. A Novel Optimal Mapping Algorithm with Less Computational Complexity for Virtual Network Embedding. *IEEE Trans. Netw. Serv. Manag.* **2018**, *15*, 356–371. [CrossRef]
98. Cao, H.; Guo, Y.; Qu, Z.; Wu, S.; Zhu, H.; Yang, L. ER-VNE: A Joint Energy and Revenue Embedding Algorithm for Embedding Virtual Networks. *IEEE Access* **2018**, *6*, 47815–47827. [CrossRef]
99. Chowdhury, M.; Rahman, M.R.; Boutaba, R. ViNEYard: Virtual Network Embedding Algorithms with Coordinated Node and Link Mapping. *IEEE/ACM Trans. Netw.* **2012**, *20*, 206–219. [CrossRef]
100. Jarray, A.; Karmouch, A. Decomposition Approaches for Virtual Network Embedding with One-shot Node and Link mapping. *IEEE/ACM Trans. Netw.* **2015**, *23*, 1012–1025. [CrossRef]
101. Shahin, A.A. Memetic Multi-Objective Particle Swarm Optimization-Based Energy-Aware Virtual Network Embedding. *Int. J. Adv. Comput. Sci. Appl.* **2015**, *6*, 1–12.
102. Li, R.; Zhao, Z.; Sun, Q.; Chih-Lin, I.; Yang, C.; Chen, X.; Zhao, M.; Zhang, H. Deep Reinforcement Learning for Resource Management in Network Slicing. *IEEE Access* **2018**, *6*, 74429–74441. [CrossRef]
103. Sun, G.; Zemuy, G.T.; Xiong, K. Dynamic Reservation and Deep Reinforcement Learning based Autonomous Resource Management for Wireless Virtual Networks. In Proceedings of the International Performance Computing and Communications Conference (IPCCC), Orlando, FL, USA, 17–19 November 2018.
104. Haeri, S.; Trajković, L. Virtual Network Embedding via Monte Carlo Tree Search. *IEEE Trans. Cybern.* **2018**, *48*, 510–521. [CrossRef] [PubMed]
105. Yao, H.; Zhang, B.; Zhang, P.; Wu, S.; Jiang, C.; Guo, S. RDAM: A Reinforcement Learning Based Dynamic Attribute Matrix Representation for Virtual Network Embedding. *IEEE Trans. Emerg. Top. Comput.* **2018**, 1–13. [CrossRef]
106. Blenk, A.; Kalmbach, P.; Zerwas, J.; Jarschel, M.; Schmid, S.; Kellerer, W. NeuroViNE: A Neural Preprocessor for Your Virtual Network Embedding Algorithm. In Proceedings of the IEEE INFOCOM 2018—IEEE Conference on Computer Communications, Honolulu, HI, USA, 16–19 April 2018; pp. 405–413.
107. Dolati, M.; Hassanpour, S.B.; Ghaderi, M.; Khonsari, A. DeepViNE: Virtual Network Embedding with Deep Reinforcement Learning. In Proceedings of the IEEE INFOCOM 2019—IEEE Conference on Computer Communications Workshops (INFOCOM WKSHPS), Paris, France, 29 April–2 May 2019; pp. 879–885.
108. Kitindi, E.J.; Fu, S.; Jia, Y.; Kabir, A.; Wang, Y. Wireless Network Virtualization with SDN and C-RAN for 5G Networks: Requirements, Opportunities, and Challenges. *IEEE Access* **2017**, *5*, 19099–19115. [CrossRef]
109. Liang, C.; Yu, F.R. Wireless Network Virtualization: A Survey, Some Research Issues and Challenges. *IEEE Commun. Surv. Tutor.* **2015**, *17*, 358–380. [CrossRef]
110. Costanzo, S.; Fajjari, I.; Aitsaadi, N.; Langar, R. DEMO: SDN-based network slicing in C-RAN. In Proceedings of the 2018 15th IEEE Annual Consumer Communications and Networking Conference (CCNC), Las Vegas, NV, USA, 12–15 January 2018; pp. 1–2.
111. Alemany, P.; Vilalta, R.; De La Cruz, J.L.; Pol, A.; Román, A.; Casellas, R.; Martínez, R.; Muñoz, R. Experimental Validation of Network Slicing Management for Vertical Applications on Multimedia Real-time Communications over a Packet/optical Network. In Proceedings of the 2019 21st International Conference on Transparent Optical Networks (ICTON), Angers, France, 9–13 July 2019; pp. 3–6.
112. Zhao, C.; Parhami, B. Virtual Network Embedding Through Graph Eigenspace Alignment. *IEEE Trans. Netw. Serv. Manag.* **2019**, *16*, 632–646. [CrossRef]
113. Pavon-Marino, P.; Izquierdo-Zaragoza, J.L. Net2plan: An Open Source Network Planning Tool for Bridging the Gap between Academia and Industry. *IEEE Netw.* **2015**, *29*, 90–96. [CrossRef]
114. Romero-Gazquez, J.L.; Bueno-Delgado, M.V.; Moreno-Muro, F.J.; Pavon-Marino, P. Net2plan-GIS: An Open-Source Net2Plan Extension Integrating GIS Data for 5G Network Planning. In Proceedings of the 2018 20th International Conference on Transparent Optical Networks (ICTON), Bucharest, Romania, 1–5 July 2018.
115. Garrich, M.; Hernández-Bastida, M.; San-Nicolás-Martínez, C.; Moreno-Muro, F.J.; Pavon-Marino, P. *The Net2Plan-OpenStack Project: IT Resource Manager for Metropolitan SDN/NFV Ecosystems*; OFC: Auckland, New Zealand, 2019.

116. Szyrkowiec, T.; Autenrieth, A.; Gunning, P.; Wright, P.; Lord, A.; Elbers, J.-P.; Lumb, A. First Field Demonstration of Cloud Datacenter Workflow Automation Employing Dynamic Optical Transport Network Resources under OpenStack and OpenFlow Orchestration. *Opt. Express* **2014**, *22*, 2595–2602. [CrossRef]
117. Yang, H.; Zhang, J.; Ji, Y.; Tan, Y.; Lin, Y.; Han, J.; Lee, Y. Data Center Service Locationlization based on Virtual Resource Migration in software Defined Elastic Optical Network. In Proceedings of the IEEE Optical Fiber Communications Conference and Exhibition (OFC), Los Angeles, CA, USA, 22–26 March 2015.
118. Hammad, A.; Aguado, A.; Kondepu, K.; Zong, Y.; Marhuenda, J.; Yan, S.; Nejabati, R.; Simeonidou, D. Demonstration of NFV Content Delivery using SDN- enabled Virtual Infrastructures. In Proceedings of the IEEE Optical Fiber Communications Conference and Exhibition (OFC), Los Angeles, CA, USA, 19–23 March 2017.
119. Diallo, T.; Beldachi, A.F.; Muqaddas, A.S.; Silva, R.S.; Nejabati, R.; Tzanakaki, A.; Simeonidou, D. Enabling Heterogenous Low Latency and High- bandwidth Virtual Network Services for 5G Utilizing a Flexible Optical Transport Network. In Proceedings of the IEEE Optical Fiber Communications Conference and Exhibition (OFC), San Diego, CA, USA, 3–7 March 2019.
120. Minami, Y.; Taniguchi, A.; Kawabata, T.; Sakaida, N.; Shimano, K. An Architecture and Implementation of Automatic Network Slicing for Microservices. In Proceedings of the NOMS 2018 - 2018 IEEE/IFIP Network Operations and Management Symposium, Taipei, Taiwan, 23–27 April 2018; pp. 1–4.
121. Costanzo, S.; Cherrier, S.; Langar, R. Network Slicing Orchestration of IoT-BeC3applications and eMBB services in C-RAN. In Proceedings of the IEEE INFOCOM 2019—IEEE Conference on Computer Communications Workshops (INFOCOM WKSHPS), Paris, France, 29 April–2 May 2019; pp. 975–976.
122. Landi, G.; Giardina, P.; Capitani, M.; Kondepu, K.; Valcarenghi, L.; Avino, G. Provisioning and automated scaling of network slices for virtual Content Delivery Networks in 5G infrastructures. In Proceedings of the ACM International Symposium on Mobile Ad Hoc Networking and Computing, Catania, Italy, 2–5 July 2019; pp. 397–398.
123. Ramanathan, S.; Tacca, M.; Razo, M.; Mirkhanzadeh, B.; Kondepu, K.; Giannone, F.; Valcarenghi, L.; Fumagalli, A. A programmable optical network testbed in support of C-RAN: a reliability study. *Photonic Netw. Commun.* **2019**, *37*, 311–321. [CrossRef]
124. Ou, Y.; Yan, S.; Hammad, A.; Guo, B.; Peng, S.; Nejabati, R.; Simeonidou, D. Demonstration of Virtualizeable and Software-Defined Optical Transceiver. *J. Lightwave Technol.* **2016**, *34*, 1916–1924. [CrossRef]
125. Ou, Y.; Davis, M.; Aguado, A.; Meng, F.; Nejabati, R.; Simeonidou, D. Optical Network Virtualisation Using Multitechnology Monitoring and SDN-Enabled Optical Transceiver. *J. Lightwave Technol.* **2018**, *36*, 1890–1898. [CrossRef]
126. Wang, S.; Bi, J.; Wu, J.; Vasilakos, A.V.; Fan, Q. VNE-TD: A Virtual Network Embedding Algorithm Based on Temporal- Difference Learning. *Comput. Netw.* **2019**, *161*, 251–263. [CrossRef]
127. Yan, Z.; Ge, J.; Wu, Y.; Zheng, H.; Li, L.; Li, T. Automatic Virtual Network Embedding based on Deep Reinforcement Learning. In Proceedings of the IEEE International Conference on High Performance Computing and Communications, IEEE International Conference on Smart City and IEEE International Conference on Data Science and Systems, Zhangjiajie, China, 10–12 August 2019; pp. 625–631.
128. Yan, Z.; Ge, J.; Wu, Y.; Li, L.; Li, T. Automatic Virtual Network Embedding: A Deep Reinforcement Learning Approach with Graph Convolutional Networks. *IEEE J. Sel. Areas Commun.* **2020**; doi:10.1109/jsac.2020.2986662. [CrossRef]
129. Zhang, H.; Zheng, X.; Tian, J. Virtual Network Mapping based on the Prediction of Support Vector Machine. In Proceedings of the 2016 8th International Conference on Information Technology in Medicine and Education (ITME), Fuzhou, China, 23–25 December 2017; pp. 853–858.
130. Alvizu, R.; Troia, S.; Maier, G.; Pattavina, A. Matheuristic With Machine-Learning-Based Prediction for Software- Defined Mobile Metro-Core Networks. *J. Opt. Commun. Netw.* **2017**, *9*, D19–D30. [CrossRef]
131. Le, V.A.; Nguyen, P.L.; Ji, Y. Deep Convolutional LSTM Network-based Traffic Matrix Prediction with Partial Information. In Proceedings of the 2019 IFIP/IEEE Symposium on Integrated Network and Service Management (IM), Arlington, TX, USA, 8–12 April 2019; pp. 261–269.

sensors

Letter

# Bragg Grating Assisted Sagnac Interferometer in $SiO_2$-$Al_2O_3$-$La_2O_3$ Polarization-Maintaining Fiber for Strain–Temperature Discrimination

Zhifang Wu [1,*], Peili Wu [2,3], Maryna Kudinova [4], Hailiang Zhang [2,3], Perry Ping Shum [2,3], Xuguang Shao [2], Georges Humbert [4], Jean-Louis Auguste [4], Xuan Quyen Dinh [2,3,5] and Jixiong Pu [1]

1 Fujian Key Laboratory of Light Propagation and Transformation, College of Information Science and Engineering, Huaqiao University, Xiamen 361021, China; jixiong@hqu.edu.cn

2 School of Electrical and Electronic Engineering, Nanyang Technological University, 50 Nanyang Avenue, Singapore 639798, Singapore; WUPE0004@e.ntu.edu.sg (P.W.); hzhang023@e.ntu.edu.sg (H.Z.); shum@ieee.org (P.P.S.); XGshao@ntu.edu.sg (X.S.); XDDinh@ntu.edu.sg (X.Q.D.)

3 CINTRA CNRS/NTU/THALES, UMI 3288, 50 Nanyang Drive, Singapore 637553, Singapore

4 XLIM Research Institute, UMR 7252 CNRS/University of Limoges, 123 Avenue Albert Thomas, 87060 Limoges, France; marynakudinova@gmail.com (M.K.); georges.humbert@xlim.fr (G.H.); jean-louis.auguste@xlim.fr (J.-L.A.)

5 Thales Solutions Asia Pte Ltd., R&T Department, 21 Changi North Rise, Singapore 498788, Singapore

* Correspondence: zfwu@hqu.edu.cn

Received: 4 August 2020; Accepted: 21 August 2020; Published: 24 August 2020

**Abstract:** Polarization-maintaining fibers (PMFs) have always received great attention in fiber optic communication systems and components which are sensitive to polarization. Moreover, they are widely applied for high-accuracy detection and sensing devices, such as fiber gyroscope, electric/magnetic sensors, multi-parameter sensors, and so on. Here, we demonstrated the combination of a fiber Bragg grating (FBG) and Sagnac interference in the same section of a new type of PANDA-structure PMF for the simultaneous measurement of axial strain and temperature. This specialty PMF features two stress-applied parts made of lanthanum-aluminum co-doped silicate ($SiO_2$-$Al_2O_3$-$La_2O_3$, SAL) glass, which has a higher thermal expansion coefficient than borosilicate glass used commonly in commercial PMFs. Furthermore, the FBG inscribed in this SAL PMF not only aids the device in discriminating strain and temperature, but also calibrates the phase birefringence of the SAL PMF more precisely thanks to the much narrower bandwidth of grating peaks. By analyzing the variation of wavelength interval between two FBG peaks, the underlying mechanism of the phase birefringence responding to temperature and strain is revealed. It explains exactly the sensing behavior of the SAL PMF based Sagnac interference dip. A numerical simulation on the SAL PMF's internal stress and consequent modal effective refractive indices was performed to double confirm the calibration of fiber's phase birefringence.

**Keywords:** lanthanum-aluminum silicate glass; polarization-maintaining fiber; fiber Bragg grating; Sagnac interferometer

## 1. Introduction

High-birefringence (Hi-Bi) fibers, a category of polarization-maintaining fibers (PMFs), refer to those having significant effective refractive index difference between their two orthogonally polarized fundamental modes [1,2]. Thanks to their special capability of maintaining the polarization state of transmitted light, PMFs play very important roles in optical communication and sensing systems.

Especially, PMFs provide unique convenience for constructing fiber-based Sagnac interferometers, which have been demonstrated with great success in optical fiber gyroscopes, comb filters and other high-accuracy measurements of current, vibration, strain, temperature and so on [3–6]. Besides, by combining with fiber gratings [7], Fabry–Perot interference [8], Mach–Zehnder interference [5], or microresonators [9], PMF-based Sagnac interferometers have been exploited for multi-parameter sensing applications as well.

Generally, there are two approaches to break the degeneracy of two fundamental modes and enlarge their effective refractive index difference considerably. The first one basically consists of designing and fabricating noncircular-shape-core fiber [10,11]. For example, elliptical-core fibers were fabricated and showed appreciable birefringence as early as 1970s [12,13]. Benefiting from the development of microstructured optical fibers (MOFs) in 1990s [14,15], lots of very high-birefringence MOFs were demonstrated with various sizes of air holes [16,17]. The second approach to generating stable birefringence in optical fibers is essentially related to elasto-optic effect. By replacing some parts of cladding with a high-expansion glass, asymmetric stress will be frozen in the fiber due to the different thermal expansion coefficients between the replacement part and the rest of cladding as the fiber is drawn and cooled down. Such stress in turn creates a significant effective refractive index difference between the core modes, which are parallel and perpendicular to the stress direction through elasto-optic effect, resulting in a linear birefringence in the drawn fiber. The areas filled with high-expansion glass are called stress-applying parts (SAPs). The classic geometry of SAPs includes bow-tie, PANDA, and elliptical inner cladding [18].

Considering the thermo-mechanical compatibility between SAPs and pure silica cladding, most SAPs are made of borosilicate glass whose thermal expansion coefficient ($\sim 1.0 \times 10^{-6}$ $K^{-1}$) is about two times higher than that of pure silica ($5.4 \times 10^{-7}$ $K^{-1}$) [19,20]. Recently, a new kind of lanthanum-aluminum co-doped silicate ($SiO_2$-$Al_2O_3$-$La_2O_3$, SAL) glass was used as SAPs in a PANDA-type PMF [20,21]. The thermal expansion coefficient of this SAL glass was calibrated to be around $5.32 \times 10^{-6}$ $K^{-1}$, which is almost ten times that of silica and five times higher than that of borosilicate glass. The fabricated SAL PMF showed comparable phase birefringence with commercial PANDA PMFs with borosilicate SAPs.

In this paper, we demonstrated for the first time the integration of Bragg grating into an SAL PMF and its sensing capabilities when it was inserted in a Sagnac interferometer. At the same time, the phase birefringence of the SAL PMF was further calibrated more precisely by calculating the difference between the Bragg peaks of two orthogonal-polarization modes. The distinct responses of the Bragg grating peaks and interference fringe were applied in the discrimination of axial strain and temperature. The underlying mechanism of the behavior of the Sagnac interference fringe was analyzed as well.

## 2. Materials and Methods

The specialty PMF used in this work was conceived and fabricated at XLIM research institute France by using powder-in-tube technology [20]. The scanned electronic microscopy (SEM) picture in Figure 1a shows that the fiber has PANDA geometry, consisting of a germania-doped silica core and two flanked stress-applied rods made of SAL glass. Quantitatively, the latter one contains 70 mol% $SiO_2$, 20 mol% $Al_2O_3$ and 10 mol% $La_2O_3$. The diameters of the core and cladding are of 8.4 μm and 140 μm, respectively. The centers of two stress rods are about 36.5 μm away from the core center, and their diameters are about 25.0 μm and 25.2 μm, respectively.

The corresponding refractive index profile of this fiber was measured by an IFA-100 Fibre Index Profiler and then plotted in Figure 1b. The core has a quasi-step-index profile and the maximum refractive index difference from that of pure silica is around 0.0062. The refractive indices of two SAL rods show parabolic-like distribution and the maximum values are about 0.0428 and 0.0504, respectively. Contrary to conventional PANDA PMFs whose stress rods are made of borosilicate, in our fiber, the refractive indices of SAPs are higher than that of the fiber core since the dopant ($La_2O_3$) has much higher refractive index than pure silica [22]. Referring to conventional PANDA PMFs, the axis parallel to the applied-stress direction is defined as "slow axis", whereas the perpendicular axis is defined as "fast axis" [1].

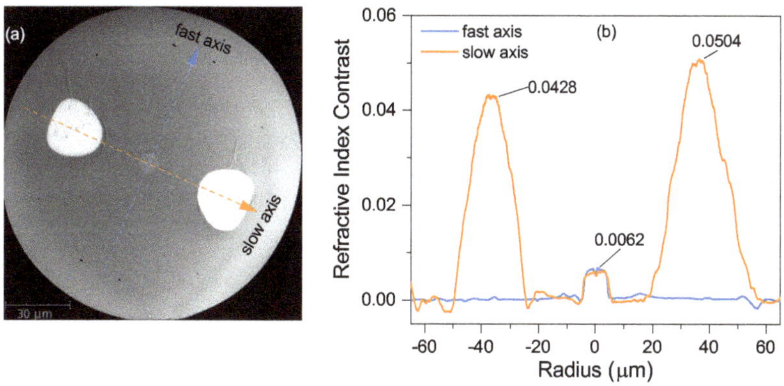


**Figure 1.** Scanning electronic microscopy picture (**a**) and corresponding refractive index profile (**b**) of the $SiO_2$-$Al_2O_3$-$La_2O_3$ (SAL) polarization-maintaining fiber (PMF).

As a high-birefringence fiber, the SAL PMF can be employed to construct a Sagnac interferometer [3,23]. Meanwhile, since the core of the SAL PMF is doped with germanium, it is also suitable for functionalization by inscribing Bragg gratings. When the SAL PMF with the inscribed Bragg grating is inserted into a Sagnac loop mirror, the Sagnac interference and Bragg grating resonance are combined together, as shown in Figure 2. It is noteworthy that the configuration is different from those in previous works [7,24], in which the grating is inscribed in standard single-mode fiber (SMF) and then connected with PMFs. The spatial distance between the grating and PMF may cause the device to respond to external variation from different places, especially in sensing mechanical quantities. In our scheme, however, the FBGs are integrated inside the SAL PMF directly. Thus, there is almost no spatial distance between the FBGs and PMF, indicating that they respond to the perturbation from same place.

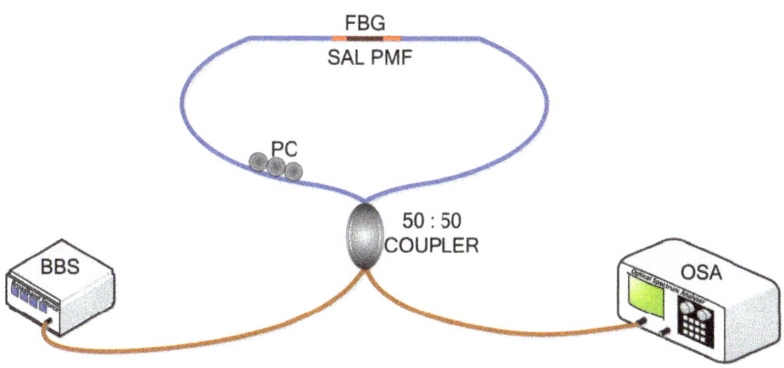


**Figure 2.** Schematic diagram of the proposed configuration and experimental setup. BBS: broadband source; OSA: optical spectrum analyzer; PC: polarization controller; FBG: fiber Bragg grating; SAL PMF: $SiO_2$-$Al_2O_3$-$La_2O_3$ polarization-maintaining fiber.

The Bragg grating in this SAL PMF was fabricated by using conventional ultraviolet (UV) laser lateral exposure method, in which the fiber was exposed to the interference fringe by scanning a phase mask with a UV laser [25]. Before being exposed to UV light, the photosensitivity of SAL PMF was enhanced via hydrogen loading process at 12 Mpa and 80 °C for one week. Next, a 56-mm-long SAL

PMF was spliced with two single-mode fibers and then connected to a broadband source (BBS, Infinon Research Broadband Source) and an optical spectrum analyzer (OSA, Yokogawa AQ6370C) through a 50:50 coupler, as shown as the schematic diagram in Figure 2. A polarization controller was inserted into the fiber loop to optimize the output interference spectrum. The UV light was generated from a frequency-doubled continuous-wave (CW) Argon laser (Coherent, INNOVA 90C) and the power was about 70 mW. The period of phase mask was 1070.20 nm, corresponding to the grating pitch of 535.10 nm. The resolution of OSA was set to 0.02 nm. Before laser scanning, an additional alignment of the refractive index axis was conducted to avoid that the core was shadowed by the stress rods. The UV laser was moved out of the mask region and focused on the SAL PMF. Since the diffraction pattern illuminated through the fast axis was significantly different from that through the slow axis, the illumination direction was aligned along the fast axis by rotating the fiber and monitoring the diffraction pattern. Finally, the Bragg grating was inscribed by moving the UV laser to scan the phase mask with the speed of 0.02 mm/s. The inscription length was 10.00 mm.

## 3. Results and Discussions

### 3.1. Transmission Spectra of Fabricated Device

After the grating inscription process finished, the transmission spectrum was measured and plotted in Figure 3. As shown, two Bragg grating peaks, corresponding to the Bragg reflections of slow-axis and fast-axis polarized modes, are embedded in the interference fringe and their central wavelengths are 1549.62 nm and 1549.27 nm, respectively. According to the phase-matching condition of fiber Bragg grating, $\lambda_{res} = 2n_{eff} \cdot \Lambda$, the effective refractive indices of slow-axis and fast-axis polarized modes ($n_{eff,s}$ and $n_{eff,f}$) can be calculated, respectively.

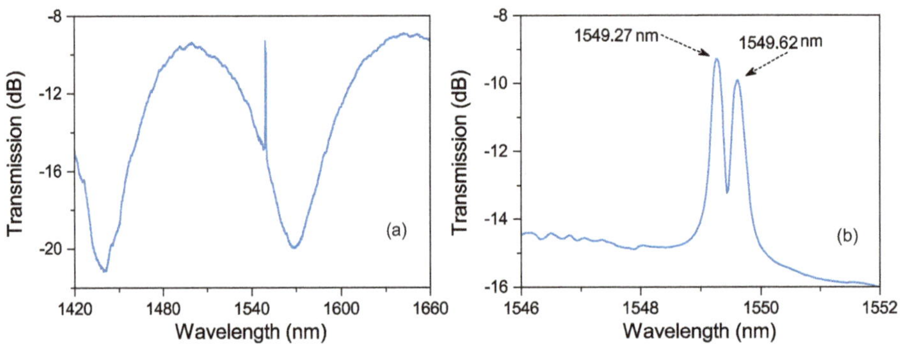


**Figure 3.** Spectra of the SAL PMF based Sagnac interference (**a**) and Bragg resonant peaks (**b**) after grating inscription.

The phase birefringence $B$, defined as the effective refractive index difference between slow-axis and fast-axis polarized modes, can be derived as follows:

$$B = n_{eff,s} - n_{eff,f} = \frac{(\lambda_{res,s} - \lambda_{res,f})}{2\Lambda}, \tag{1}$$

where $\lambda_{res,s}$ and $\lambda_{res,f}$ are the resonant wavelengths of slow-axis and fast-axis polarized modes, respectively. For the fabricated Bragg grating in the SAL PMF, $\Lambda = 535.10$ nm, $\lambda_{res,s} = 1549.62$ nm and $\lambda_{res,f} = 1549.27$ nm, the phase birefringence $B$ of this SAL PMF is characterized as $3.27 \times 10^{-4}$. It matches well with the previously reported results [20]. Moreover, this characterization result should be more precise since the bandwidth of Bragg grating peaks are much narrower than that of interference fringe. The corresponding beat length ($L_b = \lambda/B$) is around 4.74 mm, which is comparable with that of commercial PANDA PMF [1].

### 3.2. Stress and Birefringence Simulation

Since the birefringence in the SAL PMF is induced by the asymmetric thermal stress through elasto-optical effect, the internal thermal stress of the fiber and the consequent phase birefringence were numerically analyzed by using the solid mechanics and wave optics modules in COMSOL Multiphysics [26,27]. The detailed parameters of the fiber's structure and materials are listed in Table 1. In order to simplify the simulation model, the fiber's structure was defined to be axial symmetric along the fast and slow axes. Moreover, the core, SAL rods and outer cladding were all drawn by perfectly circular shape, and they were of 8.4 μm , 24.2 μm and 140 μm in diameter, respectively. A step-index profile of the refractive index was adopted. The material refractive indices of the pure silica and germania-doped silica (core region) were described by a three-term Sellmeier dispersion formula with corresponding coefficients [28]. The mole fraction of germania in fiber's core can be deduced by the measured refractive index difference of the core $\Delta n_{\text{core}}$ and the relationship between the refractive index variation and the doping concentration of germania ($C_{\text{GeO}_2}(\text{mol.\%}) = \Delta n_{\text{core}}/0.00146$) [29]. Here, $\Delta n_{\text{core}}$ equals 0.0062 according to measured index contrast shown in Figure 1b . The index of SAL glass was represented by adding a fixed value to the dispersion formula of pure silica. The fixed value was the average index difference of the SAL rods compared to the pure silica cladding. It was estimated around 0.045 based on the measured index distribution. Since the doping concentration of the fiber's core was only around 4.0 mol.%, the core was assumed to have the same Young's modulus, Poisson's ratio and relative density as those of pure silica, while those of SAL rods were much higher and cited from previous work [20]. All the materials were set with the same stress-optical coefficients.

**Table 1.** Structure and material parameters used in the simulation [20,26].

| | **Cladding** | **Core** | **SAL Rods** |
|---|---|---|---|
| Material | pure silica | $GeO_2$ doped silica | $La_2O_3$-$Al_2O_3$ doped silica |
| Diameter (μm) | 140 | 8.4 | 24.2 |
| Young's modulus (GPa) | 78.3 | 78.3 | 110.7 |
| Poisson's ratio | 0.186 | 0.186 | 0.282 |
| Relative density (kg/m$^3$) | 2203 | 2203 | 3346.7 |
| Thermal expansion coefficient (K$^{-1}$) | $5.4 \times 10^{-7}$ | $1.0 \times 10^{-6}$ | $5.32 \times 10^{-6}$ |
| First stress optical coefficient (m$^2$/N) | | $0.65 \times 10^{-12}$ | |
| Second stress optical coefficient (m$^2$/N) | | $4.2 \times 10^{-12}$ | |
| Operating temperature (°C) | | 20 | |
| Reference temperature (°C) | | 1100 | |

When the fiber was cooled down from 1100 °C (reference temperature) to 20 °C (operating temperature), the difference between the stress distribution along slow-axis and fast-axis directions was simulated and illustrated in Figure 4. As shown, the mean stress distributes anisotropically on the SAL PMF's cross section, and the mean stress near the core region is around $10^8$ Pa. Consequently, the material refractive indices become anisotropic accordingly, resulted in the breakage of the degenerateness of fiber's fundamental modes. Figure 5a,b show the simulated electric field distribution of x-polarized and y-polarized fundamental modes, denoted respectively with $LP_{01,X}$ and $LP_{01,Y}$. Their polarization states (marked with white arrows) point to the slow axis and fast axis, respectively. Their effective refractive indices decrease monotonically from wavelengths of 1100 nm to 1900 nm, while the corresponding phase birefringence increases very slightly. The simulated phase birefringence is estimated at $3.6 \times 10^{-4}$ for a wide wavelength range. It agrees basically with the experimental results

$(3.27 \times 10^{-4})$. The slight gap may be attributed to the structural mismatch between the experimental fiber and simulation model.

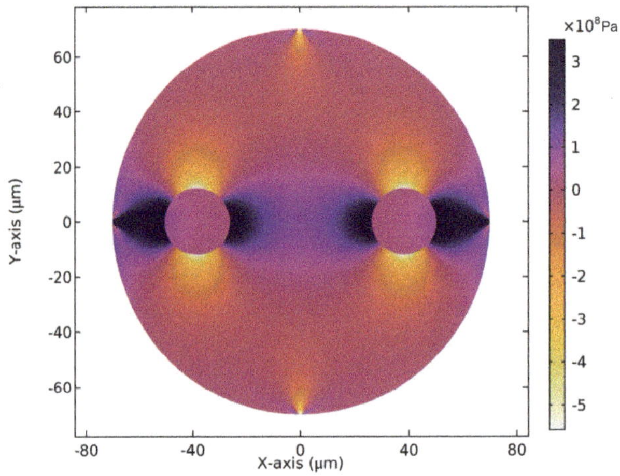


**Figure 4.** Thermal stress difference $\sigma_x - \sigma_y$ distribution of the SAL PMF cross section, where $\sigma_x$ and $\sigma_y$ are the stress distribution along the slow and fast axes, respectively.

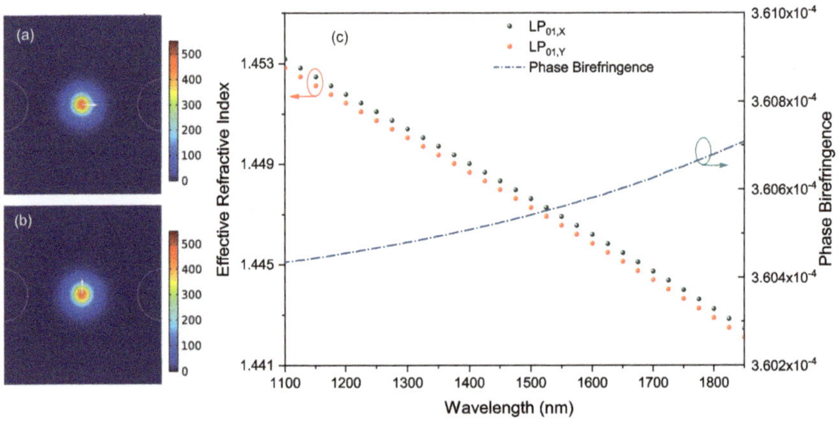


**Figure 5.** The electric field distribution of x-polarized fundamental mode (**a**) and y-polarized fundamental mode (**b**). (**c**) The effective refractive index variations of x-polarized (black sphere) and y-polarized (red sphere) modes in the wavelength range from 1100 nm to 1900 nm, and the dispersion curve of the corresponding phase birefringence (dark blue dot-dash line).

### *3.3. Sensing Performance*

The sensing performance of the proposed device was investigated as the following. Firstly, the sensor head was clamped on two translation stages whose fine resolution was limited to 0.5 μm. One stage was fixed, whereas the other one was driven by a micrometer to stretch fiber [25]. When the axial strain was increased from 0 to 1200 με, as shown in Figure 6a,b, both interference pattern and Bragg peaks shift to longer wavelengths. Moreover, it is worth noting that the red shift of two FBG peaks are very similar and the wavelength gap between two peaks becomes slightly larger with the increment of applied axial strain. It indicates that the phase birefringence of the SAL PMF is enlarged as the fiber is stretched increasingly. For PMF-based Sagnac interferometer, the relationship between the shift of interference wavelength ( $\Delta\lambda_{SI,\varepsilon}$) and applied axial strain ($\varepsilon$) can be described by

$\Delta\lambda_{SI,\varepsilon} = \lambda_{SI}(1 + P_e)\,\varepsilon$, where $P_e = \Delta B(\varepsilon)/B$ is the strain-optic coefficient of the fiber's birefringence. Since both physical length and birefringence depend on axial strain positively, the Sagnac interference spectrum will consequently shift to longer wavelengths.

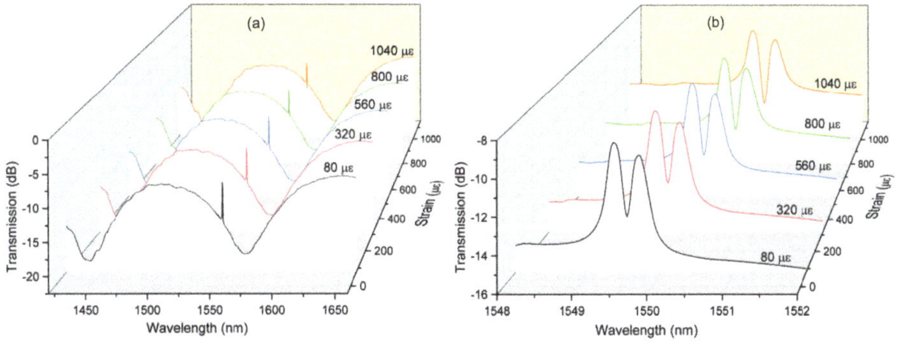


**Figure 6.** Spectral shifts of Sagnac interference dip (**a**) and Bragg resonant peaks (**b**) with varying axial strain.

By following the wavelength variations of the fast-axis Bragg peak and one interference dip, the strain sensitivities of the Bragg grating and the interference dip of the proposed device can be characterized to be $7.60 \times 10^{-4}$ nm/$\mu\epsilon$ and $1.00 \times 10^{-2}$ nm/$\mu\epsilon$, respectively, as shown as the linear fitting curves in Figure 7. Their corresponding R-squared factors of linear fitting are 0.9996 and 0.9921, respectively.

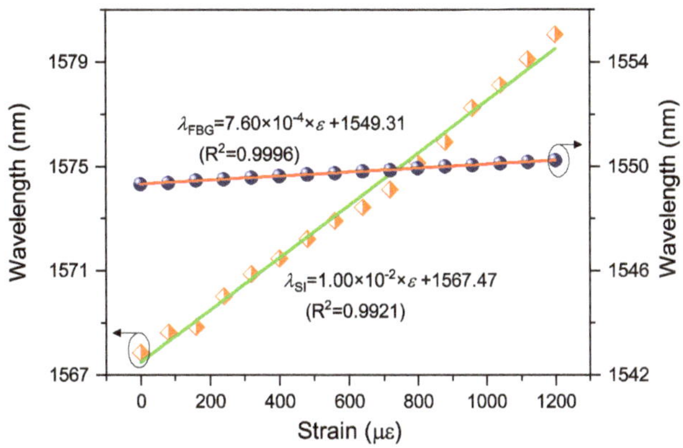


**Figure 7.** Axial-strain responses of Sagnac interference dip (orange square with green line) and Bragg peak of fast-axis polarized mode (blue sphere with red line).

Then, the sensor head was placed into a high-accuracy column oven to characterize its thermal performance. The temperature precision of this oven is 0.1 °C. A tiny tension was applied at both ends of the sensor head to keep it straight and then the temperature was increased from 25 °C to 85 °C with the step of 5 °C. The corresponding transmission spectra of the Sagnac interference and Bragg grating were recorded and illustrated in Figure 8a,b, respectively. As shown, the interference spectrum shifts to shorter wavelengths with the increment of temperature, while the Bragg peaks shift to longer wavelengths. Moreover, the gap between two Bragg peaks decreased apparently in the process of rising temperature. It means that the phase birefringence of the SAL PMF tends downwards when fiber temperature goes upwards. Similarly, the variation of interference wavelength

($\Delta\lambda_{SI,T}$) is related to fiber's thermal expansion and the birefringence change caused by thermo-optic effect, $\Delta\lambda_{SI,T} = \lambda_{SI} (\alpha + \Gamma) T$, where $\alpha$ is the thermal expansion coefficient and $\Gamma$ is the equivalent thermo-optic coefficient of birefringence. As illustrated in Figure 8b, two FBG peaks goes closer with increasing temperature. It indicates that the difference between the effective refractive indices of two polarized modes (i.e., phase birefringence) becomes smaller and the equivalent thermo-optic coefficient of birefringence $\Gamma$ should be negative. Although $\alpha$ is positive, the thermo-optic effect induced birefringence variation contributes dominantly to the shift of interference fringe [6]. Therefore, the negative response of the interference dip to temperature, shown in Figure 8a, can be well explained.

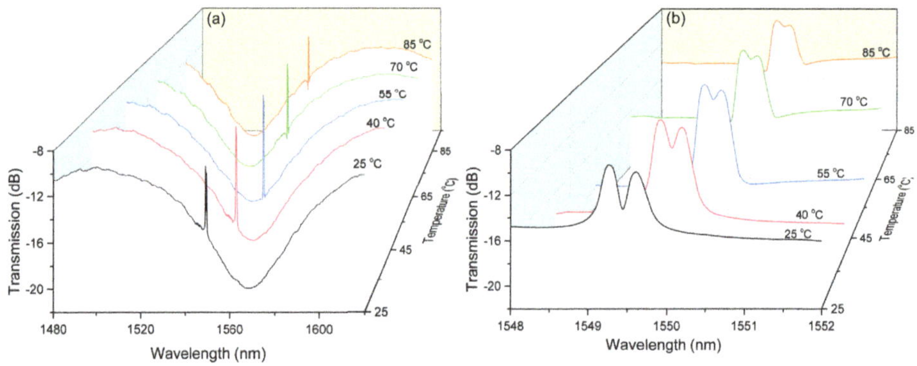


**Figure 8.** Spectral shifts of Sagnac interference dip (**a**) and Bragg resonant peaks (**b**) with varying temperature.

In parallel with the strain measurement, the temperature sensitivities of the fast-axis polarization Bragg peak and the interference dip are taken at $1.18 \times 10^{-2}$ nm/°C and $-8.57 \times 10^{-1}$ nm/°C, respectively, as shown by the linear fitting results in Figure 9. Their corresponding R-squared factors of linear fitting are of 0.9996 and 0.9921, respectively.

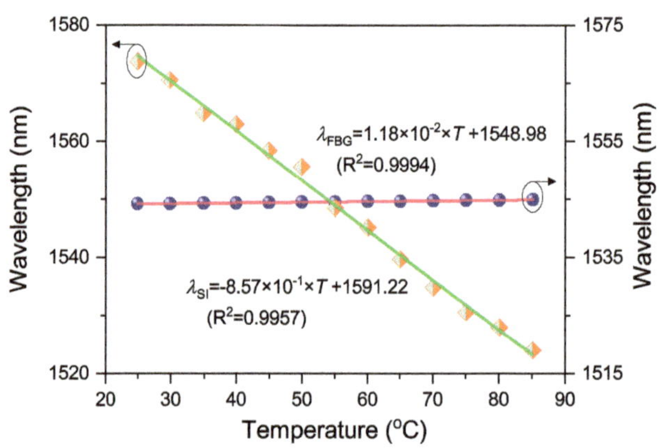


**Figure 9.** Thermal responses of the Sagnac interference dip (orange square with green line) and the Bragg peak of fast-axis polarized mode (blue sphere with red line).

On the basis of the different responses of the fast-axis polarization Bragg peak and the Sagnac interference dip to axial strain and temperature, this proposed sensor can be applied to measure strain and temperature simultaneously by using the following matrix [30–32]:

$$\begin{aligned}\begin{bmatrix}\Delta\varepsilon \\ \Delta T\end{bmatrix} &= \begin{bmatrix}S_{\varepsilon,\mathrm{SI}} & S_{T,\mathrm{SI}} \\ S_{\varepsilon,\mathrm{FBG}} & S_{T,\mathrm{FBG}}\end{bmatrix}^{-1}\begin{bmatrix}\Delta\lambda_{\mathrm{SI}} \\ \Delta\lambda_{\mathrm{FBG}}\end{bmatrix} \\ &= \begin{bmatrix}1.00\times10^{-2} & -8.57\times10^{-1} \\ 7.60\times10^{-4} & 1.18\times10^{-2}\end{bmatrix}^{-1}\begin{bmatrix}\Delta\lambda_{\mathrm{SI}} \\ \Delta\lambda_{\mathrm{FBG}}\end{bmatrix}\end{aligned} \quad (2)$$

where $\Delta\varepsilon$ and $\Delta T$ refer to strain and temperature variations; $\Delta\lambda_{\mathrm{SI}}$ and $\Delta\lambda_{\mathrm{FBG}}$ correspond to wavelength shifts of the Sagnac interference dip and the fast-axis polarization Bragg peak, respectively; $S_{\varepsilon,\mathrm{SI}}$ , $S_{T,\mathrm{SI}}$, $S_{\varepsilon,\mathrm{FBG}}$, and $S_{T,\mathrm{FBG}}$ are the strain and temperature sensitivities of Sagnac interference dip and fast-axis polarization Bragg peak, respectively.

## 4. Conclusions

In conclusion, the Bragg grating fabrication in a new type of polarization-maintaining fiber (SAL PMF) whose stress-applied rods were made of SAL glass was reported for the first time, to the best of our knowledge. Based on this SAL PMF with the integrated FBG, a Sagnac interferometer was constructed and demonstrated for the simultaneous measurement of axial strain and temperature. The strain and temperature sensitivities of the proposed device reach $1.00 \times 10^{-2}$ nm/με and $-8.57 \times 10^{-1}$ nm/°C, respectively. They are about 100 times and 85 times higher than those of normal FBG-based sensors, respectively [33]. Moreover, the Bragg peaks functioned as an important indicator for investigating the phase birefringence of the SAL PMF. Firstly, by calculating the effective refractive indices of two Bragg peaks, the SAL PMF's phase birefringence was calibrated more precisely compared with the calculation from the interference spectrum, since the bandwidths of Bragg peaks were much narrower than those of the interference fringe [20]. Secondly, the variation of Bragg peaks' separation reflected the tendency of SAL PMF's birefringence with changing axial strain or temperature as well. Thus, it explained explicitly the response behavior of the Sagnac interference dip in the axial strain and temperature sensing measurements.

**Author Contributions:** Z.W. and P.P.S. contributed to the conception and design of this work, P.W., H.Z. and Z.W. contributed to the device fabrication and data acquisition, M.K., G.H. and J.-L.A. fabricated the special SAL PM fiber, Z.W. and P.W. drafted the manuscript, X.S., X.Q.D. and J.P. advised the data analysis and manuscript revision. All authors have read and agreed to the published version of the manuscript.

**Funding:** This work is partially supported by National Natural Science Foundation of China (11774102), the Scientific Research Funds and Promotion Program for Young and Middle-aged Teacher in Science & Technology Research of Huaqiao University (ZQN-YX504, 17BS412), National Research Foundation Singapore (NRF-CRP13-2014-05), and Singapore Ministry of Education Academic Research Fund Tier 1 (MOE2016-T1-001-077).

**Acknowledgments:** We would like to acknowledge the team in XLIM Research Institute for the help with specialty fiber fabrication, and also acknowledge Raghuraman Sidharthan and Daryl Ji Ming Ho from Centre for Fiber Technology, Nanyang Technological University for characterizing the index profile of SAL PMF.

**Conflicts of Interest:** The authors declare no conflict of interest.

## References

1. Méndez, A.; Morse, T.F. *Specialty Optical Fibers Handbook*; Elsevier Academic Press: Cambridge, MA, USA, 2007; pp. 243–276. [CrossRef]
2. Noda, J.; Okamoto, K.; Sasaki, Y. Polarization-maintaining fibers and their applications. *J. Lightw. Technol.* **1986**, *4*, 1071–1089. [CrossRef]
3. Liu, Y.; Liu, B.; Feng, X.; Zhang, W.; Zhou, G.; Yuan, S.; Kai, G.; Dong, X. High-birefringence fiber loop mirrors and their applications as sensors. *Appl. Opt.* **2005**, *44*, 2382–2390. [CrossRef] [PubMed]

4. Dong, X.; Tam, H.Y.; Shum, P.P. Temperature-insensitive strain sensor with polarization-maintaining photonic crystal fiber based Sagnac interferometer. *Appl. Phys. Lett.* **2007**, *90*, 151113. [CrossRef]
5. Gouveia, C.; Chesini, G.; Cordeiro, C.M.B.; Baptista, J.M.; Jorge, P.A.S. Simultaneous measurement of refractive index and temperature using multimode interference inside a high birefringence fiber loop mirror. *Sens. Actuators B* **2013**, *177*, 717–723. [CrossRef]
6. Frazão, O.; Baptista, M.J.; Santos, L.J. Recent Advances in High-Birefringence Fiber Loop Mirror Sensors. *Sensors* **2007**, *7*, 2970. [CrossRef] [PubMed]
7. Zhou, D.P.; Wei, L.; Liu, W.K.; Lit, J.W.Y. Simultaneous measurement of strain and temperature based on a fiber Bragg grating combined with a high-birefringence fiber loop mirror. *Opt. Commun.* **2008**, *281*, 4640–4643. [CrossRef]
8. Wang, S.S.; Hu, Z.F.; Li, Y.H.; Tong, L.M. All-fiber Fabry-Perot resonators based on microfiber Sagnac loop mirrors. *Opt. Lett.* **2009**, *34*, 253–255. [CrossRef]
9. Zheng, Y.; Wu, Z.; Shum, P.P.; Xu, Z.; Keiser, G.; Humbert, G.; Zhang, H.; Zeng, S.; Dinh, X.Q. Sensing and lasing applications of whispering gallery mode microresonators. *Opto-Electronic Advances* **2018**, *1*, 180015. [CrossRef]
10. Chu, P.; Sammut, R. Analytical method for calculation of stresses and material birefringence in polarization-maintaining optical fiber. *J. Lightw. Technol.* **1984**, *2*, 650–662. [CrossRef]
11. Frazão, O.; Silva, S.O.; Baptista, J.M.; Santos, J.L.; Statkiewicz-Barabach, G.; Urbanczyk, W.; Wojcik, J. Simultaneous measurement of multiparameters using a Sagnac interferometer with polarization maintaining side-hole fiber. *Appl. Opt.* **2008**, *47*, 4841–4848. [CrossRef]
12. Ramaswamy, V.; French, W.G.; Standley, R.D. Polarization characteristics of noncircular core single-mode fibers. *Appl. Opt.* **1978**, *17*, 3014–3017. [CrossRef]
13. Adams, M.J.; Payne, D.N.; Ragdale, C.M. Birefringence in optical fibres with elliptical cross-section. *Electron. Lett.* **1979**, *15*, 298–299. [CrossRef] [PubMed]
14. Knight, J.C.; Birks, T.A.; Russell, P.S.J.; Atkin, D.M. All-silica single-mode optical fiber with photonic crystal cladding. *Opt. Lett.* **1996**, *21*, 1547–1549. [CrossRef]
15. Russell, P. Photonic Crystal Fibers. *Science* **2003**, *299*, 358–362. [CrossRef] [PubMed]
16. Suzuki, K.; Kubota, H.; Kawanishi, S.; Tanaka, M.; Fujita, M. Optical properties of a low-loss polarization-maintaining photonic crystal fiber. *Opt. Express* **2001**, *9*, 676–680. [CrossRef]
17. Steel, M.J.; Osgood, R.M. Polarization and Dispersive Properties of Elliptical-Hole Photonic Crystal Fibers. *J. Lightw. Technol.* **2001**, *19*, 495. [CrossRef]
18. Varnham, M.; Payne, D.N.; Barlow, A.; Birch, R. Analytic solution for the birefringence produced by thermal stress in polarization-maintaining optical fibers. *J. Lightw. Technol.* **1983**, *1*, 332–339. [CrossRef]
19. Ramaswamy, V.; Stolen, R.H.; Divino, M.D.; Pleibel, W. Birefringence in elliptically clad borosilicate single-mode fibers. *Appl. Opt.* **1979**, *18*, 4080–4084. [CrossRef]
20. Kudinova, M.; Humbert, G.; Auguste, J.L.; Delaizir, G. Multimaterial polarization maintaining optical fibers fabricated with the powder-in-tube technology. *Opt. Mater. Express* **2017**, *7*, 3780–3790. [CrossRef]
21. Litzkendorf, D.; Grimm, S.; Schuster, K.; Kobelke, J.; Schwuchow, A.; Ludwig, A.; Kirchhof, J.; Leich, M.; Jetschke, S.; Dellith, J.; et al. Study of Lanthanum Aluminum Silicate Glasses for Passive and Active Optical Fibers. *Int. J. Appl. Glass Sci.* **2012**, *3*, 321–331. [CrossRef]
22. Dragic, P.D.; Kucera, C.; Ballato, J.; Litzkendorf, D.; Dellith, J.; Schuster, K. Brillouin scattering properties of lanthano-aluminosilicate optical fiber. *Appl. Opt.* **2014**, *53*, 5660. [CrossRef]
23. Cui, Y.; Wu, Z.; Shum, P.P.; Dinh, X.; Humbert, G. Investigation on the Impact of Hi-Bi Fiber Length on the Sensitivity of Sagnac Interferometer. *IEEE Sens. J.* **2014**, *14*, 1952–1956. [CrossRef] [PubMed]
24. Kim, H.M.; Nam, H.; Moon, D.S.; Kim, Y.H.; Lee, B.H.; Chung, Y. Simultaneous measurement of strain and temperature with high sensing accuracy. In Proceedings of the 2009 14th OptoElectronics and Communications Conference, Vienna, Austria, 13–17 July 2009; pp. 1–2. [CrossRef] [PubMed]
25. Wu, Z.; Zhang, H.; Shum, P.P.; Shao, X.; Huang, T.; Seow, Y.M.; Liu, Y.G.; Wei, H.; Wang, Z. Supermode Bragg grating combined Mach-Zehnder interferometer for temperature-strain discrimination. *Opt. Express* **2015**, *23*, 33001–33007. [CrossRef]
26. Guan, R.; Zhu, F.; Gan, Z.; Huang, D.; Liu, S. Stress birefringence analysis of polarization maintaining optical fibers. *Opt. Fiber Technol.* **2005**, *11*, 240–254. [CrossRef]

27. Zhao, J.; Tang, M.; Oh, K.; Feng, Z.; Zhao, C.; Liao, R.; Fu, S.; Shum, P.P.; Liu, D. Polarization-maintaining few mode fiber composed of a central circular-hole and an elliptical-ring core. *Photon. Res.* **2017**, *5*, 261–266. [CrossRef] [PubMed]
28. Fleming, J.W. Dispersion in $GeO_2$-$SiO_2$ glasses. *Appl. Opt.* **1984**, *23*, 4486–4493. [CrossRef]
29. Dianov, E.M.; Mashinsky, V.M. Germania-Based Core Optical Fibers. *J. Lightw. Technol.* **2005**, *23*, 3500. [CrossRef]
30. Guan, B.; Tam, H.Y.; Tao, X.M.; Dong, X.Y. Simultaneous strain and temperature measurement using a superstructure fiber Bragg grating. *IEEE Photon. Techno. Lett.* **2000**, *12*, 675–677. [CrossRef] [PubMed]
31. Wu, Z.; Liu, Y.G.; Wang, Z.; Jiang, M.; Ji, W.; Han, T.; Li, S.; Shao, X.; Dinh, X.Q.; Tjin, S.C.; et al. Simultaneous measurement of curvature and strain based on fiber Bragg grating in two-dimensional waveguide array fiber. *Opt. Lett.* **2013**, *38*, 4070–4073. [CrossRef]
32. Zhang, H.; Wu, Z.; Shum, P.P.; Shao, X.; Wang, R.; Dinh, X.Q.; Fu, S.; Tong, W.; Tang, M. Directional torsion and temperature discrimination based on a multicore fiber with a helical structure. *Opt. Express* **2018**, *26*, 544–551. [CrossRef]
33. Kersey, A.D.; Davis, M.A.; Patrick, H.J.; LeBlanc, M.; Koo, K.P.; Askins, C.G.; Putnam, M.A.; Friebele, E.J. Fiber grating sensors. *J. Lightwave Technol.* **1997**, *15*, 1442–1463. [CrossRef]

sensors

MDPI

Letter

# Fluorinated Polyimide-Film Based Temperature and Humidity Sensor Utilizing Fiber Bragg Grating

**Xiuxiu Xu [1,2,3], Mingming Luo [1,2,3,*], Jianfei Liu [1,2,3] and Nannan Luan [1,2,3]**

1 School of Electronic and Information Engineering, Hebei University of Technology, Tianjin 300401, China; 201921902022@stu.hebut.edu.cn (X.X.); jfliu@hebut.edu.cn (J.L.); luan@hebut.edu.cn (N.L.)
2 Tianjin Key Laboratory of Electronic Materials and Devices, Tianjin 300401, China
3 Hebei Key Laboratory of Advanced Laser Technology and Equipment, Tianjin 300401, China
* Correspondence: 2019013@hebut.edu.cn

Received: 30 August 2020; Accepted: 22 September 2020; Published: 24 September 2020

check for updates

**Abstract:** We propose and demonstrate a temperature and humidity sensor based on a fluorinated polyimide film and fiber Bragg grating. Moisture-induced film expansion or contraction causes an extra strain, which is transferred to the fiber Bragg grating and leads to a humidity-dependent wavelength shift. The hydrophobic fluoride doping in the polyimide film helps to restrain its humidity hysteresis and provides a short moisture breathing time less than 2 min. Additionally, another cascaded fiber Bragg grating is used to exclude its thermal crosstalk, with a temperature accuracy of ±0.5 °C. Experimental monitoring over 9000 min revealed a considerable humidity accuracy better than ±3% relative humidity, due to the sensitized separate film-grating structure. The passive and electromagnetic immune sensor proved itself in field tests and could have sensing applications in the electro-sensitive storage of fuel, explosives, and chemicals.

**Keywords:** fluorinated polyimide film; humidity hysteresis; fiber Bragg grating

## 1. Introduction

Humidity monitoring and regulation are of great significance in pharmacy, semiconductors, and costly facility maintenance and explosive storage [1–3]. Relative humidity, defined as the ratio of vapor content in the air, describes the present moisture content compared with what the air can hold at most [4]. Mechanical psychrometers and dew-point hygrometers are limited in humidity sensing due to their low accuracy and complexity [1]. Chemical methods mainly depend on the hygroscopic reaction, which has been found to be irreversible and nonreusable [5]. Thus, capacitance and resistance are selected as measurable signs of humidity change, which are accurate, fast-responding, and small-scaled [3]. However, for safety reasons, active electronic humidity sensors prove unsuitable for electro-sensitive applications such as explosives, chemicals and fuel storage [1–4].

Optical fiber sensing, as a passive and electromagnetism-immune method, provides a safe and reliable approach to temperature and humidity monitoring for such electro-sensitive storage [3,6]. However, different mechanisms, including spectral absorption, evanescent sensing, and light scattering, show poor performance in terms of flexibility and multiplexing [7–13]. Additionally, non-uniform sensor structures, such as the Fabry-Perot cavity [7], fiber taper [8], micro-nano fiber [9], side-polished fiber [10], and U-bent fiber [11], are fragile and unfit for mass production [12–16]. Moreover, the combination of the functional materials and fiber sensors shows non-uniformity and uncertainty as well, especially when it comes to graphene oxide [17], metal oxide [18], UV gels [19], and nanoparticles [20]. For practical applications, fiber Bragg grating (FBG) combined with polymer provides a better way due to its long service and uniformity in mass production. The humidity-dependent polyimide film was first investigated by Kronenberg, revealing its potential application in relative humidity (RH) sensing [21].

Later, he demonstrated a humidity sensor with polyimide coating at the grating by measuring the moisture-induced expansion [22]. T. Yeo and K. Grattan analyzed the strain transfer between the polyimide coating and the fiber Bragg grating, and explained the relationship between humidity and wavelength shift in detail [23]. The linear sensitivity was estimated to be +4.5 pm/RH with an uncertainty of ±4% RH, while the polyimide coating resulted in humidity hysteresis and a response time over 40 min. Recently, similar attempts have been made for a higher sensitivity and a shorter breathing time, which are hard to achieve simultaneously with an ordinary polyimide film in a conventional structure. The reason lies in the tough hydrogen bonding in polyimide with a porous surface, which causes humidity hysteresis and even moisture agglomeration [24,25]. Moreover, the moisture exchange and humidity sensitivity are restricted with each other depending on the tradeoff between film thickness and surface, where it is difficult to balance the response time and accuracy. According to the aforementioned analysis, a humidity sensor with high performance still remains to be realized both in chemical modification and structural design.

In this paper, a fluorinated polyimide-film-based temperature and humidity sensor utilizing fiber Bragg grating is proposed and demonstrated. The sensor was constructed with an FBG and a 20 μm-thick fluorinated polyimide film by direct manual gluing. Moisture-induced film expansion or contraction causes an extra strain, which is transferred to the FBG and leads to a humidity-dependent wavelength shift. Compared with the coating method, the separate film-grating structure provides a large surface area for moisture exchange, an improved dynamic range, and considerable sensitivity. The –$CF_3$ modification in the monomer of the polyimide reduces its moisture capacity, restrains its humidity hysteresis, and provides a short moisture breathing time less than 2 min. Another cascaded FBG temperature sensor is used to exclude its thermal crosstalk, with a temperature accuracy of ±0.5 °C. Experimental monitoring over 9000 min revealed a relative humidity accuracy better than ±3% RH, while field testing for tobacco storage proved its stability and practicability as designed. The passive and electromagnetism-immune sensor with an accurate and fast response can be applied for the electro-sensitive storage of explosives, chemicals, and fuels.

## 2. Materials and Methods

### 2.1. Functional Material

For common polyimide films, the vapor in the air is easily captured by the hydrophilic hydrogen bonds and porous surface. The strong absorption causes a slow moisture release and may even lead to humidity hysteresis and agglomeration. Thus, chemical modification is necessary to improve the surface properties of the polyimide film. The fluorinated polyimide film was synthesized and provided by the group of Pro. L. Fan, Institute of Chemistry Chinese Academy of Sciences, and is commercially available with excellent hysteresis resistance [26]. As can be seen from the chemical formula and the molecular structure of the fluorinated polyimide shown in Figure 1a, the trifluoromethyl modification replaces the ether bond in the original aromatic monomer with a molecular weight of 630. In the detailed straight-chain polycondensation, the fluorine modification improves the hydrophobicity of the film [26,27]. With the increase in fluorine content from 15.32% to 30.16% in the monomer, the transparent film becomes an applicable candidate for a fast-response humidity sensor. Compared with a non-fluorinated film synthesized with a similar monomer with a molecular weight of 496 such as the dark one shown in Figure 1b, the trifluoromethyl acts in a moisture-proof manner and restrains its capacity from 5% to 1%, which is quantified by weight gain at saturated humidity. Thus, with such a hydrophobic modification, a small humidity hysteresis and a short breathing time can be expected when using a fluorinated polyimide film.

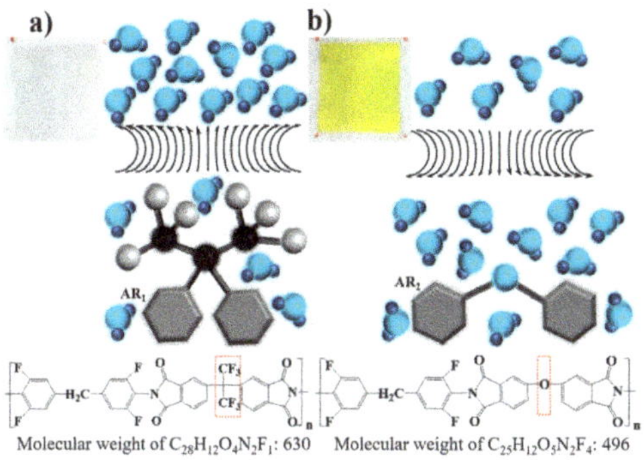


**Figure 1.** Chemical modification of the fluorinated polyimide (**a**) and non-fluorinated polyimide (**b**).

As the moisture capacity is significantly reduced in a fluorinated polyimide film, the sensitivity to humidity declines sharply along with the insufficient expansion. The conventional coating method is no longer suitable for practical use, with which it is hard to achieve a fast response, a low hysteresis, and considerable sensitivity simultaneously.

### 2.2. Sensor Structure

Thus, a separate film-grating structure was specifically designed to maintain the dynamic range and the humidity sensitivity, as a result of the large cross-sectional ratio between the functional film and fiber grating. As shown in Figure 2a, the sensing unit was fabricated with a bare FBG of 125 μm in diameter and a film of 20 μm in thickness. The FBG and the film were assembled with an aluminous convertor and clamped between two brackets of a hollow frame by direct manual gluing. The unit was pre-stretched at 2000 με to ensure an approximately linear spectral response.

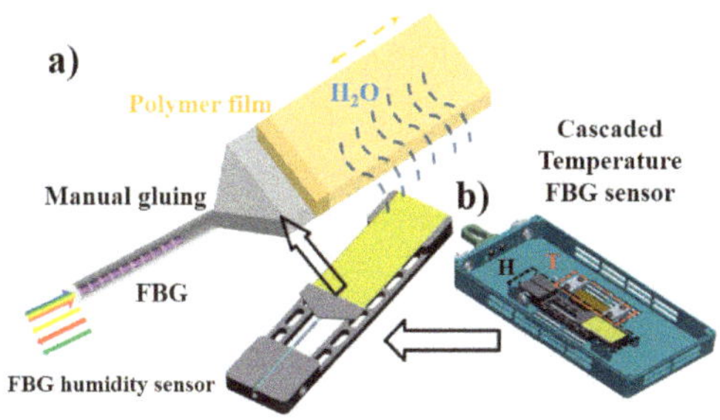


**Figure 2.** Sensing mechanism and separate film-grating structure (**a**)with another cascaded FBG temperature sensor to exclude the thermal crosstalk (**b**) in fiber Bragg grating (FBG) humidity sensor.

Once the humidity changes, the film self-rebalances by exchanging moisture with that in the air. In this way, the expansion or contraction leads to an axial strain change transferred to the FBG. Finally, the humidity variation is visualized by tracing the wavelength shift, which indicates dynamic moisture absorption and release. Moreover, Figure 2b shows another cascaded FBG temperature sensor to exclude the thermal crosstalk and extract the pure relative-humidity changes from the co-effects on the wavelength shift, as circled with the red dotted line.

### 2.3. Working Principles

The humidity sensing mechanism is further discussed below. According to the equal stress in a rigid connection as in Equation (1), the strain transferred to the FBG is strictly associated with the expansion and contraction of the polyimide film,

$$E_{\mathrm{pi}}\varepsilon_{\mathrm{pi}}W_{\mathrm{pi}}d_{\mathrm{pi}} = E_{\mathrm{G}}\varepsilon_{\mathrm{G}}\pi R_{\mathrm{G}}^2 \quad (1)$$

where $\varepsilon_{\mathrm{pi}}$ ($\varepsilon_{\mathrm{G}}$) and $E_{\mathrm{pi}}$ ($E_{\mathrm{G}}$) refer to the strain and the elastic modulus of the film (grating), respectively. $W_{\mathrm{pi}}$ and $d_{\mathrm{pi}}$ describe the transverse section of film, while $R_{\mathrm{G}}$ denotes the radius of the bare fiber. Besides, the polyimide film expansion strictly equals the contraction in the FBG, fixing an isolated length at a constant temperature in Equation (2).

$$L_{\mathrm{pi}}\alpha_{\mathrm{pi}}H(1+\varepsilon_{\mathrm{pi}}) + L_{\mathrm{pi}}\varepsilon_{\mathrm{pi}} + L_{\mathrm{G}}\varepsilon_{\mathrm{G}} = \mathrm{C}_0 \quad (2)$$

$\alpha_{\mathrm{pi}}$ and $H$ represent the hygroscopic expansion coefficient and the humidity variation, and $L_{\mathrm{G}}$ and $L_{\mathrm{pi}}$ stand for the lengths of the fiber grating and film available, respectively. After derivation calculus on Equation (2) and simplification with Equation (1), we deduce the differential sensitivity $S_H$ as the humidity-dependent wavelength shift,

$$S_H = \frac{d\lambda_G}{dH} = k\frac{d\varepsilon_G}{dH} = -k\frac{E_{\mathrm{pi}}W_{\mathrm{pi}}d_{\mathrm{pi}}}{E_{\mathrm{G}}\pi R_{\mathrm{G}}^2}\frac{1+\varepsilon_0}{H+\frac{1}{\alpha_{\mathrm{pi}}}+\frac{1}{\alpha_{\mathrm{pi}}}\frac{L_{\mathrm{G}}}{L_{\mathrm{pi}}}\frac{E_{\mathrm{pi}}W_{\mathrm{pi}}d_{\mathrm{pi}}}{E_{\mathrm{G}}\pi R_{\mathrm{G}}^2}} \quad (3)$$

where $k$ is defined as the transfer coefficient between the axial strain and wavelength shift, and $\varepsilon_0$ refers to the initialized strain in the FBG. The negative expression indicates a blue-shift direction with an alterable sensitivity, which can be enhanced by increasing the cross-sectional ratio and the length ratio between the film and FBG. A slight nonlinearity in the sensitivity is introduced at different humidity, which can be well corrected with a second-polynomial coefficient.

## 3. Results

The temperature and humidity sensors were calibrated with a highly accurate Humidity Generator (GEO Calibration, Model 2000), and their sensing performance was validated in humidity-mimicking enclosures created with different saturated solutions. As shown in Figure 3, the sensors were exposed in different humidity enclosures of 12% (LiCl), 33% ($MgCl_2$), 60% (NaBr), 75% (NaCl), and 98% ($K_2SO_4$). The wavelengths measured at humidity of 10.4%, 34.2%, 57.6%, 76.1%, and 96.2% at 25 °C were distributed near the calibration fitting curve, similar with those at other temperatures.

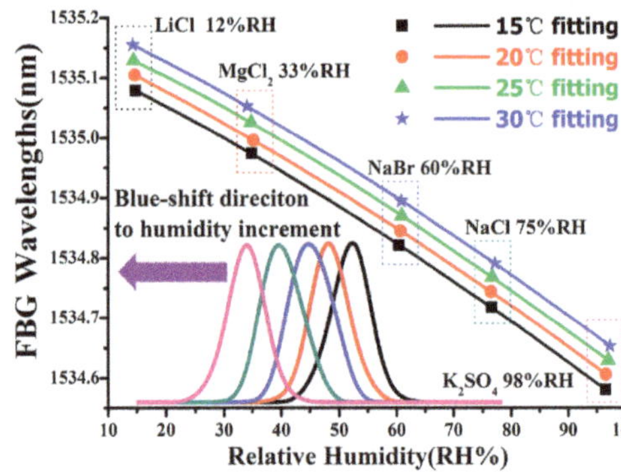


**Figure 3.** Humidity-dependent wavelength shifts with different saturated solutions and temperatures.

As assumed in theoretical analysis, the polynomial fitting curves exhibit an approximately linear dependence between the humidity and the wavelength shift. Additionally, the slight nonlinearity can be estimated with a second coefficient as the humidity varies from 12% RH to 98% RH. Its central wavelength shift towards the blue direction from 1535.1 to 1534.6 nm agrees well with the looseness of the polyimide film and decrement in axial strain. The temperature and humidity sensitivities were obtained as +5.033 pm/°C and −6.09 pm/%RH, respectively, providing opposite spectral responses to better exclude the thermal crosstalk.

The spectral response to humidity change is compared with that of a CE314 electronic sensor, and the humidity hysteresis is discussed in Figure 4. Both the FBG sensor and the electronic sensor were exposed in the same enclosures of the LiCl (12% RH) and $K_2SO_4$ (98% RH) solutions, as well as in the lab (62% RH). The red line above indicates the humidity recorded by CE314, while the blue line represents the wavelength shift determined with the FBG humidity sensor. As the humidity changes among 12% RH, 98% RH, and 62% RH, the central wavelength of the FBG exhibits a red-shift or blue-shift in response to a decrement or increment in the humidity, respectively. Due to the fluoride doping in the polyimide film, a short breathing time less than 2 min is observed with low hysteresis. Besides, the considerable sensitivity and accuracy are preserved, as a consequence of the separate film-grating structure. Thus, the humidity sensor is experimentally proved to be feasible with considerable sensitivity, a fast response, and low humidity hysteresis.

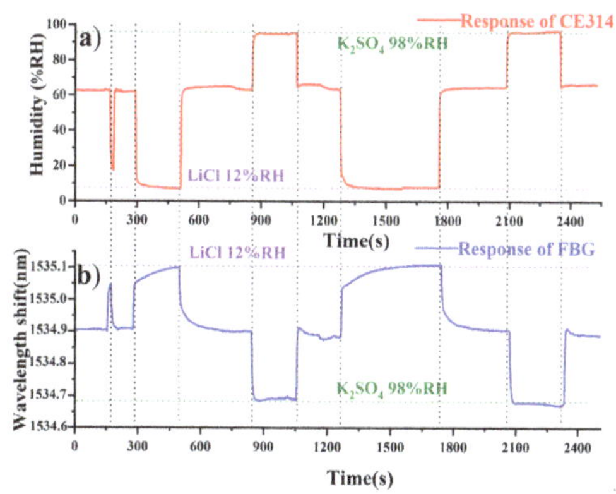


**Figure 4.** Responses to different humidity enclosures at 12% RH (LiCl), 62% RH (in the lab), and 98% RH ($K_2SO_4$) recorded by the CE314 sensor (**a**) and the FBG humidity sensor (**b**).

In the experiment as shown in Figure 5, the sensors were applied for practical monitoring after aging for one month. The signals from the Amplified Spontaneous Emission (ASE) source were modulated by temperature and humidity sensors, harvested by an Optical Spectral Analyzing (OSA) module, and demodulated with pre-recorded calibration coefficients. The temperature and relative humidity visually appeared on the display and were continuously recorded backstage.

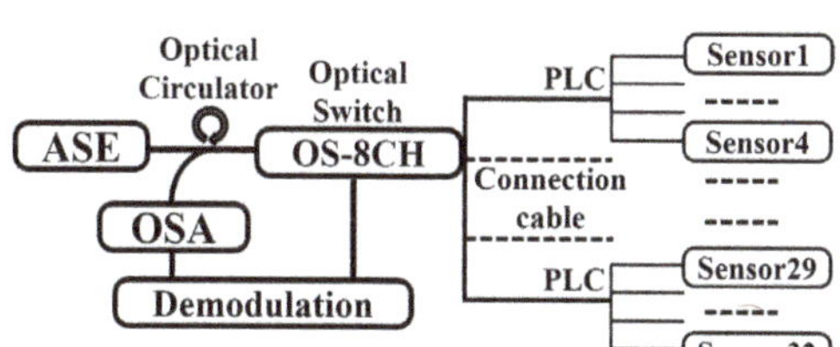


**Figure 5.** Brief illustration of the multiplexing sensing system with 8CH*4 sensors.

Figure 6 shows the temperature varying from 17 to 35 °C for the FBG sensors and an electronic sensor CE314, which is consistent between each. The deviations between the black, red, green, and blue lines demonstrate the temperature differences over 9000 min of monitoring. Mostly, the temperature deviations are well restricted within ±0.5 °C, while such errors beyond tolerance occasionally occur at the perturbations as circled by the dashed lines above.

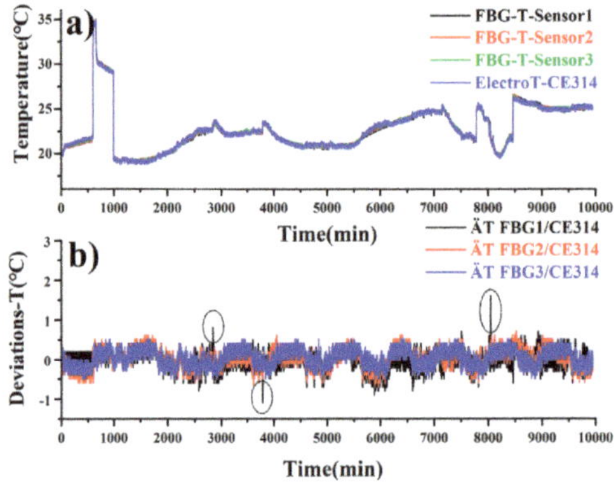


**Figure 6.** Temperature variations recorded by 3 FBG sensors and the CE314 sensor (**a**); temperature deviations between these 3 FBG sensors and the CE314 sensor (**b**).

Figure 7 shows the humidity variation aligned with the monitoring timeline of the temperature. The humidity is also demonstrated with black, red, green, and blue lines, which are consistent with each other as well. Basically, the humidity deviations are restricted within ±3% RH, while the errors mainly occur at temperature jumping.

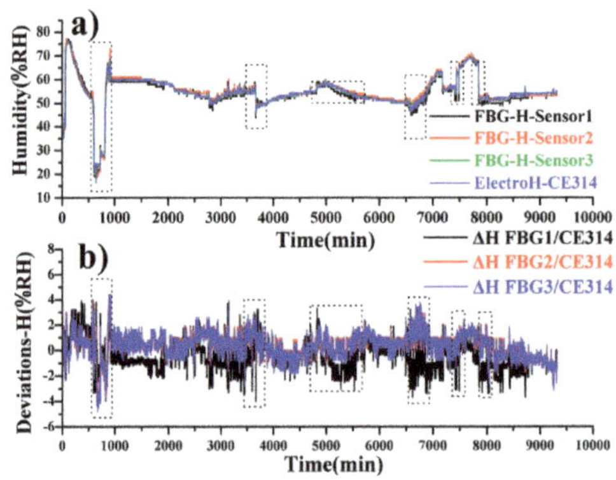


**Figure 7.** Humidity variations recorded by 3 FBG sensors and the CE314 sensor (**a**); humidity deviations between these 3 FBG sensors and CE314 sensor (**b**).

Moreover, as shown in Figure 8, the practicability was verified through field tests in a tobacco warehouse, Fujian province, Southeast of China. As the tobacco standards require, the unprocessed tobacco should be stored at 20 to 30 °C, and 65% RH to 75% RH. Three FBG temperature/humidity sensors were placed in Storage 1, Storage 2, and the outdoor area, which were connected to Channel 1, Channel 2, and Channel 3 via commercial cables with an insertion loss of about 1.1 dB. The backward optical signals were harvested by a distant interrogator and demodulated to real-time temperature and humidity. During 18 h of discontinuous monitoring over 3 days, data were recorded by the FBG sensors and are expressed with solid lines, while those from the manual inspections with CE314 are depicted with scatter points. By comparison, the FBG sensors agreed well with the electronic sensor, whether in storage or outdoors. Since the circumstance in the storage was relatively stable, the temperature varied from 27.5 to 28.4 °C and the humidity changed from 68% RH to 74% RH. Due to the airflow perturbations outdoors, the temperature varied from 28.7 to 35.3 °C and the humidity changed from 46% RH to 73% RH. The deviation of T ≤ 0.3 °C and H ≤ 2.8% RH outdoors is larger than that in storage, of T ≤ 0.2 °C and H ≤ 2.2% RH, which is also consistent with actual situations.

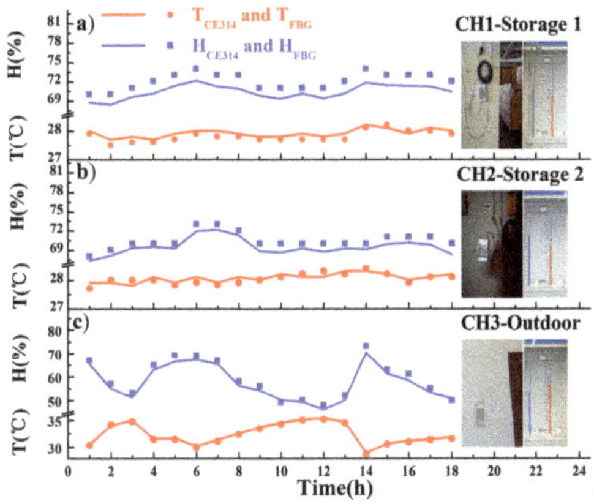


**Figure 8.** Temperature and humidity recorded by the FBG-1 and CE314 sensor at Channel1 in Storage 1 (**a**), by the FBG-2 and CE314 sensor at Channel 2 in Storage 2 (**b**), and by the FBG-3 and CE314 sensor at Channel 3 outdoors (**c**).

## 4. Conclusions

We propose and demonstrate a fluorinated-film-based FBG temperature and humidity sensor. The fluoride doping resolves the humidity hysteresis and provides a short breathing time less than 2 min. The calibrated sensors show different humidity corresponding to enclosures created with saturated solutions and good stability during 9000 min of monitoring in the lab. Additionally, the field tests prove the outstanding performance and practicability of the humidity sensors, but the consistency, size, weight, and cost of the humidity sensors can be further improved in the future, revealing potential for pharmacy, semiconductors, facility and equipment protection, and chemical and explosive storage.

**Author Contributions:** Conceptualization, X.X. and M.L.; funding acquisition, M.L., J.L. and N.L.; investigation, X.X. and M.L.; supervision, M.L., J.L. and N.L.; writing—original draft, X.X. and M.L.; writing—review and editing, X.X., M.L. and J.L. All authors have read and agreed to the published version of the manuscript.

**Funding:** Natural Science Foundation of Hebei province (Grant No. F2019202294); Natural Science Foundation of Hebei province (Grant No. A2020202013); Natural Science Foundation of Tianjin, China (Grant No. 15JCYBJC17000).

**Acknowledgments:** The authors thank the group of Lin Fan, Institute of Chemistry Chinese Academy of Sciences, for providing the fluorinated polyimide film. This work was supported in part by Jinxuan Wu and Shuang Li, attending the Hebei University of Technology.

**Conflicts of Interest:** The authors declare no conflict of interest.

## References

1. Farahani, H.; Wagiran, R.; Hamidon, M.N. Humidity Sensors Principle, Mechanism, and Fabrication Technologies: A Comprehensive Review. *Sensors* **2014**, *14*, 7881–7939.
2. Sikarwar, S.; Yadav, B.C. Opto-electronic humidity sensor: A review. *Sens. Actuators A Phys.* **2015**, *233*, 54–70.
3. Lo Presti, D.; Massaroni, C.; Schena, E. Optical Fiber Gratings for Humidity Measurements: A Review. *IEEE Sens. J.* **2018**, *18*, 9065–9074.

4. Kapic, A.; Tsirou, A.; Verdini, P.G.; Carrara, S. Humidity Sensors for High Energy Physics Applications: A Review. *IEEE Sens. J.* **2020**, *20*, 10335–10344.
5. McDonagh, C.; Burke, C.S.; MacCraith, B.D. Optical chemical sensors. *Chem. Rev.* **2008**, *108*, 400–422.
6. Lee, B. Review of the present status of optical fiber sensors. *Opt. Fiber Technol.* **2003**, *9*, 57–79.
7. Arregui, F.J.; Matias, I.R.; Cooper, K.L.; Claus, R.O. Simultaneous measurement of humidity and temperature by combining a reflective intensity-based optical fiber sensor and a fiber Bragg grating. *Sens. J. IEEE* **2002**, *2*, 482–487. [CrossRef]
8. Bariáin, C.; Matías, I.R.; Arregui, F.J.; López-Amo, M. Optical fiber humidity sensor based on a tapered fiber coated with agarose gel. *Sens. Actuators B Chem.* **2000**, *69*, 127–131.
9. Khijwania, S.K.; Srinivasan, K.L.; Singh, J.P. An evanescent-wave optical fiber relative humidity sensor with enhanced sensitivity. *Sens. Actuators B Chem.* **2005**, *104*, 217–222.
10. Gastón, A.; Pérez, F.; Sevilla, J. Optical fiber relative-humidity sensor with polyvinyl alcohol film. *Appl. Opt.* **2004**, *43*, 4127–4132. [CrossRef] [PubMed]
11. Zhang, L.; Gu, F.; Lou, J.; Yin, X.; Tong, L. Fast detection of humidity with a subwavelength-diameter fiber taper coated with gelatin film. *Opt. Express* **2008**, *16*, 13349–13353. [CrossRef] [PubMed]
12. Li, T.; Dong, X.; Chan, C.C.; Ni, K.; Zhang, S.; Shum, P.P. Humidity Sensor With a PVA-Coated Photonic Crystal Fiber Interferometer. *IEEE Sens. J.* **2013**, *13*, 2214–2216. [CrossRef]
13. Dong, X.; Li, T.; Liu, Y.; Li, Y.; Zhao, C.; Chan, C.C. Polyvinyl alcohol-coated hybrid fiber grating for relative humidity sensing. *J. Biomed. Opt.* **2011**, *16*, 1–5. [CrossRef] [PubMed]
14. Oliveira, R.; Bilro, L.; Marques, T.H.R.; Cordeiro, C.M.B.; Nogueira, R. Simultaneous detection of humidity and temperature through an adhesive based Fabry–Pérot cavity combined with polymer fiber Bragg grating. *Opt. Lasers Eng.* **2019**, *114*, 37–43. [CrossRef]
15. Shao, M.; Zang, Y.; Qiao, X.; Fu, H.; Jia, Z.-A. Humidity Sensor Based on Hybrid Fiber Bragg Grating/Abrupt Fiber Taper. *IEEE Sens. J.* **2017**, *17*, 1302–1305. [CrossRef]
16. Lewis, E.; Hernandez, F.U.; Correia, R.; Morgan, S.P.; Hayes-Gill, B.; Evans, D.; Sinha, R.; Norris, A.; Harvey, D.; Hardman, J.G.; et al. Simultaneous temperature and humidity measurements in a mechanical ventilator using an optical fibre sensor. In *Sixth European Workshop on Optical Fibre Sensors*; International Society for Optics and Photonics, SPIE: Bellingham, WA, USA, 2016. [CrossRef]
17. Jiang, B.; Bi, Z.; Hao, Z.; Yuan, Q.; Feng, D.; Zhou, K.; Zhang, L.; Gan, X.; Zhao, J. Graphene oxide-deposited tilted fiber grating for ultrafast humidity sensing and human breath monitoring. *Sens. Actuators B Chem.* **2019**, *293*, 336–341. [CrossRef]
18. Villatoro, J.; Zubia, J.; Schuster, K.; Arregui, F.J.; Elosua, C. Enhancing sensitivity of photonic crystal fiber interferometric humidity sensor by the thickness of $SnO_2$ thin films. *Sens. Actuators B. Chem.* **2017**. [CrossRef]
19. Mohamed, H.; Irawati, N.; Ahmad, F.; Ibrahim, M.H.; Harun, S.W. Optical Humidity Sensor Based on Tapered Fiber with Multi-walled Carbon Nanotubes Slurry. *Indones. J. Electr. Eng. Comput. Sci.* **2017**, *6*, 97–103. [CrossRef]
20. Massaroni, C.; Caponero, M.A.; D'Amato, R.; Lo Presti, D.; Schena, E. Fiber Bragg Grating Measuring System for Simultaneous Monitoring of Temperature and Humidity in Mechanical Ventilation. *Sensors* **2017**, *17*, 749. [CrossRef]
21. Giaccari, P.; Limberger, H.G.; Kronenberg, P. In Influence of humidity and temperature on polyimide-coated fiber Bragg gratings. In *Bragg Gratings, Photosensitivity, and Poling in Glass Waveguides*; Optical Society of America: Washington, DC, USA, 2001. [CrossRef]
22. Kronenberg, P.; Rastogi, P.K.; Giaccari, P.; Limberger, H.G. Relative humidity sensor with optical fiber Bragg gratings. *Opt. Lett.* **2002**, *27*, 1385–1387. [CrossRef]
23. Yeo, T.L.; Sun, T.; Grattan, K.T.V.; Parry, D.; Lade, R.; Powell, B.D. Characterisation of a polymer-coated fibre Bragg grating sensor for relative humidity sensing. *Sens. Actuators B Chem.* **2005**, *110*, 148–156. [CrossRef]
24. Sidhu, N.K.; Sohi, P.A.; Kahrizi, M. Polymer based optical humidity and temperature sensor. *J. Mater. Sci. Mater. Electron.* **2019**, *30*, 3069–3077. [CrossRef]
25. Xie, Z.; Yan, H.; Li, Y.; Zhao, X. A humidity fiber sensor based on both end-sides of a fiber Bragg grating coated with polyimide. *Opt. Fiber Technol.* **2020**, *57*. [CrossRef]

26. Zhao, X.; Liu, J.; Rui, J.; Fan, L.; Yang, S. Synthesis and characterization of organosoluble polyfluorinated polyimides derived from 3,3′,5,5′-tetrafluoro-4,4′-diaminodiphenylmethane and various aromatic dianhydrides. *J. Appl. Polym. Sci.* **2010**, *103*, 1442–1449. [CrossRef]
27. Huo, H.; Mo, S.; Sun, H.; Yang, S.; Fan, L. Preparation and properties of molecular-weight-controlled polyimide adhesive film. *e-Polymers* **2012**, *12*, 566–583. [CrossRef]

sensors

MDPI

Letter

# Surface Plasmon Resonance Sensor Based on Dual-Side Polished Microstructured Optical Fiber with Dual-Core

**Haixia Han [1], Donglian Hou [1], Nannan Luan [1,*], Zhenxu Bai [1], Li Song [1], Jianfei Liu [1] and Yongsheng Hu [2]**

1 Tianjin Key Laboratory of Electronic Materials and Devices, School of Electronics and Information Engineering, Hebei University of Technology, Tianjin 300401, China; 201821902029@stu.hebut.edu.cn (H.H.); 201921902013@stu.hebut.edu.cn (D.H.); baizhenxu@hotmail.com (Z.B.); songli@hebut.edu.cn (L.S.); jfliu@hebut.edu.cn (J.L.)

2 State Key Laboratory of Luminescence and Applications, Changchun Institute of Optics, Fine Mechanics and Physics, Chinese Academy of Sciences, Changchun 130033, China; huyongsheng@ciomp.ac.cn

* Correspondence: nannanluan@gmail.com; Tel.: +86-022-60438171

Received: 26 May 2020; Accepted: 8 July 2020; Published: 14 July 2020

check for updates

**Abstract:** A surface plasmon resonance (SPR) sensor based on a dual-side polished microstructured optical fiber (MOF) with a dual core is proposed for a large analyte refractive index (RI; $n_a$) detection range. Gold is used as a plasmonic material coated on the polished surface, and analytes can be directly contacted with the gold film. The special structure not only facilitates the fabrication of the sensor, but also can work in the $n_a$ range of 1.42–1.46 when the background material RI is 1.45, which is beyond the reach of other traditional MOF-SPR sensors. The sensing performance of the sensor was investigated by the wavelength and amplitude interrogation methods. The detailed numerical results showed that the proposed sensor can work effectively in the $n_a$ range of 1.35–1.47 and exhibits higher sensitivity in the $n_a$ range of 1.42–1.43.

**Keywords:** microstructured optical fiber; optical fiber sensors; refractive index sensor; surface plasmon resonance

## 1. Introduction

Surface plasmon resonance (SPR) is a unique optical property, which refers to the resonance excitation of free electron oscillations at the interface between a metallic layer and a dielectric surface under a *p*-polarized light radiation, and is extremely sensitive to variations in the refractive index (RI) of the surrounding medium [1–6]. Therefore, SPR has been exploited with a wide range of sensing applications in different fields, such as biology, chemistry, and environmental medicine [1–8]. Initially, most SPR sensors were based on the Kretschmann–Raether prism geometry, but prism-based SPR sensing devices have the disadvantages of large volume, limited mechanical reliability, high cost, unsuitability for distributed sensing, and mass production, which are difficult to adapt themselves to the development of optical communication technology [1–14].

To effectively overcome the limitations of prism-based SPR sensors, optical fiber-based SPR sensors have been developed, in which the fiber acts as a prism, coupling incident light with plasmons. Compared with prism-based devices, the optical fiber design is simpler and more flexible, which can reduce the size of the sensor to a large extent. In addition, the advantages of electromagnetic immunity, high degree of integration, mechanical stability, and in situ monitoring have made optical fiber design more and more attractive [1–11,15,16]. However, the phase-matching condition between the core mode and the surface plasmon polariton (SPP) mode of optical fiber-based SPR sensors is difficult to meet.

In principle, phase matching happens, when their effective refractive index ($n_{eff}$) values of the two modes are identical. For single-mode fibers, the $n_{eff}$ of the SPP mode is usually close to that of the adjacent analyte. For example, the $n_{eff}$ of the water is 1.33, and the $n_{eff}$ of the core mode is close to that of the core material, which for most practical materials is higher than 1.45 [15–17]. Therefore, phase matching occurs only at higher frequencies, and high frequencies limit the penetration depth of the plasmon into the analyte, which reduces the sensitivity of the sensor.

To alleviate phase-matching problems of optical fiber-based SPR sensors, microstructured optical fiber (MOF)-based SPR sensors have been widely studied owing to their design flexibility [9,12–46]. By changing the size, shape, and arrangement of air holes along the propagation direction, the $n_{eff}$ can be tuned to the anticipated values and thus can solve the phase-matching problem between the core mode and the SPP mode [9,15–17,24,25]. In addition, due to their unique structural characteristics and novel optical properties, MOF-based SPR sensors can significantly improve sensing performance. However, there exists an undetectable range of the RI detection in MOF-based SPR sensors, which limits their application in biological and chemical sensing fields. For instance, when the background material RI of an MOF is 1.45, to satisfy the total reflection condition, the upper detection limit of the analyte RI ($n_a$) is usually lower than 1.42 for low-RI MOF-SPR sensors [9,15–17,20,24,25,29], and the lower detection limit of $n_a$ is usually higher than 1.45 for high-RI MOF-SPR sensors [30–32]. In addition, the size of air holes in these sensors is in the order of micrometers, which makes them difficult to be coated with metals or filled with analytes in actual manufacturing [4,6,9,12,15,16,19,24,25].

In this paper, to overcome the above problems, we propose an SPR sensor based on a dual-side polished MOF with a dual core. The dual-side polished structure can be coated with gold films and then directly contact with analytes, which helps to reduce the manufacturing difficulty of the sensor. The dual core can reduce the impact of total reflection conditions on the optical fiber, so that the sensor can work at a wide range of $n_a$, especially at 1.42–1.46, in which range the other MOF-based SPR sensors cannot work [20–26].

## 2. Structure Design and Principle

The schematic diagram of the proposed SPR sensor based on a dual-side polished MOF with a dual core is depicted in Figure 1. This structure can be obtained by a wheel-polishing setup with a 3D mechanical platform, which can move along the X, Y, and Z directions [47]. By employing a light source and an optical spectrum analyzer to online monitor the transmission spectrum during the polishing process, the polishing position, length, and depth could be easy to set up and operate accurately via a computer program [46,47]. Gold was used as a plasmonic material to coat the polished surface, which is not as difficult as coating a gold film on the inner surface of small air holes. Here, the center-to-center distance between two adjacent air holes ($\Lambda$) was 3 μm, and the diameter of air holes ($d$) was 0.5$\Lambda$. The thickness of the gold film ($m$) was 40 nm, and the polishing depth from the fiber center to the polished surface ($h$) was 2.1$\Lambda$. The mode characteristics and the sensing performance of the proposed sensor were simulated through commercially available software COMSOL. A perfectly matched layer (PML) was added to the outer computational region, which was applied to absorb scattered light [9,10,15,16,18,19,25–29,31,36–38].

In this sensor, we used fused silica as a background material and set its RI at 1.45 [17,20,24,25,27,29,41] to detect a range of $n_a$, which the other MOF-based SPR sensors cannot realize. The RI of the air was set to 1. To achieve the dielectric constant of gold ($\varepsilon(\omega)$), we used the Drude–Lorentz model, of which the equation can be written as follows [48]:

$$\varepsilon(\omega) = \varepsilon_\infty - \frac{\omega_D^2}{\omega(\omega + j\gamma_D)} + \frac{\Delta\varepsilon \cdot \Omega_L^2}{\left(\omega^2 - \Omega_L^2\right) + j\Gamma_L\omega} \quad (1)$$

where $\varepsilon_\infty$ is the permittivity at high frequencies, $\omega$ can be interpreted as the angular frequency, $\omega_D$ and $\gamma_D$ indicate the plasma frequency and the damping frequency, respectively, $\Delta\varepsilon$ is the weighting factor, and $\Gamma_L$ and $\Omega_L$ are the spectral width and the oscillator strength of the Lorentz oscillators, respectively [10,34,36,48].

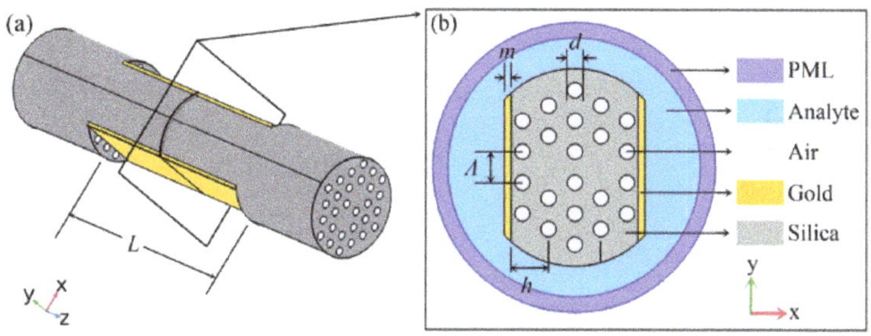


**Figure 1.** (**a**) Schematic diagram of the proposed dual-side polished microstructured optical fiber (MOF)-based surface plasmon resonance (SPR) sensor with a dual core; (**b**) cross-section of the proposed MOF-based SPR sensor.

## 3. Simulation Results and Discussion

### 3.1. Coupling Properties

Like other dual-core MOFs [37–39], the proposed sensor can support four supermodes in fundamental modes. Figure 2 shows the electric field distributions of the four supermodes at 1100 nm for $n_a$ =1.44. Insets A and B of Figure 2 represent the even mode and the odd mode in the $x$ polarized direction ($x$-even mode and $x$-odd mode), respectively. Insets C and D of Figure 2 represent the even mode and the odd mode in the $y$ polarized direction ($y$-even mode and $y$-odd mode), respectively. Here, we only investigated the coupling properties of the $x$-polarized core mode, because the $y$-polarized core mode with the electric field was parallel to the gold film surface and was not easily coupled with SPP modes [39–41].

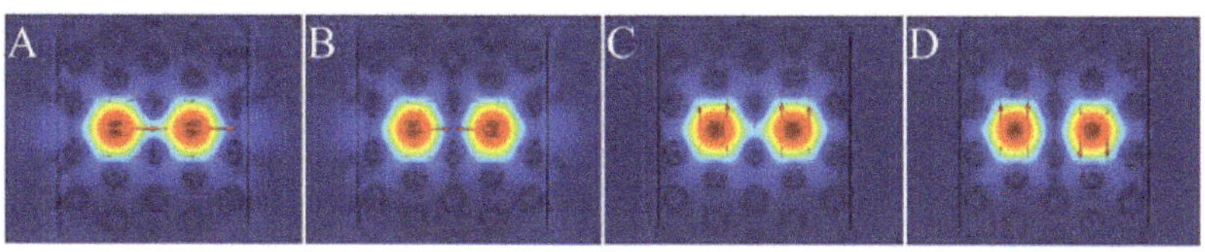


**Figure 2.** Electric field distributions of four fundamental supermodes at 1100 nm for the analyte refractive index ($n_a$) of 1.44: (**A**) $x$-even mode; (**B**) $x$-odd mode; (**C**) $y$-even mode; (**D**) $y$-odd mode (the arrows represent the direction of the electric field).

In theory, when the real parts of $n_{eff}$ (Re($n_{eff}$)) of the core mode and the SPP mode are equal, the phase-matching condition between them are satisfied. Then, the two modes will be coupled with each other, and the maximum energy transfer from the core mode to the SPP mode can be achieved [5–8,10–26,37]. Figure 3 shows the Re($n_{eff}$) curves of the $x$-even core modes and $x$-even SPP modes, the loss spectra of the $x$-even core modes, and the electric field distributions of the relevant modes for $n_a$ = 1.44, 1.45, and 1.46. The black solid line represents the Re($n_{eff}$) of the $x$-even core mode, while the red solid, dashed, and dotted lines represent the Re($n_{eff}$) of the $x$-even SPP modes at $n_a$ = 1.44, 1.45, and 1.46, respectively, as shown in Figure 3a. The blue solid, dashed, and dotted lines stand for the losses of the $x$-even core modes at $n_a$ = 1.44, 1.45, and 1.46, respectively (Figure 3b). Take $n_a$ equal to 1.44 as an example. The $x$-even core mode (inset A of Figure 3c) and the $x$-even SPP mode (inset B of Figure 3c) were coupled with each other (inset C of Figure 3c) at a wavelength of 1518 nm (point C in Figure 3a,b). At this wavelength (also called resonance wavelength), a significant loss peak appeared (see the blue solid curve in Figure 3b), which indicated the maximum energy transfer from the $x$-even core mode to the $x$-even SPP mode. The insets D, E, F, and G of Figure 3c represent the electric field distributions at points D, E, F, and G, respectively (Figure 3a,b), and they can also show the energy transfer from the $x$-even core mode to the $x$-even SPP mode at $n_a$ = 1.45 and 1.46, respectively. As shown in Figure 3, the values of resonance wavelengths were shifted from 1518 to 1533 nm and from 1533 to 1556 nm due to the variations of $n_a$ from 1.44 to 1.45 and from 1.45 to 1.46, respectively. We can observe that a tiny change of $n_a$ can lead to a significant shift of resonance wavelength. This capability can be utilized to detect the changes of the analyte RI [19,24–26,37,38].

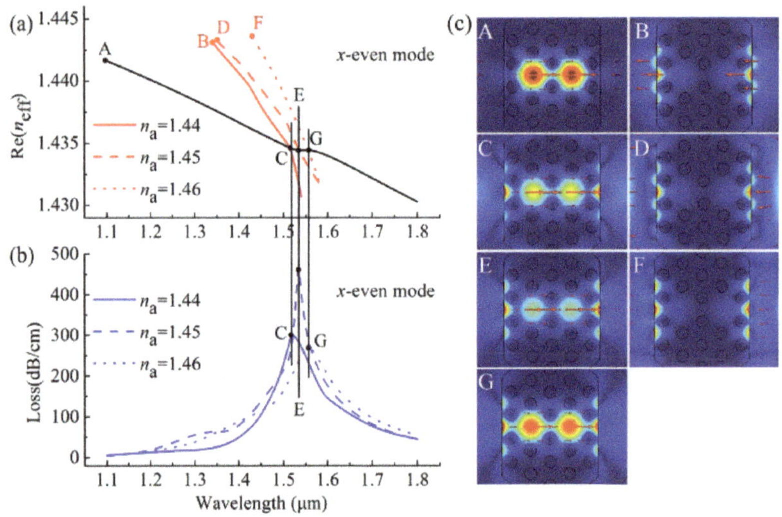


**Figure 3.** (**a**) The real parts of the effective refractive index (Re ($n_{eff}$)) curves of the $x$-even core modes (black lines) and the $x$-even SPP modes at $n_a$ = 1.44, 1.45, and 1.46 represented by the red solid, dashed, and dotted lines, respectively; (**b**) the loss spectra of the $x$-even core modes at $n_a$ = 1.44, 1.45, and 1.46 represented by the blue solid, dashed, and dotted lines, respectively; (**c**) electric field distributions of the $x$-even core mode **A** at 1100 nm, $x$-even SPP mode **B** at 1340 nm, and $x$-even core mode **C** at 1518 nm for $n_a$ = 1.44, electric field distributions of $x$-even SPP mode **D** at 1350 nm and $x$-even core mode **E** at 1533 nm for $n_a$ = 1.45, and electric field distributions of $x$-even SPP mode **F** at 1430 nm and $x$-even core mode **G** at 1556 nm for $n_a$ = 1.46.

When varying the $n_a$ from 1.44 to 1.46, the Re($n_{eff}$) curves of the $x$-odd core modes (black solid lines) and the $x$-odd SPP modes (red solid, dashed, and dotted lines) and the loss spectra (blue solid, dashed, and dotted lines) of the $x$-odd core modes, as well as the electric field distributions of the relevant modes are shown in Figure 4. Similar to the coupling properties of the $x$-even core modes, the phase-matching conditions between the $x$-odd core mode and $x$-odd SPP modes were satisfied at the wavelength of 1518 nm (point C in Figure 4a,b) for $n_a$ = 1.44, at the wavelength of 1533 nm (point E in Figure 4a,b) for $n_a$ = 1.45, and at 1556 nm (point G in Figure 4a,b) for $n_a$ = 1.46. At the resonance wavelengths, the corresponding electric field distributions are shown in insets C, E, and G of Figure 4c, when the $n_a$ values were 1.44, 1.45, and 1.46, respectively. Compared with Figure 3a,b, we found that, unlike the resonance wavelength of the $x$-even core mode shifting to longer wavelengths with $n_a$ increasing from 1.44 to 1.46, the resonance wavelength of the $x$-odd core mode moved to a shorter wavelength as $n_a$ varied from 1.44 to 1.45, whereas the resonance wavelength moved towards longer wavelengths as $n_a$ increased from 1.45 to 1.46. This unexpected peak behavior disturbs the regularity of the SPR sensor and therefore cannot be utilized to detect the changes of the analyte RI.

From Figures 3 and 4, we can see that the resonance wavelengths of the $x$-even core modes move toward longer wavelengths with increasing $n_a$, but the behavior of the resonance wavelengths of the $x$-odd core modes is not regular and cannot be used to detect the variations of the analyte RI. In the following discussion, we only consider the sensing performance of the $x$-even core modes.

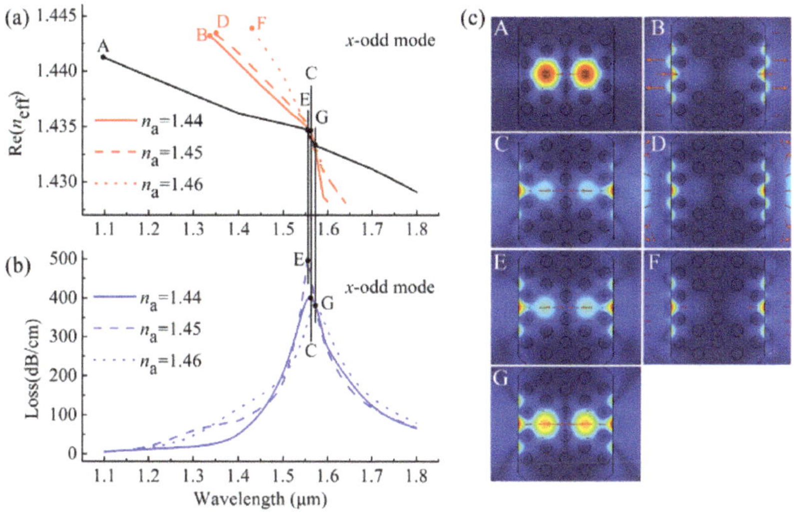


**Figure 4.** (**a**) The Re ($n_{eff}$) curves of the $x$-odd core modes (black lines) and $x$-odd SPP modes at $n_a$ = 1.44, 1.45,and 1.46 represented by the red solid, dashed, and dotted lines, respectively; (**b**) the loss spectra of the $x$-odd core modes at $n_a$ = 1.44, 1.45, and 1.46 represented by the blue solid, dashed, and dotted lines, respectively; (**c**) electric field distributions of the $x$-odd core mode **A** at 1100 nm, the $x$-odd SPP mode **B** at 1340 nm, and the $x$-odd core mode **C** at 1561 nm for $n_a$ = 1.44, electric field distributions of $x$-odd SPP mode **D** at 1350 nm and $x$-odd core mode **E** at 1556 nm for $n_a$ = 1.45, and electric field distributions of $x$-odd SPP mode **F** at 1430 nm and $x$-odd core mode **G** at 1572 nm for $n_a$ = 1.46.

### 3.2. Sensing Performance

The sensing performance of the sensor can be evaluated by wavelength sensitivity (wavelength interrogation) and amplitude sensitivity (amplitude interrogation) [5,10,15,18,26]. The wavelength sensitivity can be calculated from the following equation [5,10–15,17–20,26,37,38]:

$$S(nm/RIU) = \frac{\Delta\lambda_{peak}}{\Delta n_a} \tag{2}$$

where $\Delta\lambda_{peak}$ is the shift of the resonance wavelength and $\Delta n_a$ is the variation of $n_a$. As shown by the blue solid and dashed lines in Figure 3b, we observed $\Delta\lambda_{peak}$ of 15 nm when $n_a$ was varied from 1.44 to 1.45. According to Equation (2), the corresponding wavelength sensitivity in terms of refractive index units (RIU) was 1500 nm/RIU.

The amplitude sensitivity can be calculated at a particular wavelength. Assuming a reasonable length of the sensor was L = 1/$\alpha(\lambda, n_a)$, the amplitude sensitivity was expressed as [5,10,17,18,26,29,38]:

$$S(RIU^{-1}) = -\frac{1}{\alpha(\lambda, n_a)}\frac{\partial\alpha(\lambda, n_a)}{\partial n_a} \tag{3}$$

where $\alpha$ ($\lambda$, $n_a$) is the overall loss for a particular wavelength, $\partial\alpha(\lambda, n_a)$ is the difference between two adjacent loss spectra due to a small change in $n_a$, and $\partial n_a$ is the change of $n_a$. According to Equation (3), we plotted the amplitude sensitivity curves in Figure 5. As is shown by the blue solid curve, the maximum amplitude sensitivity of the $x$-even core mode was 72.18 $RIU^{-1}$ at 1534 nm for $m$ = 40 nm.

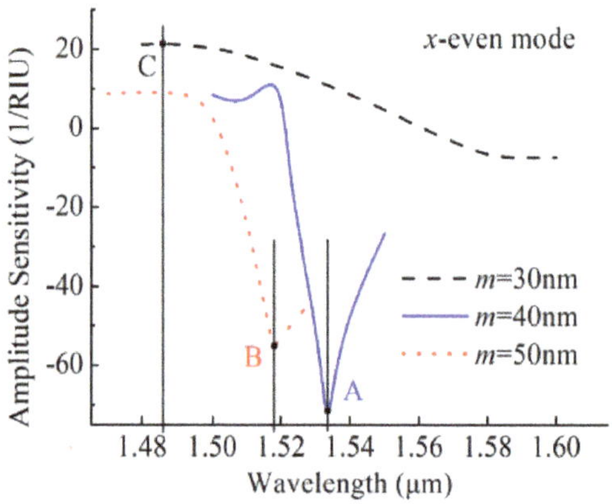


**Figure 5.** Amplitude sensitivities of $x$-even core modes for the gold film thicknesses of 30, 40, and 50 nm.

### 3.3. Gold Film Thickness

The thickness of a gold film is the most important factor that affects the SPR spectra and thus the sensing performance [8–10,12,13,15,16,18,24,33–36]. Figure 6 shows the loss spectra of the $x$-even core modes by varying $m$ for $n_a$ = 1.44 and 1.45. Comparing Figures 3b and 6, it can be evident that the resonance wavelength moved to a shorter wavelength by increasing $m$ from 30 to 50 nm in the case of $n_a$ = 1.44 and 1.45. According to Equation (2), the values of the wavelength sensitivities were 300 and 1700 nm nm/RIU for $m$ of 30 and 50 nm, respectively. The peak losses and the wavelengths influenced by varying $m$ also affected the amplitude sensitivities. Figure 5 shows the amplitude sensitivities of $x$-even core modes as $m$ varied from 30 to 50 nm. It can be found that the maximum amplitude sensitivities of 21.35 and 55.3 $RIU^{-1}$ were achieved at 1486 and 1518 nm for $m$ of 30 and 50 nm, respectively.

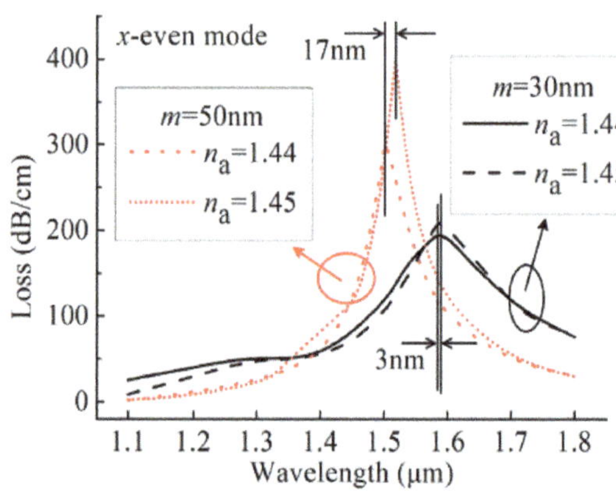


**Figure 6.** Loss spectra of the $x$-even core modes for the gold film thicknesses of 30 and 50 nm at $n_a$ of 1.44 and 1.45.

To further evaluate the performance of the designed sensor, Table 1 shows the summary of several sensing parameters including the peak wavelength, the peak loss, the wavelength sensitivity, the maximum amplitude sensitivity, and the wavelength for the maximum amplitude sensitivity of different $m$ at an $n_a$ range of 1.35–1.47. By means of a detailed investigation of theses parameters, it was found that $m$ can affect the peak wavelength and the peak loss and therefore affect the wavelength sensitivity, the maximum amplitude sensitivity, and the wavelength for the maximum amplitude sensitivity. It is worth noting that the trend of the wavelength for the maximum amplitude sensitivity was roughly the same as that of the peak wavelength, because according to Equation (3), the maximum amplitude sensitivity is related to the maximum $\partial\alpha\ (\lambda, n_a)$, which generally occurs in the vicinity of the resonance peak. The change of $m$ has a slight impact on the sensing performance in single-core MOF-SPR sensors [9,10,12,13,15,16,24,33], while it has a greater and more irregular impact on the sensing performance in the proposed dual-core MOF-SPR sensor, which indicates that this dual-core structure is more sensitive to the thicknesses of the gold film. Nevertheless, the designed dual-core MOF sensor can detect a larger RI range, and it exhibits higher wavelength and amplitude sensitivities at the $n_a$ range of 1.41–1.43, even with other gold film thicknesses.

**Table 1.** Comparison of the sensing parameters of the $x$-even core modes at the $n_a$ range of 1.35–1.47 for $m$ values of 30, 40, and 50 nm, respectively.

| | Peak Wavelength (nm) | | | Peak Loss (dB/m) | | | Wavelength Sensitivities (nm/RIU) | | | Maximum Amplitude Sensitivities (RIU−1) | | | Wavelength for the Maximum Amplitude Sensitivity (nm) | | |
|---|---|---|---|---|---|---|---|---|---|---|---|---|---|---|---|
| $n_a$ \ $m$ | 30 nm | 40 nm | 50 nm | 30 nm | 40 nm | 50 nm | 30 nm | 40 nm | 50 nm | 30 nm | 40 nm | 50 nm | 30 nm | 40 nm | 50 nm |
| 1.35 | 1107 | 1124 | 967 | 400.46 | 303.1 | 332.45 | 900 | 600 | 300 | 16.29 | 8.27 | 13.36 | 1132 | 1141 | 985 |
| 1.36 | 1116 | 1130 | 970 | 420.74 | 319.45 | 337.37 | 1000 | 700 | 400 | 19.41 | 10.69 | 17.35 | 1141 | 1146 | 989 |
| 1.37 | 1126 | 1137 | 974 | 449.8 | 343.09 | 346.69 | 1200 | 800 | 600 | 23.97 | 14.62 | 24.16 | 1152 | 1154 | 994 |
| 1.38 | 1138 | 1145 | 980 | 493.53 | 379.69 | 365.6 | 1400 | 1100 | 800 | 31.11 | 21.73 | 37.8 | 1165 | 1164 | 1001 |
| 1.39 | 1152 | 1156 | 988 | 563.82 | 442.55 | 407.66 | 1800 | 1300 | 1300 | 43.47 | 36.78 | 73.6 | 1182 | 1178 | 1012 |
| 1.40 | 1170 | 1169 | 1001 | 688.61 | 568.95 | 523.56 | 2200 | 2200 | 12000 | 67.89 | 77.88 | 100.9 | 1202 | 1198 | 1127 |
| 1.41 | 1192 | 1191 | 1121 | 946.43 | 913.32 | 748.38 | 3100 | 3100 | 4900 | 126.83 | 270.71 | 852.92 | 1230 | 1231 | 1171 |
| 1.42 | 1223 | 1225 | 1170 | 1634.6 | 2674.4 | 5094.5 | 17500 | 17500 | 20700 | 407.64 | 1479.03 | 1411.37 | 1408 | 1340 | 1377 |
| 1.43 | 1398 | 1340 | 1377 | 6944.1 | 18113 | 29443 | 18700 | 18700 | 12400 | 1262.38 | 1347.09 | 390 | 1591 | 1519 | 1503 |
| 1.44 | 1585 | 1518 | 1501 | 19434 | 30453 | 30473 | 300 | 300 | 1700 | 21.35 | 72.18 | 55.39 | 1486 | 1534 | 1518 |
| 1.45 | 1588 | 1533 | 1518 | 20922 | 48818 | 40144 | 5600 | 5600 | 600 | 42.18 | 52.2 | 54.06 | 1556 | 1533 | 1524 |
| 1.46 | 1644 | 1556 | 1524 | 16769 | 26788 | 54836 | 2600 | 2600 | 300 | 71.72 | 11.69 | 4.3 | 1482 | 1541 | 1521 |
| 1.47 | 1670 | 1565 | 1527 | 16005 | 24814 | 47015 | | | | | | | | | |

In general, the ability of detecting large RI ranges makes the designed sensor more competitive than D-shaped MOF-SPR sensors [17,21–23,32–34,38]. Compared with single-core dual-side polished MOF-SPR sensors [18,42], the designed dual-core structure can detect $n_a$ higher than the RI of the background material of the MOF. Although other structures such as inner-coated or grooved MOF-SPR sensors can detect a wide RI range [19,39,43–45], the designed dual-side polished structure has advantages that it can be readily coated with the gold films and has outside sensing channels, making it ideal for use as a real-time sensor.

## 4. Conclusions

We have proposed and numerically investigated an SPR sensor based on a dual-side polished MOF with a dual core to realize a large range of $n_a$ detection. The gold and the analyte layers were placed outside the MOF structure, which can be expected to simplify the manufacturing process. The coupling characteristics, sensing performance, and the influence of the gold film thickness of the sensor were investigated by the finite element method in the wavelength and amplitude interrogation. Since the peak behavior of the $x$-odd mode disturbed the regularity of the SPR sensor, the $x$-even mode was determined to analyze the sensing performance. The simulation results showed that the sensor could detect a large $n_a$ range covering from 1.35 to 1.47 and had higher wavelength sensitivity and amplitude sensitivity in the $n_a$ range of 1.42–1.43 when the background material RI was 1.45. The proposed sensor can overcome the defect that other traditional MOF-SPR sensors cannot work in the $n_a$ range of 1.42–1.46, which makes it exhibit great potential in biological and chemical sensing fields.

**Author Contributions:** Conceptualization, H.H. and N.L.; investigation, H.H., D.H., and N.L.; writing of the original draft preparation, H.H.; writing of review and editing, H.H., Z.B., L.S., N.L., and J.L.; supervision, N.L., Z.B., L.S., and J.L.; funding acquisition, N.L., L.S., and Y.H. All authors have read and agreed to the published version of the manuscript.

**Funding:** This research was funded by the Natural Science Foundation of Hebei Province (grant numbers: F2019202294, F2019202337, and F2019202252), the State Key Laboratory of Luminescence and Applications (grant number: SKLA-2019-07), and the Natural Science Foundation of Tianjin City (grant number: 15JCYBJC17000).

**Conflicts of Interest:** The authors declare no conflicts of interest.

## References

1. Homola, J.; Yee, S.S.; Gauglitz, G. Surface plasmon resonance sensors: Review. *Sens. Actuators B Chem.* **1999**, *54*, 3–15. [CrossRef]
2. Singh, P. SPR biosensors: Historical perspectives and current challenges. *Sens. Actuators B Chem.* **2016**, *229*, 110–130. [CrossRef]
3. Sharma, A.K.; Pandey, A.K.; Kaur, B. A review of advancements (2007–2017) in plasmonics-based optical fiber sensors. *Opt. Fiber Technol.* **2018**, *43*, 20–34. [CrossRef]
4. Zhao, Y.; Deng, Z.; Li, J. Photonic crystal fiber based surface plasmon resonance chemical sensors. *Sens. Actuators B Chem.* **2014**, *202*, 557–567. [CrossRef]
5. Aruna Gandhi, M.S.; Chu, S.; Senthilnathan, K.; Babu, P.R.; Nakkeeran, K.; Li, Q. Recent advances in plasmonic sensor-based fiber optic probes for biological applications. *Appl. Sci.* **2019**, *9*, 949. [CrossRef]
6. Rifat, A.A.; Ahmed, R.; Yetisen, A.K.; Butt, H.; Sabouri, A.; Mahdiraji, G.A.; Yun, S.H.; Adikan, F.R.M. Photonic crystal fiber based plasmonic sensors. *Sens. Actuators B Chem.* **2017**, *243*, 311–325. [CrossRef]
7. Sharma, A.K.; Jha, R.; Gupta, B.D. Fiber-optic sensors based on surface plasmon resonance: A comprehensive review. *IEEE Sens. J.* **2007**, *7*, 1118–1129. [CrossRef]
8. Klantsataya, E.; Jia, P.; Ebendorff-Heidepriem, H.; Monro, T.M.; Francois, A. Plasmonic fiber optic refractometric sensors: From conventional architectures to recent design trends. *Sensors* **2016**, *17*, 12. [CrossRef]
9. Hassani, A.; Skorobogatiy, M. Design of the microstructured optical fiber-based surface plasmon resonance sensors with enhanced microfluidics. *Opt. Express* **2006**, *14*, 11616–11621. [CrossRef] [PubMed]
10. Islam, M.S.; Sultana, J.; Rifat, A.A.; Ahmed, R.; Dinovitser, A.; Ng, B.W.; Ebendorff-Heidepriem, H.; Abbott, D. Dual-polarized highly sensitive plasmonic sensor in the visible to near-IR spectrum. *Opt. Express* **2018**, *26*, 30347–30361. [CrossRef]
11. Klantsataya, E.; François, A.; Ebendorff-Heidepriem, H.; Hoffmann, P.; Monro, T. Surface plasmon scattering in exposed core optical fiber for enhanced resolution refractive index sensing. *Sensors* **2015**, *15*, 25090–25102. [CrossRef] [PubMed]
12. Rifat, A.A.; Mahdiraji, G.A.; Chow, D.M.; Shee, Y.G.; Ahmed, Y.G.; Adikan, F.R.M. Photonic crystal fiber-based surface plasmon resonance sensor with selective analyte channels and graphene-silver deposited core. *Sensors* **2015**, *15*, 11499–11510. [CrossRef] [PubMed]
13. Liu, C.; Yang, L.; Su, W.; Wang, F.; Sun, T.; Liu, Q.; Mu, H.; Chu, P.K. Numerical analysis of a photonic crystal fiber based on a surface plasmon resonance sensor with an annular analyte channel. *Opt. Commun.* **2017**, *382*, 162–166. [CrossRef]
14. Liu, C.; Yang, L.; Liu, Q.; Wang, F.; Sun, Z.; Sun, T.; Mu, H.; Chu, P.K. Analysis of a surface plasmon resonance probe based on photonic crystal fibers for low refractive index detection. *Plasmonics* **2017**, *13*, 779–784. [CrossRef]
15. Hassani, A.; Skorobogatiy, M. Design criteria for microstructured-optical fiber based surface plasmon resonance sensors. *J. Opt. Soc. Am. B* **2007**, *24*, 1423–1429. [CrossRef]
16. Peng, Y.; Hou, J.; Huang, Z.; Lu, Q. Temperature sensor based on surface plasmon resonance within selectively coated photonic crystal fiber. *Appl. Opt.* **2012**, *51*, 6361–6367. [CrossRef]
17. Luan, N.; Wang, R.; Lv, W.; Yao, J. Surface plasmon resonance sensor based on D-shaped microstructured optical fiber with hollow core. *Opt. Express* **2015**, *23*, 8576–8582. [CrossRef]
18. Chen, N.; Chang, M.; Zhang, X.; Zhou, J.; Lu, X.; Zhuang, S. Highly sensitive plasmonic sensor based on a dual-side polished photonic crystal fiber for component content sensing applications. *Nanomaterials* **2019**, *9*, 1587. [CrossRef]
19. Shuai, B.; Xia, L.; Zhang, Y.; Liu, D. A multi-core holey fiber based plasmonic sensor with large detection range and high linearity. *Opt. Express* **2012**, *20*, 5974–5986. [CrossRef]

20. Zhao, L.; Han, H.; Lian, Y.; Luan, N.; Liu, J. Theoretical analysis of all-solid D-type photonic crystal fiber based plasmonic sensor for refractive index and temperature sensing. *Opt. Fiber Technol.* **2019**, *50*, 165–171. [CrossRef]
21. Haque, E.; Hossain, M.A.; Ahmed, F.; Namihira, Y. Surface plasmon resonance sensor based on modified D-shaped photonic crystal fiber for wider range of refractive index detection. *IEEE Sens. J.* **2018**, *18*, 8287–8293. [CrossRef]
22. Chen, X.; Xia, L.; Li, C. Surface plasmon resonance sensor based on a novel D-shaped photonic crystal fiber for low refractive index detection. *IEEE Photonics J.* **2018**, *10*, 1–9. [CrossRef]
23. Gangwar, R.K.; Singh, V.K. Highly sensitive surface plasmon resonance based D-shaped photonic crystal fiber refractive index sensor. *Plasmonics* **2017**, *12*, 1367–1372. [CrossRef]
24. Yu, X.; Zhang, Y.; Pan, S.; Shum, P.; Yan, M.; Leviatan, Y.; Li, C. A selectively coated photonic crystal fiber based surface plasmon resonance sensor. *J. Opt.* **2010**, *12*, 015005. [CrossRef]
25. Zhang, Y.; Xia, L.; Zhou, C.; Yu, X.; Liu, H.; Liu, D.; Zhang, Y. Microstructured fiber based plasmonic index sensor with optimized accuracy and calibration relation in large dynamic range. *Opt. Commun.* **2011**, *284*, 4161–4166. [CrossRef]
26. Akter, S.; Abdur Razzak, S.M. Highly sensitive open-channels based plasmonic biosensor in visible to near-infrared wavelength. *Results Phys.* **2019**, *13*, 102328. [CrossRef]
27. Zhao, L.; Han, H.; Luan, N.; Liu, J.; Song, L.; Hu, Y. A temperature plasmonic sensor based on a side opening hollow fiber filled with high refractive index sensing medium. *Sensors* **2019**, *19*, 3730. [CrossRef]
28. Liu, C.; Su, W.; Liu, Q.; Lu, X.; Wang, F.; Sun, T.; Chu, P.K. Symmetrical dual D-shape photonic crystal fibers for surface plasmon resonance sensing. *Opt. Express* **2018**, *26*, 9039–9049. [CrossRef]
29. Luan, N.; Yao, J. Surface plasmon resonance sensor based on exposed-core microstructured optical fiber placed with a silver wire. *IEEE Photonics J.* **2016**, *8*, 1–8. [CrossRef]
30. Liu, B.; Jiang, Y.; Zhu, X.; Tang, X.; Shi, Y. Hollow fiber surface plasmon resonance sensor for the detection of liquid with high refractive index. *Opt. Express* **2013**, *21*, 32349–32357. [CrossRef]
31. Luan, N.; Yao, J. High refractive index surface plasmon resonance sensor based on a silver wire filled hollow fiber. *IEEE Photonics J.* **2016**, *8*, 1–9. [CrossRef]
32. Luan, N.; Zhao, L.; Lian, Y.; Lou, S. A high refractive index plasmonic sensor based on D-shaped photonic crystal fiber with laterally accessible hollow-core. *IEEE Photonics J.* **2018**, *10*, 1–7. [CrossRef]
33. Wang, G.; Li, S.; An, G.; Wang, X.; Zhao, Y.; Zhang, W.; Chen, H. Highly sensitive D-shaped photonic crystal fiber biological sensors based on surface plasmon resonance. *Opt. Quant. Electron.* **2016**, *48*, 46. [CrossRef]
34. An, G.; Li, S.; Qin, W.; Zhang, W.; Fan, Z.; Bao, Y. High-sensitivity refractive index sensor based on D-shaped photonic crystal fiber with rectangular lattice and nanoscale gold film. *Plasmonics* **2014**, *9*, 1355–1360. [CrossRef]
35. Hautakorpi, M.; Mattinen, M.; Ludvigsen, H. Surface-plasmon-resonance sensor based on three-hole microstructured optical fiber. *Opt. Express* **2008**, *16*, 8427–8432. [CrossRef]
36. Chang, M.; Li, B.; Chen, N.; Lu, X.; Zhang, X.; Xu, J. A compact and broadband photonic crystal fiber polarization filter based on a plasmonic resonant thin gold film. *IEEE Photonics J.* **2019**, *11*, 1–12. [CrossRef]
37. Wang, F.; Sun, Z.; Liu, C.; Sun, T.; Chu, P.K. A highly sensitive dual-core photonic crystal fiber based on a surface plasmon resonance biosensor with silver-graphene layer. *Plasmonics* **2016**, *12*, 1847–1853. [CrossRef]
38. Jabin, M.A.; Ahmed, K.; Rana, M.J.; Paul, B.K.; Luo, Y.; Vigneswaran, D. Titanium-coated dual-core D-shaped sPR-based PCF for hemoglobin sensing. *Plasmonics* **2019**, *14*, 1601–1610. [CrossRef]
39. Luan, N.; Han, H.; Zhao, L.; Liu, J.; Yao, J. Opening up dual-core microstructured optical fiber-based plasmonic sensor with large detection range and linear sensitivity. *Opt. Mater. Express* **2019**, *9*, 819–825. [CrossRef]
40. Zhang, X.; Wang, R.; Cox, F.; Kuhlmey, B.; Large, M. Selective coating of holes in microstructured optical fiber and its application to in-fiber absorptive polarizers. *Opt. Express* **2007**, *15*, 16270–16278. [CrossRef]
41. Luan, N.; Lv, W.; Wang, R.; Yao, J. Surface plasmon resonance sensor based on exposed-core microstructured optical fibres. *Electron. Lett.* **2015**, *51*, 714–715. [CrossRef]
42. An, G.; Jia, P.; Liang, T.; Hong, Y.; Wang, H.; Ghaffar, A.; Xiong, J. Double-sided polished ultra-stable and ultra-sensitive optical fiber sensor. *Plasmonics* **2020**. [CrossRef]
43. Luan, N.; Yao, J. A hollow-core photonic crystal fiber-based SPR sensor with large detection range. *IEEE Photonics J.* **2017**, *9*, 1–7. [CrossRef]

44. Wang, G.; Lu, Y.; Duan, L.; Yao, J. A refractive index sensor based on PCF with ultra-wide detection range. *IEEE J. Sel. Top. Quant.* **2020**. [CrossRef]
45. Han, H.; Hou, D.; Zhao, L.; Luan, N.; Song, L.; Liu, Z.; Lian, Y.; Liu, J.; Hu, Y. A large detection-range plasmonic sensor based on an H-shaped photonic crystal fiber. *Sensors* **2020**, *20*, 1009. [CrossRef] [PubMed]
46. Wu, T.; Shao, Y.; Wang, Y.; Cao, S.; Cao, W.; Zhang, F.; Liao, C.; He, J.; Huang, Y.; Hou, M.; et al. Surface plasmon resonance biosensor based on gold-coated side-polished hexagonal structure photonic crystal fiber. *Opt. Express* **2017**, *25*, 20313–20322. [CrossRef] [PubMed]
47. Zhao, J.; Cao, S.; Liao, C.; Wang, Y.; Wang, G.; Xu, X.; Fu, C.; Xu, G.; Lian, J.; Wang, Y. Surface plasmon resonance refractive sensor based on silver-coated side-polished fiber. *Sens. Actuators B Chem.* **2016**, *230*, 206–211. [CrossRef]
48. Vial, A.; Grimault, A.S.; Macías, D.; Barchiesi, D.; Chapelle, M.L.D.L. Improved analytical fit of gold dispersion: Application to the modeling of extinction spectra with a finite-difference time-domain method. *Phys. Rev. B.* **2005**, *71*, 085416. [CrossRef]

*sensors*

MDPI

*Letter*

# Ultrafast Resolution-Enhanced Digital Optical Frequency Comb-Based BOTDA with Pump Pulse Array Coding

**Yichang Wu [1], Chengkun Yang [1,*], Jingshun Pan [1], Qi Sui [2] and Dawei Wang [1]**

1 School of Electronics and Information Technology, Sun Yat-sen University, Guangzhou 510275, China; wuych27@mail2.sysu.edu.cn (Y.W.); panjsh3@mail2.sysu.edu.cn (J.P.); wangdw9@mail.sysu.edu.cn (D.W.)

2 Southern Marine Science and Engineering Guangdong Laboratory (Zhuhai), Zhuhai 519000, China; suiqi@sml-zhuhai.cn

* Correspondence: yangchk5@mail.sysu.edu.cn

Received: 20 September 2020; Accepted: 4 November 2020; Published: 10 November 2020

**Abstract:** In this letter, a resolution enhancement and signal-to-noise ratio (SNR) improvement scheme for digital optical frequency comb (DOFC)-based Brillouin optical time-domain analysis (BOTDA) ultrafast distributed sensing employing a pump pulse array is proposed. Based on the properties of the time-invariant linear system and the cyclic revolution theorem, experimental results indicate that its spatial resolution reaches 10.24 m while the frequency uncertainty is below 2 MHz over a 9.5 km fiber. Moreover, the response time is only 209.6 μs and the temperature measurement error is less than 0.52 °C.

**Keywords:** Brillouin; ultrafast; distributed sensing; pump pulse array; resolution enhancement; SNR

## 1. Introduction

In recent years, Brillouin optical time-domain analysis (BOTDA)-based distributed optical fiber sensing has become the subject of numerous studies for its highly precise temperature and strain-distributed measurement accuracy [1–4]. For conventional BOTDA systems, the frequency difference of the pump-probe scanning process is used to reconstruct the Brillouin gain spectrum (BGS) and Brillouin frequency shift (BFS), which can be linearly mapped to temperature and strain distributions along the fiber. It normally takes a few minutes to complete the single-time measurement of the entire fiber link due to the time-consuming frequency scanning process and the trace averaging algorithm, dropping off the distributed dynamic information of the sensing fiber [5,6].

To eliminate these two major factors that consume the sensing time, several sweeping-free techniques have been proposed to realize fast and dynamic measurements, such as slope-assisted BOTDA [7,8] and ultrafast sweeping BOTDA [9,10]. Besides, digital optical frequency comb (DOFC)-based BOTDA [11–13] and chirp-chain-based BOTDA [14–16] are presented, which are able to reconstruct the BGS and locate the BFS with one single shot. The combination of DOFC-BOTDA and multiple pump pluses provides a better signal-to-noise ratio (SNR) and breaks the inherent limitation between the spatial resolution and the detection accuracy within 0.1 ms [17]. The chirp-chain-based BOTDA utilizing the Brillouin loss scheme achieves 150 km long range fiber sensing within a few seconds [14].

However, the single-shot or fast-scanning BOTDA is confronted with several questions. First, renouncement of frequency scanning and time average severely increases the random noise on the measured BGS, decreasing the SNR of the signal and precision of the frequency [12]. Furthermore, short measurement time means much less effective information, suffering poor spatial resolution, sensitivity, or short sensing range [12,17]. Besides, the multiple-pump-pulses scheme requires an identical power of pulses, which is difficult to achieve with a high peak power of pulses [17]. Some of the fast sensing schemes can provide a fine resolution or long sensing distance by a large amount of averaging, leading to a long measuring time (a few seconds) and missing the dynamic information [14].

To overcome the challenge mentioned above, this paper starts with an introduction of a new resolution and SNR enhancement scheme without lengthening the measurement time in Section 2, combining the DOFC-BOTDA with pump pulse array coding, which can cut down the frequency uncertainty while maintaining the spatial resolution. In addition, the simulation test of the proposed sensing scheme is demonstrated in Section 3, by comparing the sensing results between our proposed scheme and the conventional counterpart. In Section 4, the experimental setup and the data processing method are introduced, and the experimental performances of the proposed sensing scheme are evaluated. Finally, some conclusions and a summary are given in Section 5.

## 2. Sensing Principle

In the proposed scheme based on scanning-free BOTDA, a continuous wide-band DOFC signal is used as the probe to detect and acquire the BGS (Brillouin gain spectrum) with a single shot. The DOFC probe with single- and multiple-pump BOTDA has been proposed for high-speed sensing and dynamic measurement [11]. Figure 1 shows the sensing principle of the DOFC-based BOTDA.

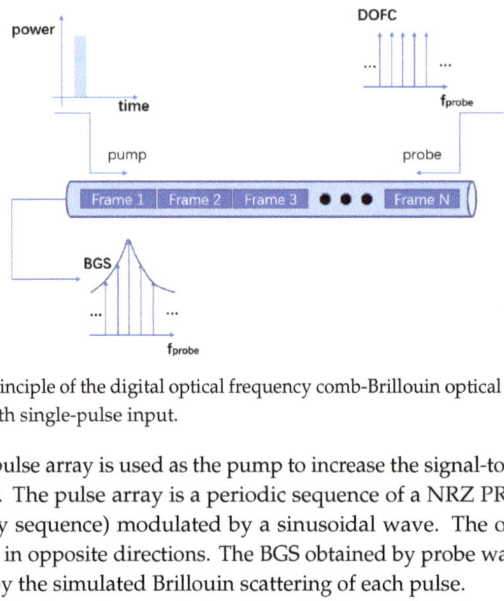


**Figure 1.** Sensing principle of the digital optical frequency comb-Brillouin optical time-domain analysis (DOFC-BOTDA) with single-pulse input.

In this paper, the pulse array is used as the pump to increase the signal-to-noise ratio and enhance the detection accuracy. The pulse array is a periodic sequence of a NRZ PRBS (Non-return-to-zero pseudo-random binary sequence) modulated by a sinusoidal wave. The optical probe and pump signals are transmitted in opposite directions. The BGS obtained by probe waves is the superposition of the BGS generated by the simulated Brillouin scattering of each pulse.

Intensity variations in the probe wave induced by stimulated Brillouin scattering can be expressed as [18]

$$\Delta I_{CW}(t,\Delta\nu) = I_{CWL}e^{-\alpha L}\left\{\exp\left(\int_{v_g t/2}^{v_g t/2+\Delta z} g_B(\xi,\Delta\nu)I_P(\xi,\Delta\nu)\mathrm{d}\xi\right)-1\right\} \quad (1)$$

where $\Delta I_{CW}$ is the intensity fluctuation measured at the near end of the fiber $z = 0$ as a function of time $t$ and frequency offset between pump and probe light $\Delta\nu$. $I_{CWL}$ is the input probe intensity at $z = L$, $L$ is the length of the fiber, $\alpha$ is the loss of optical fibers, $v_g$ is the group velocity, $g_B(\xi,\Delta\nu)$ and $I_P(\xi,\Delta\nu)$ are the Brillouin gain coefficient and pump intensity at $z = \xi$, respectively. The short simulated Brillouin scattering (SBS) interaction length taking place in BOTDA sensors leads to a very small Brillouin gain, allowing us to linearize Equation (1) as

$$\Delta I_{CW}(t, \Delta\nu) \propto \int_{v_g t/2}^{v_g t/2+\Delta z} g_B(\xi, \Delta\nu) I_P(\xi, \Delta\nu) d\xi \quad (2)$$

If $I_P(\xi, \Delta\nu)$ is no longer a single pulse, the entire interaction length should be taken into account in the upper limit of the above integration. From Equation (2), the SBS process can be regarded as a linear time-invariant, as well as frequency-invariant, system. As for conventional linear time-invariant (LTI) systems, the linear convolution theorem tells us that the product of the Fourier transform of the impulse response function and input function in the time domain is equal to the Fourier transform of the output response function. In our case of the SBS process, it can be extended to two dimensions (2D), i.e., the frequency–time domain. The 2D Fourier transform of the output response function should be the product of the Fourier transforms of the input function and impulse response function, i.e.,

$$\mathcal{F}_{2D}\{I(t, \nu)\} \cdot \mathcal{F}_{2D}\{R(t, \nu)\} = \mathcal{F}_{2D}\{O(t, \nu)\} \quad (3)$$

Input function $I$, impulse response function $R$, and output response function $O$ are functions of time $t$ and frequency $\nu$. In real applications, as sampling is discrete, a 2D function in the frequency–time domain is converted to a discrete 2D matrix. In addition, the linear convolution theorem should be replaced by the cyclic convolution theorem [19], i.e., the product of the discrete Fourier transform of the impulse response matrix and input matrix in the time–frequency domain is equal to the discrete Fourier transform of the output response matrix.

$$I_F(x_n, y_m) = \mathcal{F}_{2D}\{I(t_n, \nu_m)\}, R_F(x_n, y_m) = \mathcal{F}_{2D}\{R(t_n, \nu_m)\}, O_F(x_n, y_m) = \mathcal{F}_{2D}\{O(t_n, \nu_m)\} \quad (4)$$

$$I_F(x_i, y_j) \cdot R_F(x_i, y_j) = O_F(x_i, y_j), i = 1, 2 \ldots N, j = 1, 2 \ldots M$$

where the size of all matrices is $N * M$. In this case, the product of the matrix is the Hadamard product, as presented in Equation (4). The output response of the system is detectable, while the impulse response matrix implies that the output of the system with a single pump pulse should be like the conventional BOTDA as we expect. By designing the input response matrix $I$, detecting the output response matrix $O$, and performing the inverse operation of Equation (4), the impulse response matrix $R$ is solvable, as depicted in Figure 2. Compared to the BOTDA systems with single or several pump pulses, the quasi-continuous pump pulse array is able to improve the SNR of the system.

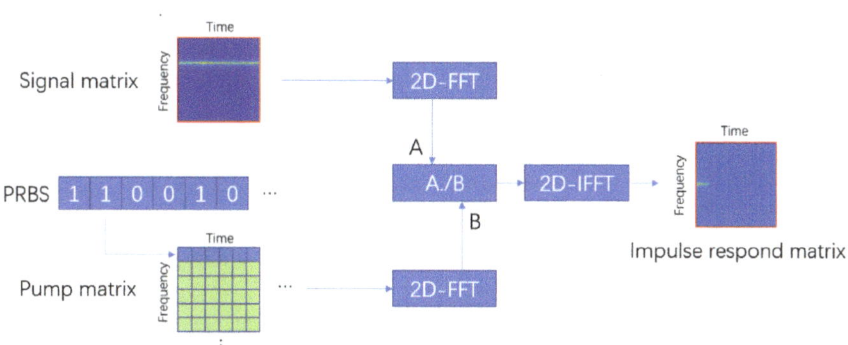


**Figure 2.** Impulse response matrix calculation procedure.

In Figure 2, two dimensions of matrices represent time and frequency. The input matrix of the system, pump matrix, is carefully designed. In this paper, we choose PRBS as the code of the pump array, while bits 1 and 0 indicate that a single pump pulse is on or off, respectively. Every pump pulse here is modulated by a sinusoidal wave with an identical frequency. Therefore, the elements of the pump matrix are 0 except those in the corresponding row of frequency, as shown in the first row of the pump matrix in Figure 2. The output matrix is constructed by splitting the received signal into independent frames, performing FFT on each frame and aligning the FFT results. After the construction of the input and output matrix, 2D-FFT is performed to change the matrix into the "frequency" domain (the frequency generated by 2D-FFT, not the real frequency). The product on the left side of Equation (4) is the Hadamard product instead of matrix multiplication. In order to recover the impulse response matrix, the inverse operation of Hadamard multiplication should be applied. Divided by the input matrix element by element, the output matrix becomes the impulse response matrix in the frequency domain. After a two-dimensional inverse fast Fourier transform (2D-IFFT) on the result, the impulse response matrix is recovered.

## 3. Simulation

We have simulated the Brillouin sensing process for a 1.6 km single-mode fiber (SMF). The length of one DOFC frame is 200 ns, corresponding to 20 m spatial length, while the frequency step is 5 MHz according to the inherent restriction imposed by the length of a DOFC frame and the frequency spacing after the FFT, i.e., the spatial and frequency resolution. Figure 3a shows the simulated impulse response matrix without noise by a 12.09 GHz sinusoidally modulated pump pulse input. Each column in this matrix represents a DOFC frame, carrying a Lorentzian-shaped BGS. The full-width at half-maximum (FWHM) of the BGS is set to 30 MHz. The typical BFS of this fiber is 10.8 GHz, resulting in the BGS being centered at 1.29 GHz on the DOFC probe. In addition, there are two irregular parts with different BFSs. One is located at 980–1000 m with a 10.96 GHz BFS, while another is at 1280–1320 m with a 10.84 GHz BFS. According to the setting frequency of the pump pulse, their central frequencies of the BGS should be located at 1.13 and 1.25 GHz, respectively.

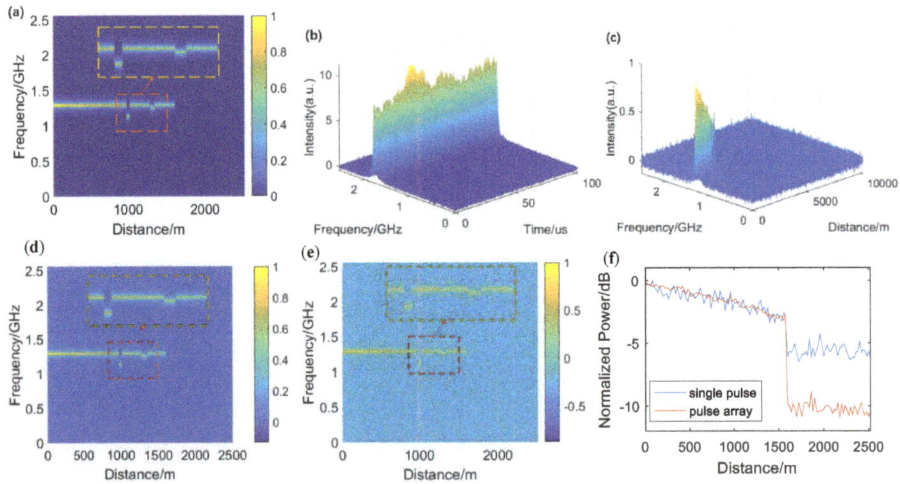


**Figure 3.** Simulation results. (**a**) Set impulse response matrix without noise, using two irregular Brillouin frequency shift (BFS) points; (**b**) simulated received output signal generated by the sensing system with pump array input; (**c**) calculated impulse response matrix; (**d**) calculated impulse response matrix with pump array input; (**e**) calculated impulse response matrix with single pump pulse input; (**f**) simulated trace for single pump pulse and pump pulse array input case.

Here, we choose PRBS9 (511 bits) as the pump array in the time domain using 11.09 GHz sinusoidal wave modulation, while bits 1 and 0 indicate that a single pump pulse is on or off, respectively. The length of each bit in the pump array is equal to a DOFC frame (200 ns in this case). The PRBS can provide a perfectly flat intensity plane without extreme points in the frequency domain after a discrete 2D Fourier transform, which can avoid the severe accuracy issue during the calculation and become robust. Without a flat intensity plane in the frequency domain, the quality of the solved impulse response matrix will deteriorate rapidly as the power of noise in the output response matrix increases. As there is only one single frequency in this pump array, all the elements of the pump matrix should be 0 except the first line that represents the 11.09 GHz sinusoidally modulated PRBS9. Although the pump array is periodic, the pump matrix only records one period as the cyclic convolution regards it as a periodic signal already. The size of the pump matrix should be expanded to 511 × 511 to ensure that the intensity remains flat after the 2D fast Fourier transform (2D-FFT).

Figure 3b shows the simulated output signal matrix of the sensing system in the presence of the pump matrix input. It is generated by overlapping the impulse response matrix that each single pulse in the pump matrix provides. As the pump array is periodic, the output signal should have the same cycle period in the time domain. Besides, white Gaussian noise is added on the output signal matrix, and the SNR is set to 20. Figure 3d,e respectively show the calculation results with the pump array input and a single pump pulse input under the same SNR. The BGS is more obvious using the pump array input than the single pump pulse input. It is an efficient way to improve the SNR of the received BGS and reduce the BFS uncertainty. Figure 3f demonstrates the comparison of the trace of the BGS peak along the fiber. Background noise is significantly suppressed using the pump pulse array input, with an SNR enhancement of about 5 dB.

## 4. Experiment, Data Processing, and Discussion

The experimental setup of the proposed BOTDA sensing system is depicted in Figure 4. A tunable continuous-wave laser (Keysight N7714A) provides a 1550 nm beam with a 100 kHz linewidth. A 90/10 optical coupler is employed to split the 1550 nm laser into two branches for the pump and probe signal.

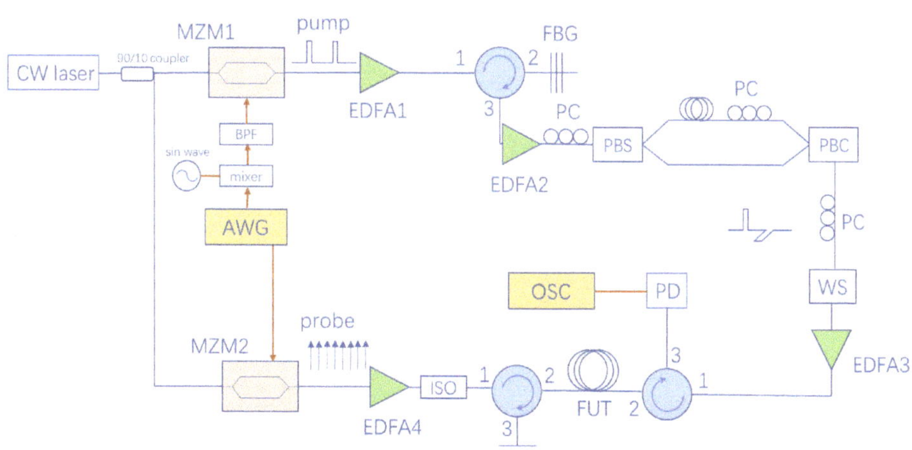


**Figure 4.** Experimental setup of the proposed BOTDA sensing system. CW laser: Continuous-wave laser; MZM: Mach–Zehnder modulator; EDFA: Erbium-doped optical fiber amplifier; FBG: Fiber Bragg grating; PC: Polarization controller; PBS: Polarizing beam splitter; WS: Wave-shaper; PBC: Polarizing beam combiner; PD: Photodiode; ISO: Optical isolator; FUT: Fiber under test; OSC: Real-time oscilloscope.

As for the pump side, an arbitrary waveform generator (AWG, Keysight M8195A, Santa Rosa, CA, US) gives an electric pulse array arranged as PRBS11 modulated with 800 MHz sinusoidal waves. Each bit in this array lasts for 102.4 ns. For the first half of a bit (about 50 ns), the signal turns to the zero level in the case of 0, whereas it turns to a sinusoidal wave in the case of 1. However, it always turns to the zero level at the latter half of the bit in order to avoid the overlapping issue of orthogonally polarized pulses mentioned later. Then, the pulse array is upconverted to 11.1 GHz by a mixer and a sinusoidal wave generated by a signal generator, and goes through a microwave bandpass filter to filter out the superfluous frequencies induced by the mixing process. The filtered pulse array is imposed on the pump light via a high-extinction-ratio Mach–Zehnder modulator (MZM, iXblue MXER-LN-20, Saint-Germain-en-Laye, France) biased at the null point. The achieved extinction ratio of carrier suppression exceeds 30 dB in order to avoid the four-wave mixing phenomenon in optical fibers. After being amplified by an erbium-doped optical fiber amplifier (EDFA), the lower sideband and remaining carrier are filtered out by a fiber Bragg grating (FBG) with 10 GHz 3 dB-bandwidth. The second EDFA amplifies the remaining upper sideband signal. To eliminate the influence of the polarization fluctuations to the SBS effect, a scheme of orthogonally polarized pump pulses is used here [20]. A polarizing beam splitter (PBS) is used to split the optical pump array into two branches with an orthogonal polarization state. Two polarization controllers (PC) are set before the PBS and one of the branches to adjust the polarization state, ensuring both branches get identical optical power. Besides, one of the branches is delayed by an SMF, in which the pump array can go for a 51 ns delay, about half the length of one bit of the PRBS array. By using a polarizing beam combiner (PBC), two branches are combined to generate an orthogonally polarized optical pump pulse array. As the effective signal only occupies the front half of the bits in the original pump array, delay and recombination will avoid the overlapping of the two orthogonally polarized optical signals. Due to the orthogonally polarized pump pulses with identical power, regardless of the direction of polarization of the probe light, the Brillouin gain is stable. Furthermore, a programmable wave-shaper is used to reshape the optical pump array and obtain a higher extinction ratio. Finally, the array is amplified by EDFA3 and launched into a fiber-under-test (FUT) of 9.5 km in length via an optical circulator. The measurement for one single test of the entire fiber link is equal to one period of the pump array, i.e., 209.6 μs.

As for the probe side, another channel of the AWG gives a periodic PRBS as a digital electrical frequency comb (DEFC) signal. Each frame of the DEFC lasts for 102.4 ns, which is as long as the length of one bit in the pump array, corresponding to a frequency spacing of 9.77 MHz and spatial resolution of 10.24 m. The light is modulated by this DEFC signal via another MZM (Fujitsu FTM7937EZ) biased at the linear operating point to generate the baseband DOFC probe signal. After being amplified by EDFA4, the probe signal passes through an optical isolator and an optical circulator, which would filter out the pump signal from the opposite side, and finally, the probe signal is sent into the FUT.

At the receiver end, the probe signal is detected by a photodiode (PD). The output signal is transmitted into a real-time oscilloscope (OSC) with a 20 GS/s sampling rate. After data acquisition, digital signal processing (DSP) is performed to solve the impulse response matrix. The DSP is shown as the flowchart in Figure 5a.

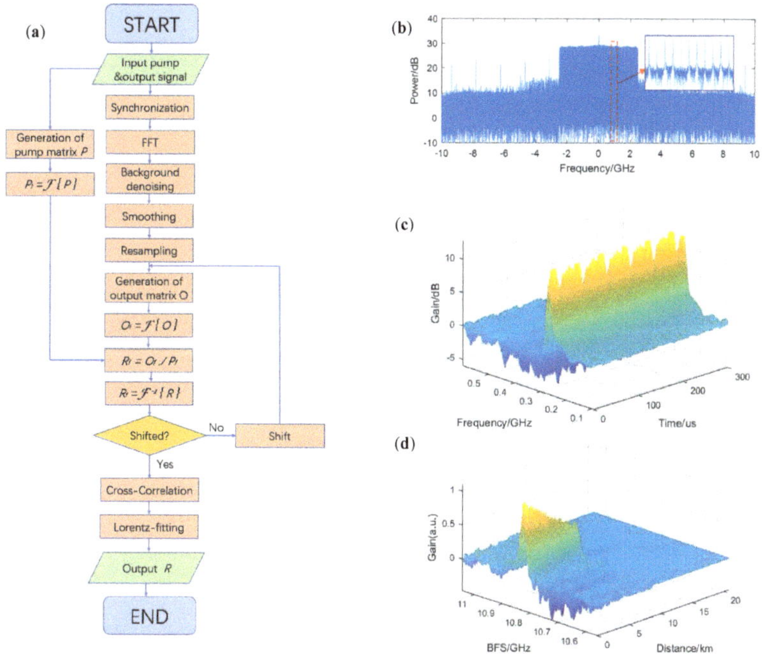


**Figure 5.** (**a**) Flowchart of digital signal processing; (**b**) DOFC signal in frequency domain; (**c**) experimental periodic received output signal from the sensing system; (**d**) calculated impulse response matrix.

The DOFC signal without the BGS in the frequency domain is shown in Figure 5b as a background reference noise. By synchronization of the received signal, the start points of all DOFC frames can be located. By dividing the continuous received signal into independent frames, fast Fourier transform (FFT) and background denoising could be used to obtain the Brillouin gain information carried by each frame. As the pump array is periodic, the experimental received output signal should also be periodic in the time dimension, as shown as Figure 5c. Then, smoothing in the frequency dimension would help to reduce the glitch on the signal. Furthermore, each frame is up-sampled by a factor of four in the frequency domain to achieve a narrower frequency spacing of 2.44 MHz. The generation of the pump matrix has been mentioned above, and the output matrix is generated by using a suitable data window onto the up-sampling signal. Here, the size of the pump matrix and the data window is $2047 \times 201$. In addition, by performing 2D-FFT on both matrices, dividing element by element, and conducting two-dimensional inverse fast-Fourier-transform (2D-IFFT) retrieval to the time–frequency (or space–frequency) domain, the impulse response matrix can be solved. In order to align the start point of the FUT to the zero point in the space dimension of the matrix, it is necessary to introduce a proper shift in the data window on the up-sampling signal. The impulse response matrix is shown in Figure 5d, where the time dimension has been changed into the space dimension (10 ns corresponds to 1 m based on the refractive index of optical fiber of 1.5) and the probe frequency has been changed into the BFS by being subtracted from the pump frequency of 11.1 GHz.

To acquire the accurate BFS, we have performed a cross-correlation algorithm and Lorentz fitting. The cross-correlation algorithm is based on the fact that Gaussian white noise is uncorrelated with the Lorentz lineshape while the effective BGS signal is strongly correlated. As the response signal is a linear combination of the BGS and noise, cross-correlation with a standard Lorentz lineshape can be used to denoise the signal [15]. As the correlation of two Lorentz shapes is still a Lorentz shape, iterative correlations can further denoise the signal. However, the correlation will broaden the width of the Lorentz shape and increase the frequency uncertainty, so we take the proper iterative time as two. Besides, by up-sampling, better results can be obtained owing to the fact that a narrower frequency spacing of a standard Lorentz lineshape would allow more effective information. Here, we perform up-sampling by a factor of four. Figure 6a shows a comparison among the signals with no cross-correlation, with a one-time correlation, and with a two-time correlation. The BGS seems asymmetric because of the beating between the upper and lower sideband of the optical probe signal. In addition, after the twice-correlation algorithm, Lorentz fitting can obtain the BFS distribution over the fiber link.

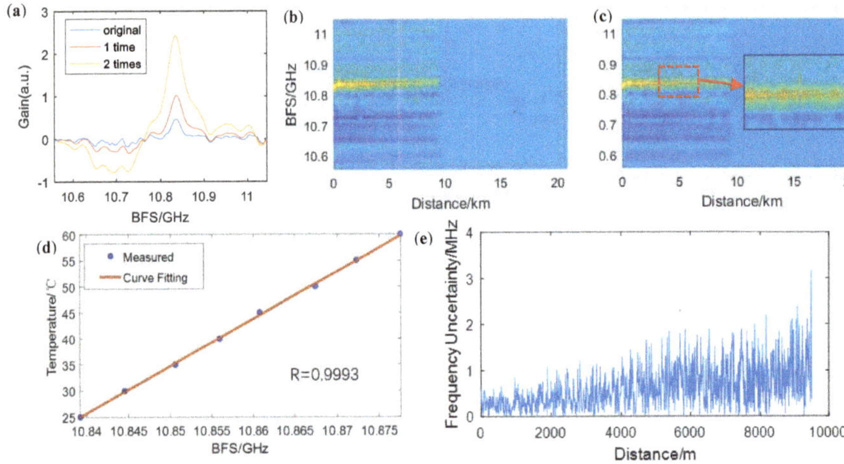


**Figure 6.** (**a**) Results of cross-correlation with Lorentz lineshape; (**b**) Brillouin gain spectrum (BGS) distribution in normal environment; (**c**) BGS distribution with a 10 m fiber section heating; (**d**) temperature-BFS measurement results; (**e**) frequency uncertainty over the fiber link.

In the temperature test configuration, a segment of 10 m long fibers at 5 km is heated using a thermostat water bath. Figure 6b shows the BGS distribution of the FUT under a room temperature of 25 °C, and Figure 6c shows the BGS distribution when the 10 m fiber section is heated up to 45 °C, where only one frame of the DOFC corresponding to the BFS of the heating part is shifted. Besides, the 10 m fiber section is heated from 25 °C to 60 °C with a 5 °C step, and the measured BFS is shown in Figure 6d. The environment temperature and BFS fit the linear relation, and the maximum deviation from the fitting curve is 0.52 °C. The temperature-BFS ratio is measured to be 0.91 °C/MHz according to the fitting curve. Figure 6e depicts the frequency uncertainty along the FUT through repeated measurement for 10 times under the same environment. The frequency uncertainty is below 1.5 MHz for the first 5 km fiber section. As for the latter 5 km fiber section, though the frequency uncertainty reaches 3.2 MHz when getting close to the end of the fiber link, it is below 2 MHz in most cases, which is much better than those resulting from the inherent restriction between the spatial resolution and frequency uncertainty when using the single pump pulse and DOFC probe, i.e., spatial resolution of 10.24 m with 9.77 MHz frequency uncertainty. As the joints of the optical fibers will provide extra frequency uncertainty, the frequency uncertainty of our proposed this system is below 2 MHz.

## 5. Conclusions

In summary, we have demonstrated an SNR improvement scheme for DOFC-based BOTDA sensing employing a pump pulse array, by which a BFS sensing time of only 209.6 μs could be achieved for a 9.5 km fiber link. By applying two-dimensional cyclic convolution and linear system theory, time as well as frequency, the input power of the system could be greatly improved to obtain better SNR. The spatial measurement resolution reached 10.24 m and the frequency uncertainty was below

2 MHz. Owing to its fast sensing speed and improved spatial resolution, our proposed sensing system is expected to find applications in FPGA-based long-time dynamic measurement.

**Author Contributions:** Y.W. and Q.S. proposed the concept and initiated the study. Y.W. carried out numerical simulations. Y.W., C.Y. and J.P. performed the experiment. C.Y., Q.S. and D.W. supervised the studies. The paper was written by Y.W. and C.Y. All authors have read and agreed to the published version of the manuscript.

**Funding:** This work was supported in part by the National Key R&D Program of China (2019YFB1803702, 2019YFA0706300), in part by the National Natural Science Foundation of China (NSFC) (61525502, 61805104) and in part by the Science and Technology Planning Project of Guangdong Province (2017B010123005, 2018B010114002).

**Conflicts of Interest:** The authors declare no conflict of interest.

## References


1. Kurashima, T.; Horiguchi, T.; Tateda, M. Distributed-temperature sensing using stimulated Brillouin scattering in optical silica fibers. *Opt. Lett.* **1990**, *15*, 1038–1040. [CrossRef] [PubMed]
2. Bao, X.; Chen, L. Recent progress in optical fiber sensors based on Brillouin scattering at university of Ottawa. *Photonic Sens.* **2011**, *1*, 102–117. [CrossRef]
3. Denisov, A.; Soto, M.A.; Thévenaz, L. Going beyond 1000000 resolved points in a Brillouin distributed fiber sensor: Theoretical analysis and experimental demonstration. *Light Sci. Appl.* **2016**, *5*, e16074. [CrossRef] [PubMed]
4. Motil, A.; Bergman, A.; Tur, M. State of the art of Brillouin fiber-optic distributed sensing. *Opt. Laser Tech.* **2016**, *78*, 81–103. [CrossRef]
5. Bernini, R.; Minardo, A.; Zeni, L. Dynamic strain measurement in optical fibers by stimulated Brillouin scattering. *Opt. Lett.* **2009**, *34*, 2613–2615. [CrossRef] [PubMed]
6. Minardo, A.; Catalano, E.; Zeni, L. Cost-effective method for fast Brillouin optical time-domain analysis. *Opt. Express* **2016**, *24*, 25424–25431. [CrossRef] [PubMed]
7. Peled, Y.; Motil, A.; Yaron, L.; Tur, M. Slope-assisted fast distributed sensing in optical fibers with arbitrary Brillouin profile. *Opt. Express* **2011**, *19*, 19845–19854. [CrossRef] [PubMed]
8. Zhou, D.; Dong, Y.; Wang, B.; Jiang, T.; Ba, D.; Xu, P.; Zhang, H.; Lu, Z.; Li, H. Slope-assisted BOTDA based on vector SBS and frequency-agile technique for wide-strain-range dynamic measurements. *Opt. Express* **2017**, *25*, 1889–1902. [CrossRef]
9. Sovran, I.; Motil, A.; Tur, M. Frequency-scanning BOTDA with ultimately fast acquisition speed. *IEEE Photonics Tech. Lett.* **2015**, *27*, 1426–1429. [CrossRef]
10. Mizuno, Y.; Hayashi, N.; Fukuda, H.; Song, K.Y.; Nakamura, K. Ultrahigh-speed distributed Brillouin reflectometry. *Light Sci. Appl.* **2016**, *5*, e16184. [CrossRef] [PubMed]
11. Jin, C.; Guo, N.; Feng, Y.; Wang, L.; Liang, H.; Li, J.; Li, Z.; Yu, C.; Lu, C. Scanning-free BOTDA based on ultra-fine digital optical frequency comb. *Opt. Express* **2015**, *23*, 5277–5284. [CrossRef] [PubMed]
12. Jin, C.; Wang, L.; Chen, Y.; Guo, N.; Chung, W.; Au, H.; Li, Z.; Tam, H.-Y.; Lu, C. Single-measurement digital optical frequency comb based phase-detection Brillouin optical time domain analyzer. *Opt. Express* **2017**, *25*, 9213–9224. [CrossRef] [PubMed]
13. Li, Z. Single-measurement BOTDA using DOFC based Brillouin phase spectrum detection. In Proceedings of the 7th International Multidisciplinary Conference on Optofluidics, Singapore, 25–28 July 2017.
14. Dong, Y.; Wang, B.; Pang, C.; Zhou, D.; Ba, D.; Zhang, H.; Bao, X. 150 km fast BOTDA based on the optical chirp chain probe wave and Brillouin loss scheme. *Opt. Lett.* **2018**, *43*, 4679–4682. [CrossRef] [PubMed]
15. Zhou, D.; Dong, Y.; Wang, B.; Pang, C.; Ba, D.; Zhang, H.; Lu, Z.; Li, H.; Bao, X. Single-shot BOTDA based on an optical chirp chain probe wave for distributed ultrafast measurement. *Light Sci. Appl.* **2018**, *7*, 32. [CrossRef] [PubMed]
16. Wang, B.; Fan, B.; Zhou, D.; Pang, C.; Li, Y.; Ba, D.; Dong, Y. High-performance optical chirp chain BOTDA by using a pattern recognition algorithm and the differential pulse-width pair technique. *Photonics Res.* **2019**, *7*, 652–658. [CrossRef]
17. Liang, Z.; Pan, J.; Gao, S.; Sui, Q.; Feng, Y.; Li, F.; Li, J.; Liu, W.; Li, Z. Spatial resolution improvement of single-shot digital optical frequency comb-based Brillouin optical time domain analysis utilizing multiple pump pulses. *Opt. Lett.* **2018**, *43*, 3534–3537. [CrossRef] [PubMed]

18. Soto, M.A.; Bolognini, G.; Di Pasquale, F.; Thévenaz, L. Simplex-coded BOTDA fiber sensor with 1 m spatial resolution over a 50 km range. *Opt. Lett.* **2010**, *35*, 259–261. [CrossRef] [PubMed]
19. Smith, J. *Mathematics of the Discrete Fourier Transform (DFT)*; Stanford University: Stanford, CA, USA, 2002.
20. Urricelqui, J.; López-Fernandino, F.; Sagues, M.; Loayssa, A. Polarization diversity scheme for BOTDA sensors based on a double orthogonal pump interaction. *J. Lightwave Tech.* **2015**, *33*, 2633–2638. [CrossRef]

**Publisher's Note:** MDPI stays neutral with regard to jurisdictional claims in published maps and institutional affiliations.

*sensors*

MDPI

*Letter*

# Numerical Analysis and Recursive Compensation of Position Deviation for a Sub-Millimeter Resolution OFDR

**Yueying Cheng [1,2,3], Mingming Luo [1,2,3], Jianfei Liu [1,2,3,*] and Nannan Luan [1,2,3]**

[1] School of Electronic and Information Engineering, Hebei University of Technology, Tianjin 300401, China; 201821902035@stu.hebut.edu.cn (Y.C.); 2019013@hebut.edu.cn (M.L.); luan@hebut.edu.cn (N.L.)

[2] Tianjin Key Laboratory of Electronic Materials and Devices, Tianjin 300401, China

[3] Hebei Key Laboratory of Advanced Laser Technology and Equipment, Tianjin 300401, China

* Correspondence: jfliu@hebut.edu.cn

Received: 30 August 2020; Accepted: 24 September 2020; Published: 27 September 2020

check for updates

**Abstract:** We analyze the source of the position deviation and propose a demodulation recursive compensation algorithm to ensure a sub-millimeter resolution in improved optical frequency domain reflectometry. The position deviation between the geometric path and optical path changes with the temperature or strain, due to the elastic-optic and thermal-optic effects. It accumulates along the fiber and becomes large enough to affect the spectral correlation between the measured and reference spectra at the fiber end. The proposed algorithm compensates for the position deviation of each measuring point and aligns the measured spectra with its reference. The numerical and experimental results both reveal that the signal-to-noise ratio of the correlation is improved doubly and a sub-millimeter spatial resolution becomes available at a 30 m fiber end. The recursive compensation algorithm helps to restrain the correlation degeneration at the fiber end and promises an effective approach to a sub-millimeter resolution in optical frequency domain reflectometry.

**Keywords:** optical frequency domain reflectometry; position deviation compensation; sub-millimeter spatial resolution

## 1. Introduction

Optical frequency domain reflectometry (OFDR), as a promising technique based on intrinsic Rayleigh scattering (RS), was firstly introduced by W. Eickhoff in 1981 [1]. Initially, OFDR was mainly used for loss and breakpoint diagnosis in optical fiber devices and networks [2,3]. The spatial resolution can be achieved at several micrometers in the frequency domain [4,5], with an expandable measuring length to tens or hundreds of meters. With the development of the narrow linewidth tunable laser source (TLS) and nonlinear phase noise compensation in OFDR, the sensing range can be promoted to dozens of kilometers [6–9]. In addition, Froggatt et al. proposed a distributed quantitative sensing method with a high spatial resolution of 0.6 cm and sensitivity of 5 micro-strains over 30 cm, utilizing spectral correlation shifts in OFDR [10,11]. This method is suitable for practical applications, where the higher spatial resolution and longer sensing range is essential, such as in temperature sensing in nuclear reactors, deformation monitoring of wind turbines, and wing skin monitoring of aircraft [12–14]. However, the similarity of RS spectra degenerates at the fiber end, which limits the spatial resolution in the long detection range. Feng K. et al. demonstrated that similarity of the RS spectra degrades in the event that a large strain is loaded. They also found highly similar characteristics of local spectra and proposed a novel method to obtain wavelength offset, relying on matching measured and local reference spectra. New evaluation coefficients of spectral similarity were illustrated, based on minimum residual sum of squares and least square, respectively [15,16]. This technology significantly helps to maintain spectral signal-to-noise ratio (SNR). They obtained a 3000 micro-strain distribution curve along a 3 m sensing fiber with a 3 mm long fiber gauge. Then, Zhao S. et al. [17] proposed spectrum registration to restrain the similarity degeneration of RS spectra and ensure a high spatial resolution, using the narrow box window to extract a local spectrum in the case of a large measurable range. The combination of spatial calibration and spectrum registration can avoid the mismatch of the fiber segments. Distributed 7000 micro-strain along a 1.2 m sensing fiber is demodulated over a 5 mm spatial resolution accurately. In addition, this team also proposed a demodulation approach based on image processing to remove the noise caused by data acquired and realize accurate measurements. This technology was verified at the spatial resolution of 0.4 mm with a 350 mm sensing distance [18]. Luo M. et al. [19] realized a spatial resolution as an order of 0.5 mm over a 25 m long sensing gauge by the compensation algorithm for position deviation. They demonstrated that position deviation accumulating along the fiber degrades the similarity of the RS spectra, due to the elastic-/thermal-optic effects. However, they did not quantitatively analyze the causes, effects, and compensation mechanisms of position deviation in detail.

In this paper, aiming at the position deviation in OFDR strain sensing system, we quantitatively analyze the influence on the system. Then, the recursive compensation mechanism for position deviation is introduced in detail. This method can align the measured spectra with their reference and preserve the similarity between the measurement and reference spectra at a high spatial resolution. The experimental results indicate that the SNR of correlation is improved doubly, and the fake peaks can be efficiently eliminated. Distributed strain is demodulated with a 5 mm spatial resolution over a 30 m long detection range finally. The homogeneity between theoretical analysis and experimental verification demonstrates that this proposed technique can satisfy the requirements of high-precision distributed sensing.

## 2. Principles

Figure 1 illustrates the schematic diagram of the dual-polarization harvesting OFDR, which consists of a TLS, a reference Michelson interferometer (RMI), a heterodyne coherent detection module, and a data acquisition-processing module. The beat signal from the RMI provides an external reference clock to resample the signals and corrects the scanning nonlinearity of the TLS. Due to the optical path difference between the fiber under test (FUT) and the local reference, the reflected light interferes with the local light and generates a position dependent signal. In the event that the axial strain or temperature changes, the optical path changes along with the length and the refractive index due to the elastic-optic and thermal-optic effects, resulting in Rayleigh spectra shift. According to the correlation calculation in the spatial domain, linear temperature and strain dependent wavelength shift can be observed. Meanwhile, position deviation is introduced in the demodulation between the measured and reference spectra as well. This causes spectral dislocation and correlation degeneration, especially at the fiber end.

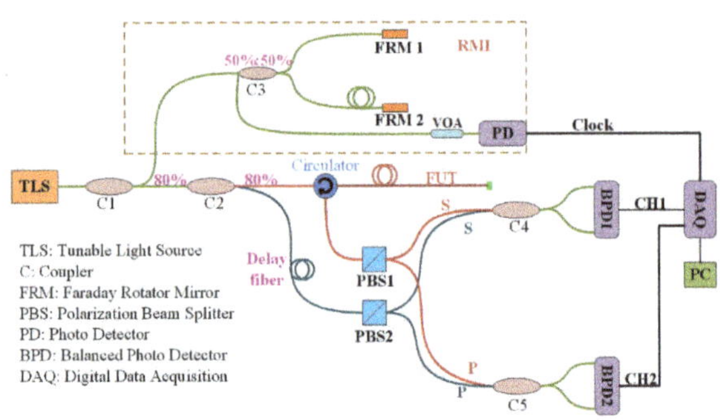


**Figure 1.** Schematic illustration of the OFDR sensing system.

The position deviation is discussed in detail, as shown in Figure 2. The orange dots represent the Rayleigh scattering center of the fiber with strain or temperature applied. It includes four equivalent measuring points, with $\Delta x$ deviation increment in optical path for each one. Besides, the position deviation accumulates along the fiber and increases to 4 $\Delta x$ through the stretched fiber segment. The effect for position deviation is similar in strain and temperature sensing systems. In this paper, the position deviation only is analyzed in the static strain sensing system. In theory, the accumulation is expressed with the coeffects on the optical path, which is described in integral form, as shown in Equation (1). Meanwhile, in the practical applications, all signals are digitized and discretized in the sum form, as shown in Equation (2).

$$X_{deviation} = \int_0^x (\kappa_{eoc} \cdot \Delta S_{dx} + \kappa_{toc} \cdot \Delta T_{dx}) dx \tag{1}$$

$$X_{deviation} = \sum_0^i (\kappa_{eoc} \cdot \Delta S_{x_i} + \kappa_{toc} \cdot \Delta T_{x_i}) \Delta x_i \tag{2}$$

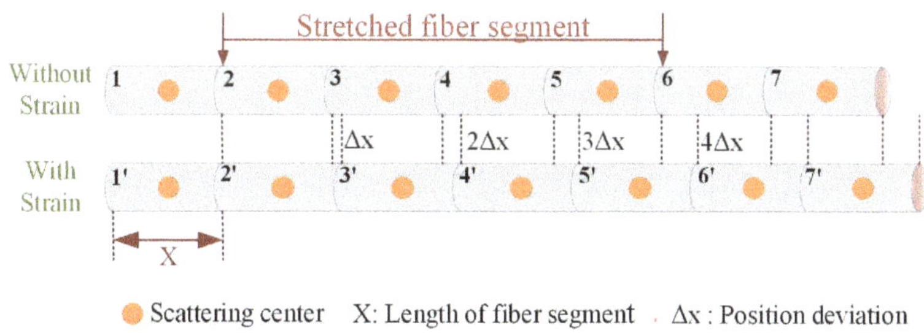


**Figure 2.** Distribution of scattering center along sensing fiber before and after strain.

The $\kappa_{eoc}$ and $\kappa_{toc}$ are the elastic-optic and thermal-optic coefficient, respectively; $\Delta S_{x_i}$ and $\Delta T_{x_i}$ refer to the strain and temperature variations on a $\Delta x_i$ long fiber segment.

As shown in Figure 3, different axial strain stages are loaded along the 30 m long FUT. The slope of the accumulation in position deviation (brown line) strictly agrees with the strain distribution (olive line), which achieves 0.491 mm at the fiber end. The deviation seems negligible to the whole sensing fiber, but it approaches the 0.5 mm gauge length of the sensor and causes spectral dislocation, which indicates the difficulty of achieving a high spatial resolution at the fiber end.

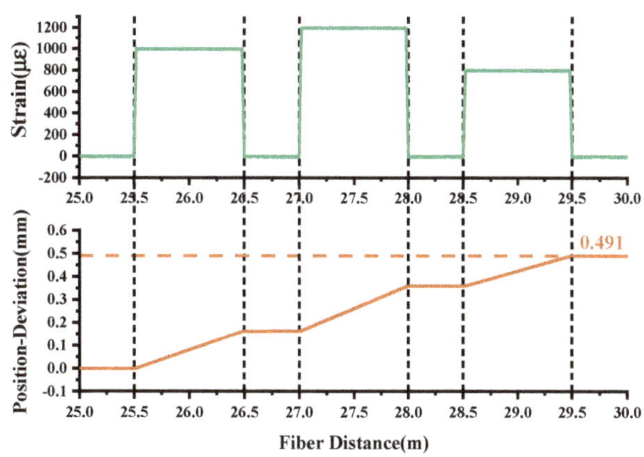


**Figure 3.** Accumulation of position deviation along sensing fiber after axial strain loaded.

For different spatial resolutions, the spectral dislocation caused by the 0.491 mm position deviation is approximately 25% (at 2 mm), 50% (at 1 mm), and 62.5% (at 0.8 mm). The SNR is used as an indicator to quantify the correlation between the measured and reference spectrum. As depicted in Figure 4a, with the increasing deviation between the measured and the reference spectra, the characteristic peak gradually submerges into the noise. The threshold of SNR, ensuring signal contrast, was defined as 4.59 dB ($e^{-1}$ times of the SNR with no deviation). Figure 4b shows that the SNR of the correlation declines with the position deviation increment. In the event that the deviation reaches ~60% of the sensor gauge length, the SNR cannot satisfy the threshold, and the position deviation should be compensated to eliminate the fake peaks in correlation.

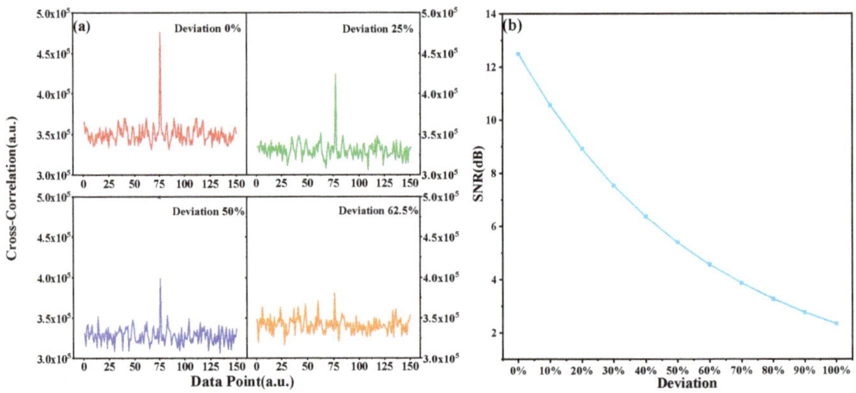


**Figure 4.** Influence of position deviation on spectral similarity. (**a**) SNR of the characteristic peak to quantify the strength of the cross-correlation. (**b**) SNR decreasing with position deviation.

For a typical OFDR sensing system with limited data capacity, a higher spatial resolution means a weaker cross-correlation and a lower tolerance for spectral dislocation. Therefore, to achieve a higher spatial resolution at a longer fiber end, the compensation for position deviation is supposed to rematch the measured spectrum to its reference and ensure the spectral correlation with a recursive algorithm.

OFDR can be regarded as the cascaded fiber Bragg gratings (FBGs) with random period. Once the strain or temperature is applied on the so-called FBGs, the central wavelength $\lambda_B$ shows a strain or temperature dependent wavelength shift. By analogy with the FBG, the elastic- and thermal- coeffects on the wavelength shift are expressed by Equation (3).

$$\frac{\Delta(nx_i)}{nx_i} \approx \frac{\Delta n \cdot x_i}{nx_i} + \frac{n\Delta x_i}{nx_i} = \frac{\Delta n}{n} + \frac{\Delta x_i}{x_i} \approx \frac{\Delta\lambda}{\lambda_c} \tag{3}$$

where $n$ and $x_i$ refer to the refractive index of fiber core and the fiber segment of $i_{th}$ sensor. The $\Delta n$ and $\Delta x_i$ represent the refractive index and optical path changes, respectively, while $\lambda_c$ and $\Delta\lambda_c$ refer to center wavelength and wavelength shift. Therefore, the position deviation can be calculated with previous $i$ sensors using integral Equation (4) and discrete Equation (5), respectively.

$$X_{deviation}^{i+1} = \int_0^{l_0} \frac{\Delta\lambda_i}{\lambda_{ic}} n dx \tag{4}$$

$$X_{deviation}^{i+1} = \sum_0^{i} \frac{\Delta\lambda_i}{\lambda_{ic}} n\Delta x_i \tag{5}$$

During the compensation process, the starting point of the measured spectrum is aligned with its reference, with the correlation well improved. According to the recursive algorithm analyzed above, the compensation for position deviation makes a high spatial resolution possible at the fiber end.

Assuming no deviation at the initial position of the stretched fiber segment. The position deviation of the second measuring point is corrected by the first point. With recursive compensation, the position deviation at the measuring point $i+1$ is corrected by previous $i$ points, and the position deviation is corrected from one fiber segment to another, from the beginning to the end. The logical process based on the recursive compensation algorithm for position deviation is illustrated in Figure 5. Firstly, the spectrum at the $x_i$ fiber segment is selected and extracted by a sliding window, where the position deviation is corrected with the previous $i-1$ sensors. Then, the wavelength offset of the current sensor $\Delta\lambda$ is obtained by correlation demodulation, which can be transferred into optical path variation by $\Delta\lambda/\lambda_c$ Utilizing Equation (5) to calculate the position deviation accumulated to the $x_i+1$ fiber segment, correlation demodulation is then performed again to get accurate strain value. Finally, the above process is repeated until the last fiber segment. Experimental verification is executed to prove the effectiveness of the above recursive compensation algorithm.

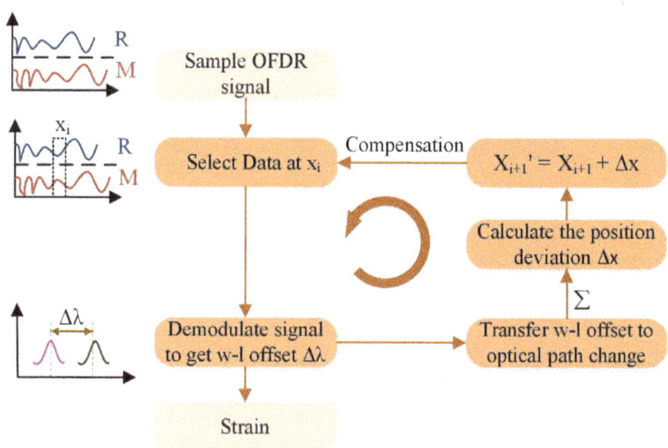


**Figure 5.** Process of the recursive solution for position deviation to obtain strain distribution.

## 3. Results and Analysis

Figure 6 shows the dependence of the spectral shift on strain at a single measuring point. The blue and red lines in Figure 6a,b are the cross-correlation respectively referring to the measured spectrum without/with strain. Figure 6a is obtained by direct solution in time domain, and Figure 6b is calculated utilizing convolution of frequency domain. Whatever method is used, the blue peaks are located at the origin in the middle, while the red peaks show the offset towards the positive spectral range. By comparing Figure 6a,b, the direct solution to the cross-correlation increases the computation and complexity of the system. Besides, a large baseline, as seen in Figure 6b, also affects the signal contrast and peak-seek accuracy. Thus, we use the convolution in frequency domain to obtain wavelength shift.

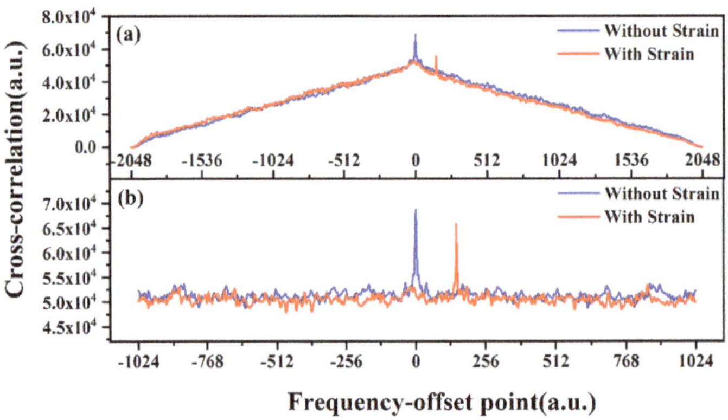


**Figure 6.** The dependence of the wavelength shift on strain at a single measuring point. (**a**) Cross-correlation in time domain; (**b**) cross-correlation in frequency domain.

The recursive compensation algorithm for position deviation is demonstrated with a static strain verification in the experiment. With massive data harvested from the experiment in frequency domain, the distributed spectra are mapped to their positions determined by the time delay. Particularly, the measured and reference spectra of the fiber segment at 21.24 m are selected as a verification for the position-deviation compensation algorithm. The SNR of the correlation peak is used as an indicator to describe the spectral correlation at different spatial resolution at 2, 1, and 0.5 mm. According to the theoretical analysis above, a higher spatial resolution means excessive data segmentation and insufficient spectral reconstruction. Thus, the same position deviation becomes more obvious and leads to a serious degeneration in correlation at a higher spatial resolution. By comparing Figure 7a,b, the position deviation causes correlation degeneration at a spatial resolution of 2 mm. Nevertheless, such position deviation is not large enough to depress the spectral contrast and affect the peak-seek accuracy. After compensation, the SNR of correlation is improved doubly. For a higher spatial resolution at 1 mm, the same deviation leads to significant correlation degeneration in Figure 7c. Even outliers are recorded as well by mistake due to low contrast during the peak-seek, unless the deviation is well corrected and the fake peaks are removed, as in Figure 7d. For extreme cases at a 0.5 mm spatial resolution, the spectral correlation is totally lost in Figure 7e, leaving the strain demodulation impossible at such a high resolution. However, with the position-deviation compensation in Figure 7f, the characteristic peak appears where it should be. Since a higher spatial resolution means a narrower sliding window, the data capacity of correlation calculation becomes smaller. In the event that the data capacity approaches its limit, improvement can hardly be achieved in SNR at high spatial resolution. Experimentally, the recursive algorithm is proven to be effective through the comparison of correlation SNR with and without the position-deviation compensation. Moreover, the alignment between the measured and reference spectra promises the possibility of a high spatial resolution at the fiber end.

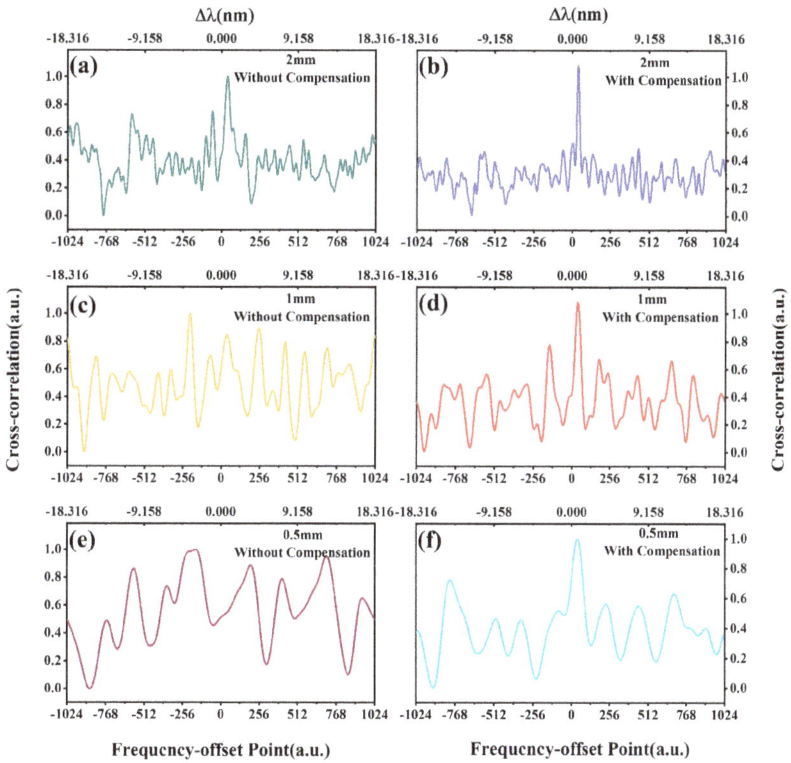


**Figure 7.** Cross-correlation result of RS spectra before and after position deviation correction: (**a**,**b**) 2 mm spatial resolution; (**c**,**d**) 1 mm spatial resolution; (**e**,**f**) 0.5 mm spatial resolution.

A static strain is axially applied at the end of a 30 m long optical fiber in an experimental verification. Figure 8 shows the strain distribution curves along an optical fiber segment at different spatial resolutions. The correlation SNR still remains considerable at 2 mm spatial resolution, while it decreases down to the background noise at 0.5 mm spatial resolution. In this case, the fake peaks randomly appear anywhere along the correlation curve, and the fake strain is obtained with the fake peak shift. Once the deviation becomes significant at a 0.5 mm spatial resolution, the errors randomly scatter near the strain distribution curve in Figure 8b. With the recursive compensation, the random errors are effectively suppressed and the restored strain distribution curve is close to that at 2 mm spatial resolution. Thus, the recursive algorithm for position deviation compensation is proven to be effective to restrain the correlation degeneration and improve SNR of the characteristic peak. The approach to a sub-millimeter spatial resolution is theoretically analyzed and demonstrated with a verification of 0.5 mm. Besides, the reorganization of 0.5 mm fiber segment is practically achieved in the experiment, referring to the experimental results in [19].

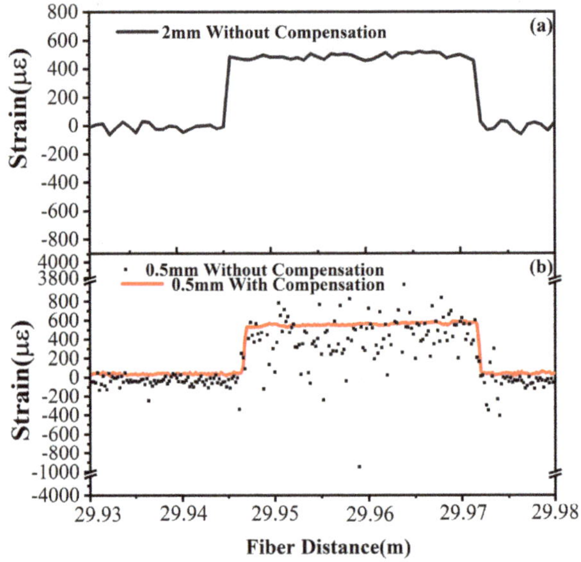


**Figure 8.** Strain distribution at the fiber end before and after position deviation correction: (**a**) 2 mm spatial resolution; (**b**) 0.5 mm spatial resolution.

## 4. Conclusions

We quantitatively analyze the position deviation and demonstrate that it degrades the correlation between the reference and measured spectra. A recursive compensation algorithm is proposed to realize a distributed sensor with higher spatial resolution and precise real-time measurement. This method helps maintain recognizable SNR of correlation by aligning the measured spectra with their reference. Moreover, the more obvious characteristic peak can be located with an accurate strain at the fiber end. With compensation, the distributed strain curve along the 30 m long fiber, at 0.5 mm spatial resolution, approaches that at 2 mm. Besides, the proposed technique can be further developed when combined with parallel computing to increase data capacity and reduce inversion distortion and applied to embedded sensors in the composite materials.

**Author Contributions:** Conceptualization, Y.C. and M.L.; funding acquisition, M.L. and N.L.; investigation, M.L. and J.L.; supervision, Y.C., M.L., J.L. and N.L.; writing—original draft, Y.C.; writing—review and editing, Y.C., M.L. and J.L. All authors have read and agreed to the published version of the manuscript.

**Funding:** Natural Science Foundation of Hebei province (Grant Nos. F2019202294 and A2020202013); Natural Science Foundation of Tianjin City (Grant No. 15JCYBJC17000).

**Acknowledgments:** This work was supported in part by Huiru Xue attended in Tianjin University of Technology and Ruiming Qi and Chao Li in Hebei University of Technology.

**Conflicts of Interest:** The authors declare no conflict of interest.

## References

1. Eickhoff, W. Optical frequency domain reflectometry in single-mode fiber. *Appl. Phys. Lett.* **1981**, *39*, 693–695. [CrossRef]
2. Ghafoori-Shiraz, H.; Okoshi, T. Optical-fiber diagnosis using optical-frequency-domain reflectometry. *Opt. Lett.* **1985**, *10*, 160–162. [CrossRef] [PubMed]
3. Ghafoori-Shiraz, H.; Okoshi, T. Fault location in optical fibers using optical frequency domain reflectometry. *J. Lightware Technol.* **1986**, *4*, 316–322. [CrossRef]

4. Dolfi, D.W.; Nazarathy, M.; Newton, S.A. 5-mm-resolution optical-frequency-domain reflectometry using a coded phase-reversal modulator. *Opt. Lett.* **1988**, *13*, 678. [CrossRef] [PubMed]
5. Tsuji, K.; Shimizu, K. Coherent optical frequency domain reflectometry for a long single-mode optical fiber using a coherent lightwave source and an external phase modulator. *Photonics Technol. Lett.* **1995**, *7*, 804–806. [CrossRef]
6. Liu, Q.; Fan, X.; He, Z. Time-gated digital optical frequency domain reflectometry with 1.6-m spatial resolution over entire 110-km range. *Opt. Express* **2015**, *23*, 25988. [CrossRef] [PubMed]
7. Baker, C.; Lu, Y.; Song, J.; Bao, X. Incoherent optical frequency domain reflectometry based on a Kerr phase-interrogator. *Opt. Express* **2014**, *22*, 15370–15375. [CrossRef] [PubMed]
8. Ding, Z.; Yao, X.S.; Liu, T.; Du, Y.; Liu, K.; Han, Q.; Meng, Z.; Jiang, J.; Chen, H. Long Measurement Range OFDR Beyond Laser Coherence Length. *IEEE Photonics Technol.* **2013**, *25*, 202–205. [CrossRef]
9. Geng, J.; Spiegelberg, C.; Jiang, S. Narrow linewidth fiber laser for 100-km optical frequency domain reflectometry. *IEEE Photonics Technol.* **2005**, *17*, 1827–1829. [CrossRef]
10. Kreger, S.T.; Gifford, D.K.; Froggatt, M.E.; Soller, B.J.; Wolfe, M.S. High Resolution Distributed Strain or Temperature Measurements in Single- and Multi-Mode Fiber Using Swept-Wavelength Interferometry. In Proceedings of the Optical Fiber Sensors, Cancun, Mexico, 23–27 October 2006.
11. Froggatt, M.; Moore, J. High-spatial-resolution distributed strain measurement in optical fiber with rayleigh scatter. *Appl. Opt.* **1998**, *37*, 1735–1740. [CrossRef] [PubMed]
12. Wada, D.; Igawa, H.; Tamayama, M.; Kasai, T.; Arizono, H.; Murayama, H.; Shiotsubo, K. Flight demonstration of aircraft fuselage and bulkhead monitoring using optical fiber distributed sensing system. *Smart Mater Struct.* **2018**, *27*, 025014. [CrossRef]
13. Sang, A.K.; Froggatt, M.E.; Gifford, D.K.; Kreger, S.T.; Dickerson, B.D. One Centimeter Spatial Resolution Temperature Measurements in a Nuclear Reactor Using Rayleigh Scatter in Optical Fiber. *IEEE Sens. J.* **2008**, *8*, 1375–1380. [CrossRef]
14. Kirikera, G.R.; Shinde, V.; Schulz, M.J.; Sundaresan, M.J.; Hughes, S.; Van Dam, J.; Nkrumah, F.; Grandhi, G.; Ghoshal, A. Monitoring Multi-Site Damage Growth During Quasi-Static Testing of a Wind Turbine Blade using a Structural Neural System. *Struct Health Monit.* **2008**, *7*, 157–173. [CrossRef]
15. Feng, K.P.; Cui, J.W.; Jin, Y.; Sun, X.; Jiang, D.; Dang, H.; Niu, Y.Z.; Tan, J.B. Enhancement of the Performance and Data Processing Rate of an Optical Frequency Domain Reflectometer Distributed Sensing System Using A Limited Swept Wavelength Range. *Sensors* **2018**, *18*, 3480. [CrossRef] [PubMed]
16. Feng, K.; Cui, J.; Jiang, D.; Dang, H.; Jin, Y.; Sun, X.; Niu, Y.; Tan, J. Improvement of the strain measurable range of an OFDR based on local similar characteristics of a Rayleigh scattering spectrum. *Opt. Lett.* **2018**, *43*, 3293–3296. [CrossRef] [PubMed]
17. Zhao, S.; Cui, J.; Suo, L.; Wu, Z.; Zhou, D.; Tan, J. Performance Investigation of OFDR Sensing System With a Wide Strain Measurement Range. *J. Lightwave Technol.* **2019**, *37*, 3721–3727. [CrossRef]
18. Zhao, S.; Cui, J.; Wu, Z.; Tan, J. Accuracy improvement in OFDR-based distributed sensing system by image processing. *Opt. Lasers Eng.* **2020**, *124*, 105824. [CrossRef]
19. Luo, M.; Liu, J.; Tang, C.; Wang, X.; Kan, B. 0.5 mm spatial resolution distributed fiber temperature and strain sensor with position deviation compensation based on OFDR. *Opt. Express* **2019**, *27*, 35823. [CrossRef] [PubMed]

MDPI
St. Alban-Anlage 66
4052 Basel
Switzerland
Tel. +41 61 683 77 34
Fax +41 61 302 89 18
www.mdpi.com

*Sensors* Editorial Office
E-mail: sensors@mdpi.com
www.mdpi.com/journal/sensors

www.ingramcontent.com/pod-product-compliance
Ingram Content Group UK Ltd.
Pitfield, Milton Keynes, MK11 3LW, UK
UKHW060124010826
14089UKWH00006B/136

9 783036 511832